Ecological Assessment of Environmental Degradation, Pollution and Recovery

ISPRA COURSES ON CHEMICAL AND ENVIRONMENTAL SCIENCE

A series devoted to the publication of courses and educational seminars given at the Joint Research Centre, Ispra Establishment, as part of its education and training programme.

Published for the Commission of the European Communities, Directorate-General Information Market and Innovation.

Volumes already published

- APPLICATIONS OF MASS SPECTROMETRY TO TRACE ANALYSIS
  Edited by S. Facchetti

- ANALYTICAL TECHNIQUES FOR HEAVY METALS IN BIOLOGICAL FLUIDS
  Edited by S. Facchetti

- OPTICAL REMOTE SENSING OF AIR POLLUTION
  Edited by P. Camagni and S. Sandroni

- MASS SPECTROMETRY OF LARGE MOLECULES
  Edited by S. Facchetti

- REGIONAL AND LONG-RANGE TRANSPORT OF AIR POLLUTION
  Edited by S. Sandroni

70294
ITT

£82.45
40052079

MEDWAY CAMPUS LIBRARY

This book is due for return or renewal on the last date stamped below, but may be recalled earlier if needed by other readers.
Fines will be charged as soon as it becomes overdue.

1 5 DEC 1997

1 3 JUL 1998

2 3 APR 1999

filing terms

294
4
DEVELOPMENT NATURAL
S INSTITUTE LIBRARY
11/89

Vol No.
Copy No.

577.4
RAV

LIBRARY
SEAS DEVELOPMENT
RESOURCES INSTITUTE
ENTRAL AVENUE
ATHAM MARITIME
CHATHAM
KENT ME4 4TB

the
UNIVERSITY
of
GREENWICH

Ecological Assessment of Environmental Degradation, Pollution and Recovery

Lectures of a course held at the Joint Research Centre, Ispra (Italy)
12–16 October 1987

Edited by

O. Ravera

Joint Research Centre, Ispra, Italy

Published for the Commission of the European Communities
by
ELSEVIER Amsterdam – Oxford – New York – Tokyo   1989

 Commission of the European Communities
Joint Research Centre, Ispra (Varese), Italy

Publication arranged by:
Directorate-General information Market and Innovation
Luxembourg

Published under licence by:
Elsevier Science Publishers B.V.
Sara Burgerhartstraat 25
P.O. Box 211, 1000 AE Amsterdam
The Netherlands

Distributors for the United States and Canada:

Elsevier Science Publishing Company inc.
655, Avenue of the Americas
New York, NY 10017, U.S.A.

© ECSC, EEC, EAEC, Brussels and Luxembourh, 1989

EUR 12054 EN

Legal Notice

Neither the Commisssion of the European Communities nor any person on behalf of the Commission is responsible for the use which might be made of the following information.

ISBN 0-444-87361-9

All rights reserved. No part of this publication may be reproduced, stored in a retrieval system or transmitted in any form or by any means, electronic, mechanical, photocopying, recording or otherwise, without the prior written permission of the publisher, Elsevier Science Publishers B.V./ Physical Sciences & Engineering Division, P.O. Box 330, 1000 AH Amsterdam, The Netherlands.

Special regulations for readers in the U.S.A. – This publication has been registered with the Copyright Clearance Center Inc. (CCC), Salem, Massachusetts. Information can be obtained from the CCC about conditions under which photocopies of parts of this publication may be made in the USA. All other copyright questions, including photocopying outside of the USA, should be referred to the publisher.

Printed in The Netherlands

PREFACE

The structure and functions of an ecosystem continuously change creating new conditions which, in turn, modify the ecosystem metabolism and its exchanges of energy and material with the adjacent ecosystems. This is the basic concept of the succession theory. The ecological succession represents the historical development of the community and its importance in ecology may be compared to that of the evolution in general biology. The time-scale of the succession varies with several factors; the size of the ecosystem and the average life-span of the individuals composing the community are the most important. For example, in few days the planktonic community shows so great changes as those that occurred in forests during some centuries. The ecosystems evolving according to the classical model of succession described in text-books are relatively few. Indeed, this ideal succession may be deeply modified by natural (for example: floods, sea level variations) and anthropogenic stresses. Man may accelerate the succession rate (for example: cultural eutrophication), prevent it (for example: conservation of natural environments at a prefixed state) or produce a regression in its trend (for example: clearing woods). The nutrient enrichment of the water bodies due to anthropogenic activities accelerates the succession towards the eutrophic state and the effluent treatment techniques aim to regress the lake evolution from the eutrophic state to the oligotrophic one. Every ecosystem tends to increase its organization level utilizing its exceeding resources in the successional process and, consequently, the output of resources from the ecosystem results in a progressive reduction. Therefore, to exploit the resources of an ecosystem, its succession must be stopped at a relatively young stage (for example: at prairy stage) or, if the successional stage is too advanced, the system must be regraded to a preceding stage (for example, deforestation). Agriculture and aquiculture are based on these principles. Changes in the community cause alterations in the physical environment, as changes in the latter influence the community that, in turn, modifies the physical environment. A stress may be absorbed by the ecosystem without producing significant alteration in its structure and functions. This ability is the measure of the inertia of the ecosystem. If the exposure time and/or the intensity of the stress overcomes a certain threshold - variable with the ecosystem and the nature of the stress - the ecosystem, after a certain time, results to be damaged. Some changes in the structure and functions of the ecosystem are considered as indices of its degradation. For example: an ecosystem is degraded when its rates are unbalanced (for example: the production-rate greatly exceeds the mineralization of the organic matter), its functions are anomalous (for example: an excess of oxygen is released from phytoplankton), some species have been eliminated or have attained too high a population density. The causes of the ecosystem degradation may be natural (for example: volcanic eruption, climatic changes) as well as anthropogenic (for example: pollution, alteration of the hydrological and geomorphological characteristics of land, transfer of species from one area to another, community simplification by eliminating weeds, parasites and predators). The recovery should theoretically return the ecosystem to its original state, but we know very little about the natural pre-human impact state of the greater part of the ecosystems. In practice, restoration consists in regressing the succession of an ecosystem to a prefixed state; for example: to recover an eutrophic lake to a mesotrophic state is commonly judged acceptable. Restoration of an ecosystem is always difficult and costly to be obtained, and a complete recovery may often be impossible because some components of the ecosystem may have been eliminated during its degradation. Generally, abolished the cause of the degradation, the ecosystem spontaneously recovers, if a sufficient period of time is considered. For example, several lakes have been restored by decreasing or abolishing the nutrient load from their watershed, cause of their degradation. Several ecosystems may recover at rates imperceptible to man and, consequently, opportune measures must be applied. The attention to the restoration problems arises from the relevant damages produced by man to the aquatic and terrestrial ecosystems. The recovery of the degraded ecosystem is necessary for saving natural resources, preserving physical environments and communities, maintaining species and preventing the loss of genotypes.

In conclusion, if it is evident that man cannot survive without modifying natural ecosystems and creating new ones (for example: reservoirs, vineyards), it is fundamental that the influences causing the ecosystem degradation must be accurately assessed to keep them at a minimum. We need more information to predict the vulnerability of the ecosystem to damages and its response to the measures to restore it. Many studies have been carried out on ecosystem degradation and its causes, but comparatively few on ecosystem recovery and, particularly, on the mechanisms producing it. As a consequence, greater attention must be paid to this important problem.

With the above-mentioned considerations in mind, it seemed opportune to organize the international course "Ecological assessment of environmental degradation, pollution and recovery" on 12-16 October, 1987, at the Joint Research Centre(JRC), C.E.C., Ispra (Varese), Italy, in the framework of its Education and Training Programme, realized by the so-called "ISPRA COURSES". This advanced course was coordinated by the undersigned and given in the form of a series of lectures presented by 12 well known experts in their specific field. The course was structured according to the following topics: (a) terrestrial and aquatic ecosystem concept; (b) structure, functions and evolution of the ecosystem in relation to the natural and anthropogenic influences; and (c) concept of stress, assessment and restoration of terrestrial and aquatic ecosystems.

These general concepts were developed in a series of lectures taking into account the ecological principles and environmental management. For the various aspects of the environmental problems, the state-of-the-art, the principles of restoration techniques, the results obtained by their application and the research needs to acquire a better knowledge of the ecological processes, were discussed. The lectures were illustrated by several case studies concerning forests, lakes, reservoirs, rivers, soil and the interrelations between air and terrestrial and aquatic ecosystems. This book contains the lectures presented at this course, reviewed by the authors. Thus, it concerns only part of the extensive subject of degradation and recovery of ecosystems.

The first part of the book deals with the new basic concepts of terrestrial and aquatic ecosystems and, particularly, the meaning of ecosystem evolution.

The concept of stress in physical, biological and ecological systems are compared and discussed in detail. The importance of natural and anthropogenic stresses in aquatic and terrestrial ecosystems is illustrated. Negative perturbation, called stress, is distinguished from the concept of subsidy, which is a modification leading to amelioration of the system. In addition, to quantify the stress in a steady-state ecosystem, the advantages and disadvantages of biochemical indicators are discussed. Some selected examples are given to illustrate the concepts of natural and man-made catastrophies, the anthropogenic impact on environmental conditions and the chronic stress produced by man and natural fluctuations.

The assessment of various types of stress in terrestrial and aquatic ecosystems is treated and illustrated with some examples.

Lake restoration is discussed from the theoretical and practical points of view. The techniques for ameliorating lake ecosystems can be grouped into two parts: (1) methods limiting the external load of nutrients; (2) methods directly applied to the lake. The latter consists in physical, chemical and biological techniques. These methods, except sediment removal, do not decrease the nutrient concentration in the water, but may reduce to a minimum the noxious effects of eutrophication. The more commonly used strategies are described and some suggestions as to their choice given. In addition, how some types of lakes react to the restoration techniques is illustrated by very clear examples.

One chapter deals with the biomanipulation of aquatic food-chain to improve the state of the eutrophicated lakes. This restoration technique is based on the concept that eutrophication, essentially, consists in a decrease of efficiency of the energy transfer along the pelagic food-web. With this method, the populations of planktivorous fish are reduced in size or abolished, with a consequent increase of the population density of filter-feeder zooplankton which control the phytoplankton populations. The results obtained are encouraging, but there is a need for more research in this field, because our quantitative information on pelagic food-web dynamics is rather scarce.

Degradation and recovery of an aquatic ecosystem are very complex processes. Consequently, by field research it is practically impossible to quantify the actual effects produced by stresses and to establish a relationship between the concentrations of various pollutants in the environment and their effects at population and community level. The results obtained from laboratory experiments cannot be extrapolated to natural ecosystems. During the last year, interesting and useful information has been acquired by studies carried out with "enclosure" method, called also limnocorral, mycrocosm and mesocosm. This technique represents a compromise between field investigation and laboratory experiments. With this method, the natural or artificial recovery of a stressed enclosed ecosystem may be followed. In one chapter the "enclosure" method, its advantages and disadvantages are described, and some examples illustrate the results obtained with this technique.

The ecological assessment of degradation and recovery of river ecosystems from pollution is the topic of another chapter. The introduction is based on essential and interesting questions, such as: "To what extent have we progressed during the last 100 years from the early ecological assessment of spatial degradation and recovery downstream of sewage discharges?". Such questions represent the discussion line of the lecture. In addition, from the introduction to the conclusions, the chapter is subdivided into three parts: (1) methods of ecological assessment; (2) pattern of degradation and recovery; and (3) selected case studies of the recovery of rivers from pollution. The first part is a critical review of commonly used techniques (for example:

diversity, biotic and functional indices, multivariate analysis, biotests). The second part describes the pattern of the degradation and recovery of river ecosystems and gives some suggestions for their investigation. The third part illustrates four selected cases of river recovery. In the conclusion, the author suggests a re-examination of research priorities and surveillance procedures. In addition, the strengths and weaknesses in both, knowledge of ecosystem behaviour and scientific support for environmental encouragement, are discussed. In spite of our sophisticated researches on aquatic pollution, there is a need for more information in order to better understand the degradation and recovery processes.

In two chapters the importance of soil pollution and its recolonization is treated. Heavy metals and other persistent micropollutants are the toxic substances considered. The author underlines that studies on soil pollution are very scarce, if compared with those on water and air, and, only in very recent times and in some countries, facilities for this research have been significantly improved. The author describes the effects of pollutants on the soil fauna and various mechanisms to achieve resistance to the pollutants. It is interesting that these mechanisms (for example: excretion, oxidation of metals to less toxic oxides) are common both to microorganisms and soil fauna. One population, living in a polluted environment, may survive because more resistant than others to certain dangerous substances; on the other hand, the same population may be less resistant to other types of stress. This characteristic is called "cost adaptation" and is clearly illustrated. In addition, the bioavailability of organic micropollutants and its relevance for bioaccumulation and bioconcentration are discussed. Other important topics treated by the author are: the biodegradation and consequent production of metabolites, the role of microflora, microfauna, physical and chemical factors in the persistent chemical degradation and soil sanitation. Some case studies illustrate the text.

The concepts on which the lecture on the effects of air pollution in terrestrial ecosystems is based, is very interesting. According to the author, all alterations of the communities and the chemical properties of the physical environment are the result of changes in the material balance. From this point of view, the ecosystem may be described in terms of storage and flow-rates of chemicals, which constitute the organisms as well as the soil. After this statement, the author lists a series of air pollutants according to their ecological role and discusses the effects of acidification in terrestrial ecosystems. Soil is considered an inert matrix as well as a sink and a source for pollutant deposition. Particular attention is paid to the effects of forest acidification. The enormous problems in improving the quality of soil polluted by air deposition are discussed in detail.

After having traced a short history of the anthropogenic acidification of water bodies, the author identifies the most important cause of acid deposition in fossil fuels. The chemical effects of acid deposition, in relation to geological and pedological conditions of the area, are discussed. The concept of dose-effect relationship and the internal resistance of an aquatic ecosystem to change (that is: its buffer capacity) are discussed. Several biological effects for the most important taxa are illustrated and the relative importance of aluminium compounds and pH are discussed and, in particular, the effects produced in fish. To reduce the effects of acid deposition in freshwater, two strategies may be applied: to reduce emissions, which act on the cause, or to apply liming, which reduces the symptoms of acidification. According to the author, liming is a temporary remedy and to abolish the effects of acid deposition, the emission must be reduced.

The last chapters deal with two interesting problems which may assume great importance in limited areas: the effects of mining on the ecosystem and the consequences of heavy metal pollution in freshwater bodies.

Air, water and soil may be polluted by mining activities, with alteration in freshwater and terrestrial ecosystems. The author briefly illustrates the history of non-ferrous mining in Great Britain. Once the mining activities cease or decrease, the recovery of soil is very slow and about 50 to 100 years are necessary to obtain a more or less complete restoration of the rivers.

The last chapter deals with the ecological effects of heavy metals and the recovery of the polluted ecosystems. The various sources of pollution, the biogeochemical cycles, the modification of the physico-chemical form of the metal, the bioconcentration and biomagnification are discussed. The metal transfer along the food-chain and the relationship between the concentration of the metal in the medium and the uptake by organisms are described. The importance of biological monitoring and bioassays is focused upon. In addition, some suggestions are listed to protect and recover polluted ecosystems.

The text is complemented with numerous figures and tables. Many references are reported at the end of each chapter. We express our gratitude to the authors who worked so hard in preparing their manuscript; to Dr. Misenta, Manager of Education and Training Programme, who fully supported the Course and the book project, and to his staff, who excellently organized the course. Many thanks are due to Dr. de Bernardi and Dr. R. Baudo (Istituto Italiano di Idrobiologia, CNR, Pallanza (Novara), Italy) for the generous help in planning the programme of the course. We are also indebted to Miss Lorenza Giannoni for her help in collaborating in the preparation of the book.

Oscar RAVERA

CONTENTS

Preface v
List of Contributors xiii

1. The evolution of terrestrial ecosystems
   F. di Castri 1
2. Ecological succession and evolution of freshwater ecosystems
   I. Ferrari and V.U. Ceccherelli 31
3. Concept of ecosystem stress and its assessment
   R.D. Gulati 55
4. Concept of stress and recovery in aquatic ecosystem
   R.D. Gulati 81
5. Concept of stress, natural and anthropogenic, in terrestrial ecosystems
   G.H. Kohlmaier 121
6. Dynamic equilibria in material systems and their response to perturbations
   G.H. Kohlmaier, A. Janecek, M. Lüdeke, J. Kindermann, K. Siebke and F. Badeck 137
7. Ecological assessment of the degradation and recovery of rivers from pollution
   R.W. Edwards 159
8. Biomanipulation of aquatic food chains to improve water quality in eutrophic lakes
   R. de Bernardi 195
9. Lake ecosystem degradation and recovery studied by the enclosure method
   O. Ravera 217
10. The impact of heavy metals on terrestrial ecosystems: Biological adaptation through behavioural and physiological avoidance
    H. Eijsackers 245
11. Persistent chemicals in soil ecosystems and optimization of biological degradation processes
    H. Eijsackers and P. Doelman 261
12. Air pollution effects in terrestrial ecosystems and their restoration
    B. Ulrich 275
13. Air pollution effects on aquatic ecosystems and their restoration
    A. Henriksen 291

14. Mining effects on ecosystems and their recovery
     B.E. Davies 313
15. Heavy metal pollution and ecosystem recovery
     R. Baudo 325
Index 353

LIST OF CONTRIBUTORS

F. Badeck	Institut für Physikalische und Theoretische Chemie, Johann Wolfgang Goethe Universität, Niederurseler Hang, 6000 Frankfurt 50, Federal Republic of Germany
R. Baudo	CNR - Istituto Italiano di Idrobiologia, Largo V. Tonolli 50-52, 28048 Verbania Pallanza, Italy
R. de Bernardi	CNR, Istituto Italiano di Idrologia, Largo V. Tonolli 50-52, 28048 Verbania Pallanza, Italy
F. de Castri	Centre d'Ecologie Fonctionnelle et Evolutive, CEPE Louis Emberger/C.N.R.S., B.P. 5051, 34033 Montpellier Cédex, France
V.U. Ceccherelli	Istituto di Zoologia, Università di Ferrara, Ferrara 44100, Italy
B.E. Davies	School of Environmental Science, University of Bradford, Bradford, West Yorkshire BD7 1DP, England
P. Doelman	Research Institute for Nature Management, P.O. Box 9201, 6800 HB Arnhem, The Netherlands
R.W. Edwards	Institute of Science and Technology, University of Wales, Cardiff and Welsh Water Authority, Cardiff, Wales, U.K.
H. Eijsackers	Programme Office, The Netherlands Integrated Soil Research Programme, P.O. Box 37, 6700 AA Wageningen, The Netherlands
I. Ferrari	Istituto di Zoologia, Università di Ferrara, Ferrara 44100, Italy
R.D. Gulati	Limnological Institute, Rijksstraatweg 6, 3631 AC Nieuwersluis, The Netherlands
A. Henriksen	Norwegian Institute for Water Research, P.O. Box 33, Blindern, 0313 Oslo 3, Norway
A. Janecek	Institut für Physikalische und Theoretische Chemie, Johann Wolfgang Goethe Universität, Niederurseler Hang, 6000 Frankfurt 50, Federal Replubic of Germany

J. Kindermann   Institut für Physikalische und Theoretische Chemie,
                      Johann Wolfgang Goethe Universität, Niederurseler Hang,
                      6000 Frankfurt 50, Federal Republic of Germany

G.H. Kohlmaier   Institut für Physikalische und Theoretische Chemie,
                      Johann Wolfgang Goethe Universität, Niederurseler Hang,
                      6000 Frankfurt 50, Federal Republic of Germany

M. Lüdeke   Institut für Physikalische und Theoretische Chemie,
                      Johann Wolfgang Goethe Universität, Niederurseler Hang,
                      6000 Frankfurt 50, Federal Republic of Germany

O. Ravera   Department of Physical and Natural Sciences, Commission
                      of the European Communities, Joint Research Center -
                      Ispra Establishment, 21020 Ispra (Varese), Italy

K. Siebke   Institut f?ur Physikalische und Theoretische Chemie,
                      Johann Wolfgang Goethe Universität, Niederurseler Hang,
                      6000 Frankfurt 50, Federal Republic of Germany

B. Ulrich   Institute of Soil Science and Forest Nutrition,
                      University of Göttingen, 13 Büsgenweg 2, 34 Göttingen,
                      Federal Republic of Germany

Chapter 1

THE EVOLUTION OF TERRESTRIAL ECOSYSTEMS

F. DI CASTRI

1.1 INTRODUCTION

Few topics of modern ecology are as controversial as the one dealing with the evolution of ecosystems. First of all, the ecosystem in itself is a debatable unit for ecology, or -at least- the interpretation of what an ecosystem is differs from one research worker to another. Secondly, evolution is a rather "uncomfortable" concept to be precisely defined in this context. Its definition could range from a vague notion of organic irreversible change to a more stringent interpretation involving the heritability and transmissibility of genetic information. *Sensu stricto*, an ecosystem, which includes both biotic elements carrying through a genetic information and abiotic components where the meaning of "information" is of a different nature, could not evolve.

Furthermore, there is often a confusion between words such as "dynamics" and "development" of ecosystems, which are by and large synonymous of ecological succession, and the word "evolution". This embraces a much longer lapse of time, and the building-up of emerging and irreversible properties that an ecosystem will keep and "transmit" through time under given disturbance regimes. To a certain extent, ecological successions (ecosystem dynamics and development) should be considered as progressive and concomitant steps -at a shorter temporal scale- of the overall process of ecosystem evolution.

In addition, treating in depth the various concepts covered by the expression "evolution of ecosystems" would imply opening the Pandora Box of true or false dichotomies of ecology and evolution, such as gradualism and punctuationalism, neo-darwinism and new synthesis, stochasticity and determinism, equilibrium and non-equilibrium, linearity and non-linearity, reductionism and holism (refs. 1-2-3-4).

Finally, this expression represents -to a certain extent- a kind of tautology or circular reasoning: ecosystems can exist only if they evolve, and they can evolve only if they exist. Indeed, tautologies are widely spread in evolutionary and ecological theories (ref. 5), together with use, abuse and misuse of analogies. They can be of a heuristic value, in so far as scientists are well aware when they are merely dealing with tautologies and analogies.

It is amazing that the two terms "ecosystems" and "evolution of ecosystems" are the core of the debate of the so-called "creationist ecology" (ref. 6), that has a few followers mostly in the United States. For creationists, the ecosystem is an organism *sensu stricto*, but this is precisely the base of their argumentation for denying the possibility of its evolution.

Taking into account debates, doubts, gaps of knowledge and other limitations, one is tempted to develop this topic in posing a number of questions. As a matter of fact, this is what I do in my lecturing at the University, and questions are as follows: Do ecosystems exist? Are they of a cybernetic nature? Are there emerging properties at the ecosystem level? Are there well defined boundaries for ecosystems? Can ecosystems be considered as units of selection? How has an ecosystem been built-up? Do ecosystems evolve? Do ecosystems transmit "inherited" characteristics? What are the evolutionary constraints as applied to ecosystems? Do ecosystems converge? How far do ecosystems optimize resources? To what extent do ecosystems resist invasion by alien species? Stochasticity or determinism in ecosystem development and evolution? Are new selective forces emerging in ecosystem evolution?

The above questions should not be considered as being of a merely rhetorical nature. They reflect present uncertainties; some of them textually reproduce the title of published papers (refs. 7-8). The core of this chapter will be built-up around a few of these questions. Nevertheless, it is not pretended here to provide definite answers to any of them. The purpose is rather to highlight present controversies, gaps of knowledge, and failures in our theoretical background. A selected and rather comprehensive bibliography could help those interested in these issues to deepen their understanding, out of the limited scope of this chapter.

1.2 DO ECOSYSTEMS EXIST?

It is not worth discussing here the origin of the term "ecosystem", and the different definitions that have been given. Furthermore, the term was coined by Tansley in 1935 (ref. 9), long after the concept had been extensively applied. In addition, many synonymous have existed before and afterwards. McIntosh (ref. 10) has comprehensively discussed in 1985 the history of the ecosystem concept and of ecosystem ecology.

In accordance with the title of this chapter, the most applicable definition should be that of Patten (1975) (ref. 11), which is as follows: "The ecosystem can be taken to consist of biotic and abiotic components that change and evolve together, and the term ecosystem implies a unit of coevolution.". Patten further elaborated this concept in 1978 and 1982 (refs. 12-13).

It is not easy to produce proofs or disproofs to the hypotheses backing this definition. Only a testable experimental approach on ecosystem studies could provide more concrete elements in this respect (ref. 7). However, specific experiments on ecosystem evolution would cover such long time periods that it is very difficult to design and implement them.

To define from the very beginning and for once the **working hypothesis** that will underlie all this chapter, it should be said that there is only infinitesimally small probability that mutualistic relations between and among species, and functional properties such as energy transfer and nutrient recycling, could occur at the ecosystem level solely by chance, and in the absence of ecosystem patterns of regulation. In any event, postulating that ecosystem attributes do exist has shown to have a value from both heuristic and operational viewpoints.

This preliminary discussion on the reality of the ecosystem may surprise those ecologists who are used to considering it as the very unit of ecology. However, unless one accepts to live in a kind of Panglossian world, it should be stressed that many ecologists do not even perceive the need to explore and recognize the ecosystem approach in their research on population dynamics, niche partitioning and overlap, or even succession of communities. In addition, understanding limits and potentials for extrapolating research results from one ecosystem to another also requires a more critical apprehension of the nature of ecosystems.

1.3 ARE ECOSYSTEMS OF A CYBERNETIC NATURE?

Polemics about this question, that were latent during several years, raised more explicitly after a paper by Engelberg and Boyarsky (ref. 14) postulating that ecosystems are of a noncybernetic nature. This also implies that there is no information network linking different parts of an ecosystem, and that control mechanisms are missing within ecosystems.

The "defense" of ecosystems (ref. 15) was taken through various responses published by the American Naturalist, where the cybernetic nature of ecosystems was strongly advocated (refs. 16-17). Patten and Odum (ref. 16), in particular, supported the viewpoint that information networks do exist in ecosystems, although they are necessarily more diffused than those built in man-made automatic control systems. In addition, a system does not need to be goal-oriented (e.g., teleologically guided) to be of a cybernetic nature (ref. 16).

Cybernetic considerations are deeply embedded in the system approach of ecology. Species assemblages (and species diversity), food chains and also nutrient cycling processes within the ecosystem are taken as information networks (refs. 18-19-20-21). Particular examples of communication mechanisms in planktonic populations are illustrated by Margalef (ref. 22).

Here again, it is not easy to produce experimental testable evidences. The working hypothesis (or postulate) mentioned above as a base for this chapter should be reiterated. It is highly improbable that successional processes and resilience of ecosystems occur merely by chance, and disregarding intrinsic mechanisms of regulation.

1.4 ARE THERE EMERGING PROPERTIES AT THE ECOSYSTEM LEVEL?

Emerging properties of an ecosystem can be defined as those appearing at this given hierarchical level, and that are not detectable by the simple analysis of properties of the component populations and communities. Admittedly, this implies accepting the postulate that ecosystems exist. With a somewhat tautological thinking, the very existence of emerging properties would provide indirect evidences on the reality of the ecosystem level.

It would be again very difficult to believe that decomposition processes, transfer of energy from one step of the trophic web to another, progressive concentration of chemicals at the top of the food chains appear by chance, because of individualistic properties of different populations, and not because of synergistic interactions at the ecosystem level. All the problem of community and ecosystem complexity and stability can only be tackled on the ground of emerging interactions (refs. 23-24), including aspects of resource availability (ref. 25). Reality of structural patterns is further discussed by May (ref. 26).

Even from an evolutionary viewpoint, coevolution processes (refs. 27-28), interspecific competition (ref. 29), regulation of species diversity (refs. 21-30), and -in general- mutualistic relations (ref. 31), are to be considered as emerging properties.

1.5 ARE THERE WELL DEFINED BOUNDARIES FOR ECOSYSTEMS?

This is yet another controversial question as related to the ecosystem concept, mostly from an operational research viewpoint. From the one side, without boundaries there could be no objects nor biological entities. From the other side, boundaries of an ecosystem largely depend on the prospective of the research worker, and of his working hypothesis. What is seen by a researcher as a whole ecosystem, for instance a termite nest with all the interactions with other species and the recycling taking place there, could well be seen by another scientist as a small part of a tropical forest ecosystem.

As for the information network, the boundaries of an ecosystem are more diffused than those of other living systems (e.g., an individual), or non-living systems (a robot or an automated industrial plant). In addition, boundaries have to be envisaged as a dynamic process within a given landscape where different ecosystems are interacting (refs. 32-33).

Accordingly, to a large extent, ecosystems boundaries can only be conceived if one accepts the hierarchical patterns of nestedness, that is to say "the property of a hierarchical system of having entities of smaller scale enclosed within those of a larger scale" (ref. 34).

This is totally consistent with the recent definition given to the ecotones as a "zone of transition between adjacent ecological systems, having a set of characteristics uniquely defined by space and time scales, and by the strength of the interactions between adjacent ecological systems." (ref. 35). Such a boundary can be envisaged at any scale, from a few centimeters to thousands of kilometers, from a successional to an evolutionary time, according to the scale of a given disturbance, but also following the working hypothesis and the problem to be tackled by a research worker.

1.6 CAN ECOSYSTEMS BE CONSIDERED AS UNITS OF SELECTION?

The meaning of this question is not whether or not ecosystems serve as an indispensable framework for selection processes and the evolution of species and populations. This is unquestionable, in my view, and it has been brightly illustrated by Hutchinson (ref. 36) in his book "The Ecological Theater and the Evolutionary Play". Rather it is meant whether or not ecosystems are in themselves entities subject to selection processes that would control their own "ecosystemic" evolution.

The fact that there is more than one unit of selection, and that selection acts on a hierarchy of units smaller and larger than the individual, from nucleotides and chromosomes to populations and species, is more and more accepted at the present stage of evolutionary and ecological theories (refs. 37-38). The hierarchical level of communities, at least, can be formulated as a natural functional unit of selection (ref. 39).

As regards the units of selection, reference should be made to some papers whose authors do not necessarily share the same approach (refs. 40-41-37-38-39). In particular, Gouyon and Gliddon (ref. 38) include ecosystems as one of the avatars subject to biological evolution through mechanisms of competition, predation, symbiosis and succession. However, the information content within ecosystems is not only provided by "a set of sets of gene pools of the various populations" (ref. 38), but also by biotic-abiotic interactions that cannot be interpreted *sensu stricto* through the genetics of information.

These aspects will be further developed in this chapter. Nevertheless, it can be anticipated that large-scale phenomena such as extreme events (ref. 42), climatic change (refs. 43-44-45) and, in general, large-scale modifications of disturbance regimes, can be considered as acting directly on ecosystems as selective forces (see also Vrba, refs. 46 and 47). Accordingly, but with great caution, ecosystems may be taken too as units of selection.

1.7 DO ECOSYSTEMS EVOLVE?

Before attempting to define the possible signification of "evolution of ecosystems", it is paradoxically needed to state *a contrario* what is not covered by this expression, at least within the framework of this chapter.

First of all, ecosystem evolution cannot match a classical genetic definition of evolution, such as the one given by Sewall Wright as "a process of cumulative change in the *heredities* characteristic of species" (ref. 48). Evolution of ecosystems does comprise genetic evolution of species, but also includes other characteristics which derive from biotic and abiotic interactions taking place at higher levels of organization.

Secondly, the numerous studies on the origin, history, evolution and replacement of floras and faunas (refs. 49-50-51-52-53) according to climatic or tectonic change, such as research based on geofloras (ref. 54) or palynological records (ref. 55), should not be considered as equivalent to those on ecosystem evolution. All historical events covered by these studies have been main causes for building-up ecosystems and promoting their evolution, but this research is not intended to provide explanations on how emergent and irreversible properties coalesced within an evolving ecosystem.

Thirdly, as already mentioned in the introduction, the development of ecosystems on a successional basis only covers a short lapse of time within the long temporal frame of the evolution of ecosystems. Nevertheless, it should be stressed how important it is to understand ecosystem dynamics at a given (and shorter) time scale, to try exploring the mode of possible evolutionary patterns of ecosystems.

The succession of terrestrial ecosystems is a far-reaching topic that cannot be discussed in depth in this chapter. After the pioneering work of Clements (ref. 56) with his deterministic approach, and the opposite views of Gleason (refs. 57-58-59), who adopted an individualistic concept, the theories of Odum and Margalef (refs. 18-16-19-22) based on thermodynamics and information theory have had a strong impact. They have still to be considered of high value, at least heuristically speaking. Odum's model of ecosystem development (ref. 18) is rarely applied, at present, to studies of ecological succession of terrestrial commmunities, but still has many followers concerned with succession of freshwater ecosystems.

All these theories have -quite understandably- been challenged or enriched by many research workers. Ecological succession has become one of the most flourishing and stimulating fields of modern ecology, and a variety of approaches is being developed (refs. 60-61-62-63-23-64-65-29-66-25-67).

Referring now more specifically to the evolution of ecosystems, it is intended to further discuss in this chapter the processes that lead, over a geological time scale, to irreversible change of ecosystem patterns (resilience, allocation of resources within the ecosystem, recycling of nutrients, clustering of species assemblages, modifications of the physical environment as regards microclimate and soil conditions, thus increasing microsites heterogeneity). These processes constitute in themselves strong selective forces for species evolution within the ecosystem.

The driving forces for ecosystem evolution are the environmental conditions that provide some determinism towards the optimization of available resources, and the disturbance regime that appeals to adaptive responses at the ecosystem level.

The background condition for such evolution is represented by the available species and gene pools existing *in situ*, at a given moment, or immigrating from neighbouring places, or arriving on account of latitudinal shifts (ref. 68) of ecosystems, as a consequence of climatic change.

Accordingly, evolution of ecosystems can be envisaged as a blending of environmental necessity leading to some optimization of resources and of historical chance because of different philogenetic origins of the "founding species" of the ecosystem. In other words, a very intricate interplay between deterministic and stochastic factors is the *leit motiv* of this evolution.

It is not conceivable that all species existing at a given moment and in a given place have all the genetic possibility of reacting in the same way -and with similar interactions among them- to given environmental conditions. Therefore, an ideal optimization is never achieved, but there can only be an "optimization of the possible", the possible being determined by genetic potentials and by changes in disturbance regime.

The evolution of ecosystems, given the complexity and irreversibility of factors involved, is not a merely iterative process. Analogically speaking, evolution is not an industrial-like production line of identical objects.

Rather, it can be seen as a kind of natural "bricolage", that is to say pottering about and handicrafting by selective forces on heterogeneous components, thus allowing to progressively reach as good a functional unit as possible.

To what extent are there "inherited" and "transmissible" characteristics in ecosystem evolution? Of course, there is no transmission at once, by reproduction, of all ecosystem properties from one ecosystem to the following one. Transmissibility in ecosystem evolution is not a discrete process, but a continuous one (or an extremely intricate blend of discrete processes). Species and populations of the ecosystem transmit genetic information, including that concerned with adaptive features to such ecosystem. A genetic transmission of inherited characteristics can also be envisaged for species assemblages of a mutualistic nature. System properties of resilience, energetic efficiency, nutrient recycling, etc. depend -after all- on interactive patterns of the component species themselves. Other "transmissions" are not of a genetic nature, such as given modifications of soil structure which have lead to the habitat progressively constructed by the very action of species living in soil (which -conversely- are further conditioned by its selective forces), or microclimatic conditions which are also the result of species metabolism, interaction and stratification. At all events, it should be stressed that the meaning of "inheritance", as applied to ecosystem evolution, has homologies and analogies, but also unquestionable differences, in comparison with the simply genetic transmissibility of inherited characteristics.

Admittedly, it would be wiser to refer to evolution of communities rather than to evolution of ecosystems. At least, in this case, only biotic components would be involved, and relations with genetics would be easier to be established. This is what has been done by several authors who have discussed -from different angles- aspects of evolution of communities (refs. 69-20-70-71-72-73). In addition, it may be wrong to place communities and ecosystems on the same hierarchical scale of biological organization (ref. 74).

Adding an abiotic and energetic component, through the ecosystem concept, implies the need to take into account -for exploring evolutionary processes at the ecosystem level- not only considerations based on biological-genetic evolution (from the molecular to the species level), but also approaches derived from thermodynamics (refs. 75-76-19-77-63-3).

In spite of this genetically heterodox interpretation of the word evolution, it is now easier at least debating on ecosystem evolution, owing to the present stage of the science of evolution. To the classical neo-darwinistic gradualism, new approaches are being added on the ground of punctuated equilibria or punctuationalism (refs. 78-79), macroevolutionary trends (refs. 46-80), consideration of hierarchy of evolutionary causes (refs. 81-41-37-39-82), as well as new syntheses (refs. 83-84-85-4), such as the turnover-pulse hypothesis (ref. 47). Within the framework of these new approaches, the existence of larger selection units like ecosystems is more widely accepted. There is, at least, an increased recognition that species evolution could not be understood in the absence of an ecosystemic framework or disregarding the impact of environmental change. Accordingly, present evolutionists may be classified (ref. 47) -as regards their approach- along a gradient ranging from "internalists" (those ignoring almost completely the role of "external" environmental conditions in species and phyletic evolution) to "strong environmentalists" (those denying any possibility of evolution in the absence of environmental change). Environmentalist approaches *sensu lato* seem to prevail now.

The relevance of biotic crises is also highlighted (refs. 86-87), particularly in relation to climatic change (refs. 43-44-45) or extreme events (ref. 42). Some consideration is already being given to the possible occurrence of a man-made biotic crisis, due to the increasingly pervasive human influence (man has become a major geological and evolutionary agent), and in particular to the forthcoming global change of climate (refs. 88-42-89-90).

Making an analogy involving recent developments in ecology and evolution, it could be said that processes of ecological succession have their counterpart in gradualistic speciations, while ecosystem evolution matches with the evolutionary mode of the turnover-pulse hypothesis. In my view, these theories should not be considered as being controversial, but are more likely to be complementary in scope. Adopting a temporal scale, ecological succession and gradualism could be "nested", respectively, within ecosystem evolution and punctuationalism (or turnover-pulses).

Several authors have already approached the question on whether or not ecosystems evolve (refs. 91-92-93-94-95-96-97). Some of them also have drawn attention to the need of caution, to avoid to be catched by false analogies.

Among possible evidences of evolution of ecosystems discussed by these research workers, a few ones will be mentioned here, as follows: the fact that -geologically speaking- there is a kind of saturation of taxa within a given ecosystem type, following periods of intense interchange, as it happened during the connections between North America and South America through the isthmus of Panama (ref. 92); the fact that in all ecosystems there seems to be a threshold as regards the number of trophic levels (no more than 5 or 6) in the food webs (ref. 92); the fact that in most ecosystems there seems to be a similar distribution of component species by size classes (ref. 92); the fact that the ecosystem vulnerability to invasion by alien species may be related to evolutionary constraints (ref. 98); and -above all- the fact that there are convergent evolutionary patterns and processes at the ecosystem level, and that these patterns may be predictable considering subtle environmental discriminators (ref. 99). In particular, evolutionary determinants leading to grassy or woody savannas have been highlighted and discussed (refs. 8-100). Convergence and divergence of ecosystems will be further developed in the next point of this chapter.

Evolution of ecosystems cannot be understood if due account is not given to environmental disturbances and biotic responses, as they act at different scales of time and space. This evolution is determined by sequences of discontinuous events, whose effects are progressively incorporated (ref. 35) in the functioning and dynamics of complex and nested ecological systems. Fig. 1 is intended to point out, as an indispensable background consideration, that

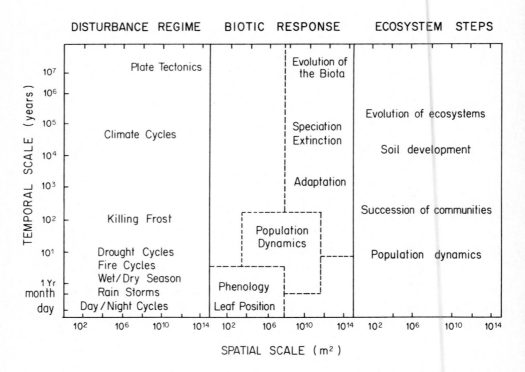

Fig. 1. Temporal and spatial scales of disturbance regimes and biotic responses as determinants of a stepwise evolution of terrestrial ecosystems, nesting biological entities (populations and communities through their own dynamics) and abiotic components (soils, through pedogenesis).
After Delcourt and Delcourt (ref. 51), largely modified.

all events and processes shown here are scale-dependent. They have, therefore, to be consistently framed in their proper scale of perception.

Fig. 2 attempts to show some steps in the evolution of ecosystems, when facing change of their disturbance regime. The starting point (step 1) -at the bottom of Fig. 2- is a so-called climax ecosystem. After the emergence of a new disturbance regime, adaptive responses at the ecosystem level

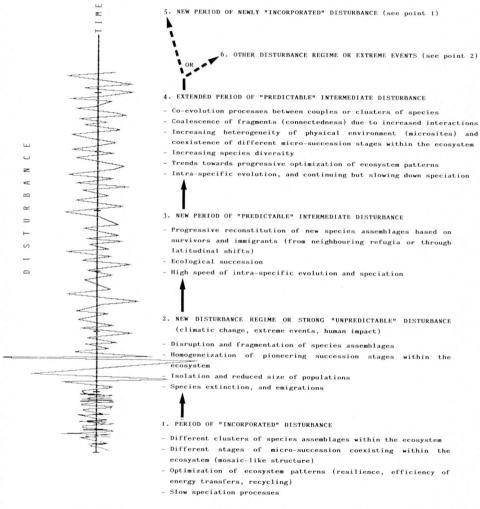

Fig. 2. Steps in the evolution of an ecosystem, following change of its previous disturbance regime. From steps 2 to 5, trends are towards increased complexity and heterogeneity, greater connectedness of component parts, emergence of ecosystem properties leading to better optimization of resources, assemblage of new co-evolutionary linkages, and change of speciation rates.

progressively take place, thus incorporating the new disturbances within patterns of ecosystem functioning and organization. Conceptually speaking, this figure is based on approaches deriving from non-equilibrium dynamics, discontinuity and non-linear reactions to disturbance. A kind of "evolutionary climax" would not be -therefore- a consequence of a long period of environmental stability, but rather of a stepwise adaptation to a given disturbance regime.

Steps of Fig. 2 have some relation with results of palynological analyses. However, the disturbance regime depicted here is a merely hypothetical one, constructed by computer programme. Arrows only indicate change in time of subsequent steps, and not linearities. Steps 2 and 3 correspond to a phase of great turbulence and large unpredictable oscillations. For instance, the fragmented and smaller populations of species pre-existing in step 1 are confronted with the large populations of the incoming species, which characterize the pioneering stage of an ecological succession.

Fig. 2 shows some resemblance to that of Lewin (ref. 101), who depicted phases of species turnover in the evolution of bovids and hominids in Southern Africa, synchronous with paleoclimatic changes.

How is an ecosystem built-up? In addition to species remaining *in situ*, new species are incorporated -while others are lost- through latitudinal shifts of vegetation (refs. 68-99). This is examplified in Fig. 3, which shows a latitudinal gradient of some 1000 km in Chile, from perarid to perhumid mediterranean-type climates and ecosystems.

Throughout northward and southward expansion and retreating of plants and animals, relicts and refugia -more xerophilous or more hygrophilous than the average ecosystem type- persist thanks to the heterogeneity of topography and mesoclimates. This is sketchily illustrated in Fig. 4. In accordance with climatic cycles, xerophilous species expand from arid refugia towards other ecosystem types, during periods of increased aridity. Conversely, hygrophilous species are invasive in periods of wetter climate. All these processes facilitate a species turnover within ecosystems, in geological times. In addition to *in situ* speciation, this turnover provides new opportunities for renewed species assemblages. Clusters (or networks) of these assemblages are at the basis of new structural and functional patterns of an evolving ecosystem organization.

Other examples of what can be referred to as "species foundation effect" on ecosystem evolution can be taken from environmental change in the Amazonian basin, with special emphasis on the expansion and retreating of refugia (refs. 103-104).

Determinants for ecosystem dynamics and evolution, as applied to mediterranean-climate sclerophyllous ecosystems, are shown in Fig. 5 (ref. 99). Change could happen at a small spatial scale where the topographical heterogeneity is so great as to provide all ranges of environmental possibilities (Fig. 4). Change also occurs at a larger latitudinal scale, following climatic gradients (Fig. 3).

As regards temporal scales, change can embrace a relatively short period, after a sudden disturbance and successive ecosystem recovery. This corresponds to an ecological succession. If changes are at a much longer time scale, thus incorporating different cycles of ecological succession, this should be referred to as ecosystem evolution. It should be stressed that man is playing an increasingly dominant role as selective force on species evolution and ecosystem dynamics, mainly through the fragmentation and reconstitution of landscapes (ref. 33).

The above discussion on ecosystem evolution refers in particular to terrestrial ecosystems. Consideration has also been given to evolution at the

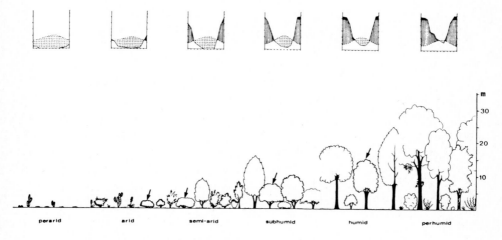

Fig. 3. A north-south latitudinal gradient in Chile (from 27° to 39° S), in correspondence to six mediterranean-climate types from perarid to perhumid. The climatic trends are schematically depicted by means of Gaussen-Walter diagrams. The plant formations are as follows : perarid = subdesert; arid = open shrublands; semi-arid = dense shrublands and matorral; subhumid = sclerophyllous woodlands and forests; humid = mesophilous woodlands and deciduous forests; perhumid = hygrophilous forests.
The particular position of *Lithraea caustica* is pointed out by arrows. This species is present in almost all plant formations, showing different life forms, from chamaephyte to shrub and tree.
During climatic fluctuations, this gradient has repeatedly shifted northwards and southwards, leaving behind species in hygric and xeric refugia in accordance to heterogeneous topographic features.
After di Castri, 1981 (ref. 99), modified.

level of ecosystem in marine environments (ref. 105). Furthermore, patterns of terrestrial and marine ecosystems have been compared (ref. 106).

Hypotheses on the evolution of terrestrial ecosystems, somewhat similar to those expressed in this chapter, have been advanced by Blandin and others (refs. 93-107-108). They put particular emphasis on climatic cycles as well as on the role of species assemblages called "coenons". As already pointed out (ref. 101), synchronous increase or decrease of the rate of speciation in different regions, in accordance with wide-ranging climatic cycles, have been put in evidence as regards evolution of mammals.

It should be stressed that evolutionary steps outlined in Fig. 2 are to be considered as sketchy generalizations. In particular, not all different components of an ecosystem react to change of the disturbance regime with the same mode and rate. There seems to be some inertia or dephasing of ecosystems (or parts of ecosystems) in the response to environmental change (ref. 109).

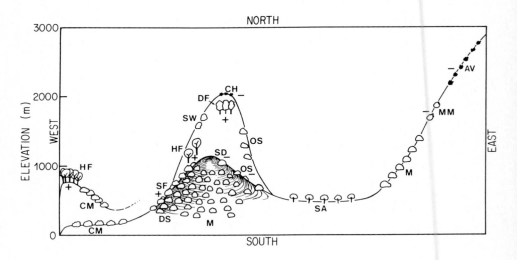

Fig. 4. West-east transect in Central Chile from the coastal terraces and hills, through the Coastal Range and the Central Valley, up to the Andean Cordillera. Hygric (+) and xeric (-) refugia exist because of atmospheric (condensation of marine fog and clouds) and topographic (slope and canyon effects) conditions.
Plant associations are practically the same as those shown in the latitudinal gradient of Fig. 3: SD (subdesert), OS (open shrubland), DS (dense shrubland), M (matorral), CM (coastal matorral), MM (montane matorral), SW (sclerophyllous woodland), SF (sclerophyllous forest), DF (deciduous forest), HF (hygrophilous forest). In addition, formations essentially controlled by edaphic, phreatic and/or anthropic conditions (SA: *Acacia caven* savanna) as well as altitudinal formations (CH: chamaephytes and herbs. AV: alpine vegetation) are represented.
After Rundel, 1981 (ref. 102), largely modified and redesigned.

This happens because of intrinsic buffering properties of ecosystems, but also -or mostly- because of the historical weight of species which have originated under different climatic conditions or disturbance regimes. For instance, trees and shrubs living in the mediterranean-climate regions of Southern-Western Australia, that is to say, under prevailing summer-drought constraints, still keep a phenological rhythm that rather corresponds to a tropical-like climate (with summer precipitations), as a kind of ancestral linkage to the environmental conditions under which their phyletic radiation took place (ref. 110).

In spite of all the caution that has inspired this chapter in dealing with such a debatable issue, it can be said, at least, that ecosystems are "historical entities", in so far as they are partly determined by their history and show sensitivity to the initial conditions.

According to the temporal scale of resolution, one can be more impressed by chance and chaos as controlling ecosystem dynamics (if events are observed on a short-term perspective), or by determinism and order when a long scale of perception is adopted to apprehend, for instance, the end result of

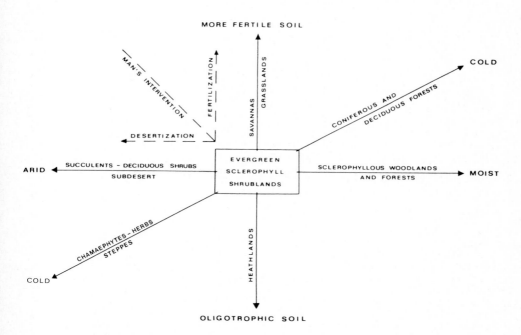

Fig. 5. Determinants of the mediterranean-type ecosystems shown in Figs. 3 and 4, in environmental gradients controlled by aridity, low temperatures and soil nutrients. The parallelism with some human impacts, as factors determining ecosystem dynamics, is outlined in the upper left of the figure. After di Castri, 1981 (ref. 99), modified.

ecosystem convergence. As a matter of fact, both chance and necessity, chaos and determinism, are inextricably acting at any given moment (refs. 76-75-1). It is worth wondering if -after all- the classical controversies between Clements' and Gleason's schools of thought (refs. 56-57-58-59) should not be attributable to mere problems of research and observational scales.

It is less important to question whether or not ecosystems -in their dynamics and evolution- behave like "individuals". Taking into account the looseness and diffuseness of ecosystems as regards boundaries and mechanisms of control, they are not "an organism". Above all, they do not "reproduce" like individuals. However, if a "drifting series of correlated events" -in physical terms- corresponds to a process, it can be said that the evolution of ecosystems constitutes "a reproduction of processes".

Nevertheless, analogies should not be pushed too far. Ecological successions (e.g., ecosystem development) can be considered as nested in ecosystem evolution, and -to a certain extent- they show "recapitulation" features. In other terms, in the evolution of ecosystems, large scale expressions of developmental constraints are detectable. However, it would not be advisable to step over a semantic threshold, and to refer to ontology (development) and phylogeny (evolution) of ecosystems. It would be still too easy to point out supposed phylogenetic sequences from tundras to taigas, from grasslands to temperate forests, or from tropical shrublands and savannas to rainforests.

Do ecosystems evolve? The reply can be mildly affirmative, in so far as it involves the evolution of hierarchically nested ecological systems (ref. 111): nested in **space** and as regards levels of organization, by considering disaggregations in communities and populations or aggregations in landscapes and biomes; nested in **time**, with due account taken of intraspecific variability, speciation and extinction, emerging of coevolutionary and mutualistic patterns, coalescence of species assemblages and networks (food webs), and increasing heterogeneity of soil and microclimate through "action and reaction" loops of abiotic and biotic components of the ecosystem.

1.8 DO ECOSYSTEMS CONVERGE?

Some hints or indirect evidence to attempt replying to questions posed earlier in this chapter can be provided by observations and studies on ecosystem convergence. It is undeniable that, all around the world, and in similar climatic zones (ref. 112), there are grossly comparable ecosystem types (e.g., deserts, steppes, tundras, taigas, deciduous forests, grasslands, tropical savannas, tropical and temperate rainforests, etc.). These are the so-called biomes.

It would be -again- highly improbable that these geographic-ecological convergences had occurred only by chance, and not because of similar selective forces or climatic determinants acting at the ecosystem level.

Incidentally, convergence of species and ecosystem types deeply impressed explorers at the time of the Great Discoveries (1500 A.D.). They had no hesitation to give to "new" plant formations the same name already used for similar formations in their native countries.

Nevertheless, abuses have been made in giving too deterministic a nature to such convergences. In fact, when these "convergent ecosystems" are studied in a more analytical way, and at a greater level of resolution, dissimilarities sharply increase (ref. 99).

In the light of earlier discussions in this chapter, it may be said straight off that convergence occurs because there are similar selective determinants (the "environmental necessity"), and divergence because of different phylogenetic pools available (the "historical chance"). Nevertheless, problems of convergence should be tackled with greater nuance.

For instance, similarities cannot merely be attributable to evolutionary convergence, but also to common phylogenetic origin of dominant species of ecosystems, as well as to similar human impacts having started in early historical periods. This implies a careful biogeographical analysis, ranging from plate tectonics (ref. 113) to human history (ref. 98).

It should also be stressed that component species of a given ecosystem can have had a different phylogenetic history and have been subjected to different selective forces and disturbance regimes. A typical sclerophyllous woodland ecosystem of Central Chile can be constituted, for instance, by soil organisms of paleoantarctic origin, shrubs and trees of neotropical origin, herbs having evolved *in situ* (or introduced in historical times, as invaders, from the Mediterranean Basin), and epigeous invertebrates having penetrated from subdesertic neighbouring ecosystems. Still, there is connectedness between components having such a different history, and the system has reached as great a resilience and resource optimization as possible.

Finally, the degree of convergence between intercontinental ecosystems may largely vary when different parts of the ecosystems are taken into account (e.g., vegetation, soil communities, bird guilds, etc.).

To tackle the problem of convergent and divergent evolution of ecosystems, analysis can be made on temporal and spatial (diachronic) comparisons. A combination of both approaches is highly desirable.

Fig. 6 takes up most of the steps illustrated in Fig. 2. The difference lies in the fact that Fig. 6 refers to a climatic cycle, that is to say, that there are comparable disturbance regimes at the beginning and at the end of the evolutionary period. Ecosystem properties at the closing cycle are not identical to those of the initial step, in spite of the fact that they respond to similar disturbance. First of all, there is a large stochastic component in the ecosystem response to a given disturbance. In other words, it is dealt with non-linear dynamics that may give room to varying responses and adaptive patterns when facing a similar selective force. Secondly, during the intermediate steps between the beginning and the end of the climatic

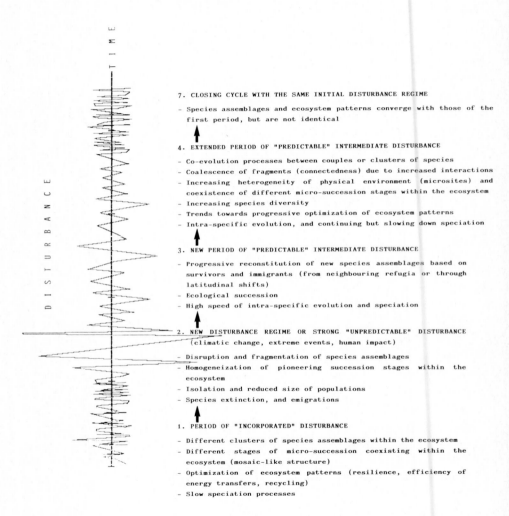

Fig. 6. Steps of ecosystem evolution within a climatic cycle, where disturbance regimes of step 1 and 7 are comparable. Ecosystem properties at the end of the cycle converge with those of step 1, but are not identical. This occurs due to the stochastic component of system response to disturbance, and to change of species composition during the intermediate steps.

cycle, new species have joined or left the ecosystem, new species have appeared, while others have become extinct; above all, new species assemblages have been established, thus creating new networks of interactions within the ecosystem. The above statements are somewhat backed by the scientific evidence of pollen profiles, showing that species composition and ecosystem features are never absolutely the same at the time of similar peaks of climatic cycles.

Therefore, it could be postulated that, given an infinite period of parallel disturbance regimes and selective forces acting on different ecosystems, the degree of convergence between them should approach that of an identical functioning, but such identity will never be totally achieved.

Looking at Figs. 2 and 6, similar patterns of ecosystem succession ("development") and ecosystem evolution come out. As already pointed out, patterns are on a different temporal scale, since ecosystem evolution comprises a whole series of successional cycles. To a certain extent, ecological succession represents a kind of recapitulation of the overall process of ecosystem evolution, since sequences leading to successive species assemblages and networks follow a comparable mode.

The overall aspect of ecosystem convergence and divergence has been approached since the early 70s, through intercontinental comparisons of mediterranean-climate ecosystems (refs. 114-115). These ecosystems have been selected in an attempt to answer the question on "whether very similar physical environments, acting on phylogenetically dissimilar organisms in different parts of the world, will produce structurally and functionally similar ecosystems" (ref. 114). The advantage of these ecosystems as "biological model" is that they are (and they have always been) largely disjunct in space and time. In other words, they correspond to regions of analogous climate, but with native plants and animals of different biogeographic origin. It would be out of the context of this chapter to give detailed conclusions on these intercontinental comparisons that are still going on, and that have given rise to a dozen of books and hundreds of publications. Only a few of them are quoted here (refs. 114-115-99-116-117).

At all events, evidences of ecosystem convergence have been demonstrated in features of primary production, species diversity, niche partitioning of various plant and animal groups, and resource allocation of comparable life forms.

Evolutionary divergences have also been outlined for these ecosystem types, for instance, as regards soil communities (ref. 118) and bird communities (ref. 119). They are primarily attributable to the weight of historical (phylogenetic) heritage. Indeed, divergences are quite foreseeable in accordance with the steps of ecosystem evolution discussed earlier in this chapter.

Moreover, the "noise" represented by factors such as phylogenetic commonalities, ecosystem resistance to change, and -above all- the now overwhelming human impact makes it more difficult to recognize "real" evolutionary convergences, as opposed to mere similarities or coincidences.

The degree of similarity between the five mediterranean-climate regions of the world is depicted in Fig. 7. Only some of these similarities are due to evolutionary convergence. It should be stressed again how intricate it is to determine convergences, even as related to these ecosystems that have been traditionally considered physiognomically as the most similar ones, despite their wide geographical separation.

Convergence of ecosystems has also been demonstrated in marine environments (refs. 120-121-122). Admittedly, singling out these convergences is even more difficult for oceanic ecosystems, because disjunction cannot be

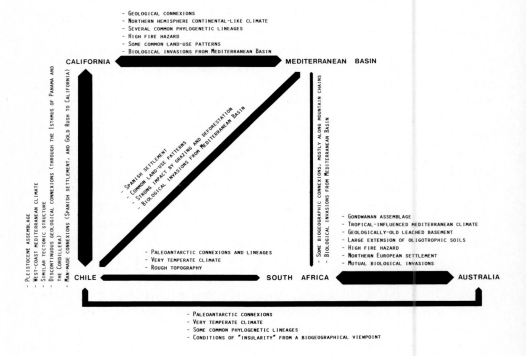

Fig. 7. Similarities among the five regions of the world with mediterranean-type climates, due to geological and topographic features, evolutionary convergence, phylogenetic commonalities, and parallel human impacts. The degree of similarity is proportional to the thickness of connecting bars. After di Castri, 1981 (ref. 99), largely modified and redesigned.

defined as sharply as for emerged lands, and differences in disturbance regimes are more subtle.

1.9 HOW FAR DO ECOSYSTEMS OPTIMIZE RESOURCES?

Resource allocation and optimization can also be envisaged at the ecosystem level. For instance, a fire-prone mediterranean shrubland ("garrigue") invests mostly in the root system and below-ground biomass, while investment of a tropical rainforest on poor leached soil concentrates on the above-ground biomass (and on development of superficial rootlets-mycorrhizae mutualistic relations to speed up nutrient cycling). Optimization can also refer to features leading to a greater ecosystem stability. Cody (ref. 123) has reviewed the overall problem of optimization in ecology, Watt (ref. 124) that of efficiency in metabolic organization (and the ecosystem shows a metabolic functioning), while Loucks (ref. 20) has discussed the evolution of efficiency properties.

Consideration has already been given to the limitations of the optimization process within ecosystem development and evolution, because of stochasticism, changes of disturbance regime, and weight of historical factors. It should be underlined that -at all events- there is not a kind of "block evolution" acting on the entire ecosystem. Therefore, different parts of an ecosystem may show different degrees of optimization. This is also due to the fact that ecosystem components can be dephased in time as regards their adaptation to present selective forces and resources from a given initial situation.

Accordingly, and with due consideration given to past constraints and opportunities, a natural ecosystem does not necessarily represent the best possible optimization of available resources.

It is not -therefore- impossible for man to increase optimization of ecosystem functioning, by modifying species composition, introducing new life forms (e.g., a shrub layer in arid steppic ecosystems), and -in general- acting on patterns of resource allocation among different ecosystem components. This process can be defined as the "domestication" of ecosystems by man, since there are several analogies with the process of plant and animal species domestication, which fundamentally consists in changing resource allocation patterns by selection of breeds.

There is another indirect evidence that not all natural ecosystems are at the best of their optimization in response to a given disturbance regime. This evidence is provided by studies on the invasion by alien species -intentionally or inadvertently introduced by man-, and on ecosystem resistance to such invasion. Several natural ecosystems are very vulnerable to invasion by aliens, thus demonstrating that their connectedness is not tightly shaped, and that new species can find a place within the ecosystem organization (ref. 98). This is, in particular, the case of ecosystems with an "insularity syndrome" (e.g., small islands, biogeographically separated zones as Australia and the southernmost parts of South America and Africa, etc.). Put in other terms, in the evolutionary process of building-up these ecosystems, a reduced amount of available pieces (e.g., founding species) were present, because of biogeographical barriers. Having had less initial "opportunities", the final state of optimization may be less advanced than that of ecosystems having been historically "open" to receive species with different phylogenetic background and biogeographical origin.

1.10 CONCLUDING REMARKS

Ecosystems, as compared with other biological units which unquestionably evolve, show several peculiarities (ref. 111). First, their biological components are individuals, populations, multi-species assemblages and communities, that have become part of the ecosystem at different time periods and according to different phylogenetic histories. Second, the diverse species assemblages (or "networks") have been and still are subject to different selective forces and disturbance regimes. Third, ecosystem boundaries are very loosely defined, and strongly fluctuate in space and time. Fourth, and unlike all lower ecological units, ecosystems include both a biotic and an abiotic component, connected by flows of energy and matter.

Under these conditions, it is not a rhetoric question to debate whether ecosystem evolve or whether they are merely individualistic epiphenomena (ref. 8).

In spite of that, there are several indirect evidences that ecosystems are evolving units, while the meaning of evolution cannot be strictly the same as that of genetic evolution. These evidences are based on constraints on the number of trophic levels of the food chains as applied to all ecosystem types, on comparable patterns in all ecosystems as regards size classes of component animal species (ref. 92), and on the development of mechanisms of regulation of species diversity and resource allocation (refs. 21-22).

Furthermore, studies on convergence (and divergence) of spatially and temporally disjunct ecosystems provide evidence of evolutionary dynamics, where environmental necessity and historical chance are inextricably blended (refs. 99-111). In addition, the degree of resistance (or non-resistance) of ecosystems to invasion by alien species may give rise to hypotheses on the connectedness and resource optimization of a given ecosystem, in relation to its evolutionary constraints and opportunities (ref. 98).

Accepting (or not accepting) the reality of ecosystems as evolving units also has practical implications. These implications include the need of a systemic approach and a multi-level research design for tackling problems such as the accumulation of pesticides and pollutants along food chains, or the building-up of more productive -or more robust- ecosystems (e.g., through agroforestry) by changing patterns of resource allocation. Above all, understanding limits and potentials for extrapolating research results from one ecosystem to another requires a critical apprehension of the nature of ecosystems.

To further advance towards an evolutionary view of ecosystems, and to get more hints to face the several questions posed in this chapter, a blend of two theoretical approaches should be achieved: hierarchical theory and theory of dissipative structures (refs. 63-75), where chaos and order, stochasticism and determinism are relevant modes of ecosystem evolution.

In addition, more effort should be made to plan experimental research on ecosystem dynamics and evolution. This research can only be conceived by combining the evolutionary ecology approach, based on population biology and community dynamics (and information networks), with the energy-budget functional approach based on energy and matter flows (refs. 1-125-126). Linkages should also be increased with molecular biology and thermodynamics, on the one extreme, and with atmospheric and geological sciences, on the other extreme.

It remains to be stressed that the term ecosystem is still too ambiguously adopted to coalesce all different subdisciplines of ecology (ref. 127). Up to now, ecosystems have too often been studied practically excluding the other units of ecology.

Hierarchical theory (refs. 34-127-128-32) and the notion of scale (refs. 129-130) have been the *leit motiv* of this chapter. More than 20 % of the publications quoted here have scales and hierarchies as their prime focus (refs.1-32-34-35-37-38-39-41-45-47-51-53-63-74-81-82-90-97-98-99-111-121-122-126-127-128-129-130). This reflects an undeniable change in present trends of ecology and evolution.

Hierarchical theory has already proved its value -at least from heuristic and semantic viewpoints- for strengthening linkages between the somewhat conflicting subdisciplines of ecology, and helping the construction of a unitary scientific logic. Fig. 8 illustrates in a very schematic way some possible linkages. They have been further developed operationally in ecological research, and even in institutional restructuring (ref. 130). Looking at a higher hierarchical level in spatial and temporal scales allows apprehending relevance and constraints of the problem under study. Descending towards lower hierarchical levels provides ways and means to single out explanatory mechanisms. This "zooming up and down" approach that allows decomposing ecosystems into their nested parts, as well as integrating

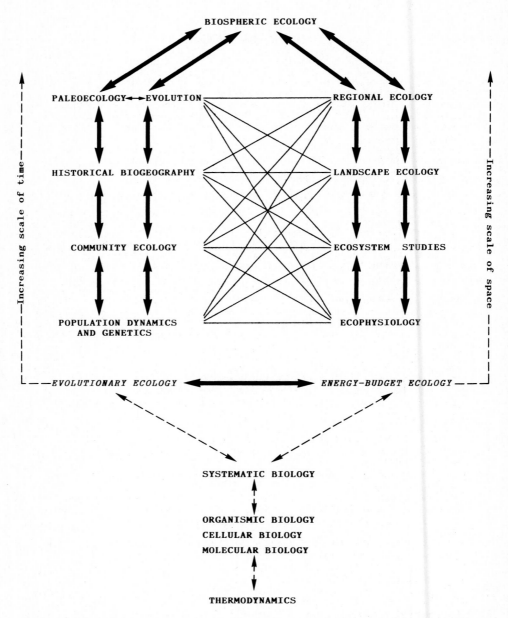

Fig. 8. The engine of ecological subdisciplines put at work by strengthening linkages along and across hierarchical scales. From di Castri, 1987 (ref. 1).

ecosystems in larger units from landscapes up to the biosphere, seems to be a rewarding method to rehabilitate the notion of ecosystem and to enhance the credibility of ecology as a science (ref. 130).

REFERENCES

1. F. di Castri, Towards a common language from molecular biology to biospheric ecology?, Biology International, Special Issue 15 (1987) 3-9.
2. J.A. Wiens, On understanding a non-equilibrium world: Myth and reality in community patterns and processes, in D.R. Strong Jr., D. Simberloff, L.G. Abele and A.B. Thistle (Editors), Ecological Communities: Conceptual Issues and the Evidence, Princeton Univ. Press, Princeton, New Jersey, 1984, pp. 439-457.
3. D. Cowan, Non-linear phenomena in physics and biology, Science, 217 (1982) 624-625.
4. G.L. Stebbins and F.I. Ayala, Is a new evolutionary synthesis necessary?, Science, 213 (1981) 967-971.
5. R.H. Peters, Tautology in evolution and ecology, Am. Nat., 110 (1976) 1-12.
6. S. Rice, Creationist Ecology?, Bull. Ecological Society of America, 67 (1986) 8-10.
7. C.F. Jordan, Do ecosystems exist?, Am. Nat., 118 (1981) 284-287.
8. B. Walker, Can ecosystems evolve or are they merely epiphenomena?, in E.S. Vrba (Editor), Species and Speciation, Transvaal Museum Monograph N° 4, Pretoria, 1985, pp. 173-176.
9. A.G. Tansley, The use and abuse of vegetational concepts and terms, Ecology, 16 (1935) 284-307.
10. R.P. McIntosh, The background of ecology. Concept and Theory, Cambridge Univ. Press, Cambridge, 1985, pp. 1-383.
11. B.C. Patten, Ecosystem as a co-evolutionary unit: A theme for teaching systems ecology, in G.S. Innis (Editor), New directions in the analysis of ecological systems. Part I, Society for Computer Simulation, La Jolla, California, 1975, pp. 1-8.
12. B.C. Patten, Systems approach to the concept of environment, Ohio J. Sci., 78 (1978) 206-222.
13. B.C. Patten, Environs: Relativistic elementary particles for ecology, Am. Nat., 119 (1982) 179-219.
14. J. Engelberg and L.L. Boyarsky, The noncybernetic nature of ecosystems, Am. Nat., 114 (1979) 317-324.
15. R.L. Knight and D.P. Swaney, In defense of ecosystems, Am. Nat., 117 (1981) 991-992.
16. B.C. Patten and E.P. Odum, The cybernetic nature of ecosystems, Am. Nat., 118 (1981) 886-895.
17. S.J. McNaughton and M.B. Coughenour, The cybernetic nature of ecosystems, Am. Nat., 117 (1981) 985-990.
18. E.P. Odum, The strategy of ecosystem development, Science, 164 (1969) 262-270.
19. R. Margalef, La Biosfera, entre la termodinámica y el juego, Ed. Omega, Barcelona, 1980, pp. 1-236.
20. O.L. Loucks, Evolution of diversity, efficiency, and community stability, Am. Zoologist, 10 (1970) 17-25.
21. J.H. Connell and E. Orias, The ecological regulation of species diversity, Am. Nat., 903 (1964) 399-414.
22. R. Margalef, Communication of structure in planktonic populations, Limnol. Oceanogr., 6 (1961) 124-128.

23   A.J. Gray, M.J. Crawley and P.J. Edwards (Editors), Colonization, Succession and Stability, Blackwell Scient. Publ., Oxford, 1987, pp. 1-482.
24   S.L. Pimm, The complexity and stability of ecosystems, Nature, 307 (1984) 321-326.
25   P.M. Vitousek and L.R. Walker, Colonization, succession and resource availability: Ecosystem-level interactions, in A.J. Gray, M.J. Crawley and P.J. Edwards (Editors), Colonization, Succession and Stability, Blackwell Scient. Publ., Oxford, 1987, pp. 207-223.
26   R.M. May, An overview: Real and apparent patterns in community structure, in D.R. Strong Jr., D. Simberloff, L.G. Abele and A.B. Thistle (Editors), Ecological Communities: Conceptual Issues and the Evidence, Princeton Univ. Press, Princeton, New Jersey, 1984, pp. 3-16.
27   P.R. Ehrlich and P.H. Raven, Butterflies and plants: A study in coevolution, Evolution, 18 (1964) 586-608.
28   D.H. Janzen, When is it coevolution?, Evolution, 34 (1980) 611-612.
29   M. Huston and T. Smith, Plant succession: Life history and competition, Am. Nat., 130 (1987) 168-198.
30   M. Rejmánek, Perturbation-dependent coexistence and species diversity in ecosystems, in P. Schuster (Editor), Stochastic phenomena and chaotic behaviour in complex systems, Springer-Verlag, Berlin, 1984, pp. 220-230.
31   P. Lavelle, Associations mutualistes avec la microflore du sol et richesse spécifique sous les tropiques: l'hypothèse du premier maillon, C.R. Acad. Sc. Paris, 302 (1986) 11-14.
32   D.I. Urban, R.V. O'Neill and H.H. Shugart Jr., Landscape Ecology: A hierarchical perspective can help scientists understand spatial patterns, BioScience, 37 (1987) 119-127.
33   R.L. Burgess and D.M. Sharpe (Editors), Forest island dynamics in man-dominated landscapes. Ecological Studies 41, Springer-Verlag, New York, 1981, pp. 1-310.
34   S.N. Salthe, Evolving Hierarchical Systems, Their Structure and Representation, Columbia Univ. Press, New York, 1985, pp. 1-343.
35   F. di Castri, A. Hansen and M. Holland (Editors), A new look at ecotones: Emerging international projects on landscape boundaries, Biology International, Special Issue 17 (1988), in press.
36   G.E. Hutchinson, The Ecological Theater and the Evolutionary Play, Yale Univ. Press, New Haven, 1965, pp. 1-139.
37   E.S. Vrba and N. Eldredge, Individuals, hierarchies and processes: Towards a more complete evolutionary theory, Paleobiology, 10 (1984) 146-171.
38   P.H. Gouyon and C. Gliddon, The genetics of information and the evolution of avatars, in G. de Jong (Editor), Population Genetics and Evolution, Springer-Verlag, Berlin, 1988, pp. 119-123.
39   J. Damuth, Selection among "species": A formulation in terms of natural functional units, Evolution, 39 (1985) 1132-1146.
40   R.C. Lewontin, The units of selection, Ann. Rev. Ecol. Syst., 1 (1970) 1-18.
41   N. Eldredge, Information, Economics, and Evolution, Ann. Rev. Ecol. Syst., 17 (1986) 351-369.
42   T.M.L. Wigley, Impact of extreme events, Nature, 316 (1985) 106-107.
43   E.T. Degens, H.K. Wong and S. Kempe, Factors controlling global climate of the past and the future, in G.E. Likens (Editor), Some perspectives of the major biogeochemical cycles, Wiley, Chichester, 1981, pp. 3-24.
44   A. Berger, J. Imbrie, J. Hays, G. Kukla and B. Saltzman (Editors), Milankovitch and Climate. Understanding the response to astronomical forcing, D. Reidel Publ. Co., Dordrecht, 1984, pp. 1-510.
45   N.A. Mörner and W. Karlén (Editors), Climatic Changes on a Yearly to Millennial Basis, D. Reidel Publ. Co., Dordrecht, 1984, pp. 1-667.
46   E.S. Vrba, Macroevolutionary trends: New perspectives on the roles of adaptation and incidental effect, Science, 221 (1983) 387-389.

47　E.S. Vrba, Environment and evolution: alternative causes of the temporal distribution of evolutionary events, South African Journal of Science, 81 (1985) 229-236.
48　S. Wright, Statistical genetics in relation to evolution, Introduction, Hermann, Paris, 1939, pp. 5-6.
49　P.H. Raven, The evolution of Mediterranean floras, in F. di Castri and H.A. Mooney (Editors), Mediterranean Type Ecosystems. Origin and Structure. Ecological Studies 7, Springer-Verlag, Berlin, 1973, pp. 213-224.
50　H.R. Delcourt, P.A. Delcourt and T. Webb III, Dynamic and plant ecology: The spectrum of vegetational change in space and time, Quaternary Science Reviews, 1 (1983) 153-175.
51　P.A. Delcourt and H.R. Delcourt, Long-term Forest Dynamics of the Temperate Zone, Ecological Studies 63, Springer-Verlag, New York, 1987, pp. 1-439.
52　S. Pignatti, Evolutionary trends in Mediterranean flora and vegetation, Vegetatio, 37 (1978) 175-185.
53　J. Blondel, From biogeography to life history theory: a multithematic approach illustrated by the biogeography of vertebrates, Journal of Biogeography, 14 (1987) 405-422.
54　D.I. Axelrod, History of the mediterranean ecosystem in California, in F. di Castri and H.A. Mooney (Editors), Mediterranean Type Ecosystems. Origin and Structure. Ecological Studies 7, Springer-Verlag, Berlin, 1973, pp. 225-277.
55　J.P. Suc, Origin and evolution of Mediterranean vegetation and climate in Europe, Nature, 307 (1984) 429-432.
56　F.E. Clements, Plant succession, Carnegie Inst. Wash. Publ., 242 (1916) 1-512.
57　H.A. Gleason, The structure and development of the plant association, Bull. Torrey Bot. Club, 44 (1917) 463-481.
58　H.A. Gleason, The individualistic concept of the plant association, Bull. Torrey Bot. Club, 53 (1926) 7-26.
59　H.A. Gleason, Further views on the succession concept, Ecology, 8 (1927) 299-326.
60　W.H. Drury and I.C.T. Nisbet, Succession, The Arnold Arbor. J., 54 (1973) 331-368.
61　J.H. Connell and R.O. Slatyer, Mechanisms of succession in natural communities and their role in community stability and organization, Am. Nat., 111 (1977) 1119-1144.
62　P.M. Vitousek and W.A. Reiners, Ecosystem succession and nutrient retention: A hypothesis, BioScience, 25 (1975) 376-381.
63　C.S. Holling, Simplifying the complex: The paradigms of ecological function and structure, European Journal of Operational Research, 30 (1987) 139-146.
64　H.H. Shugart, A Theory of Forest Dynamics. The ecological implications of forest succession models, Springer-Verlag, New York, 1984, pp. 1-278.
65　D. Tilman, The resource-ratio hypothesis of plant succession, Am. Nat., 125 (1985) 827-852.
66　S.T.A. Pickett, S.L. Collins and J.J. Armesto, Models, Mechanisms and Pathways of Succession, The Botanical Review, 53 (1987) 335-371.
67　J. Lepart and J. Escarre, La succession végétale, mécanismes et modèles: Analyse bibliographique, Bull. Ecol., 14 (1983) 133-178.
68　P. Dansereau, Biogeography - An ecological perspective, Ronald Press, New York, 1957, pp. 1-392.
69　R.H. Whittaker and G.M. Woodwell, Evolution of natural communities, in J.A. Wiens (Editor), Ecosystem structure and evolution, Oregon State Univ. Press, Corvallis, 1972, pp. 137-159.
70　E.R. Pianka, Latitudinal gradients in species diversity: a review of concepts, Am. Nat., 100 (1966) 33-46.

71 A.R. Templeton and E.D. Rothman, Evolution in heterogeneous environments, Am. Nat., 108 (1974) 409-428.
72 E.C. Olson, Community evolution and the origin of mammals, Ecology, 47 (1966) 291-302.
73 A.J. Boucot, Community evolution and rates of cladogenesis, Evol. Biol., 11 (1978) 545-655.
74 J.A. MacMahon, D.L. Phillips, J.V. Robinson and D.J. Schimpf, Levels of biological organization: An organism-centered approach, BioScience, 28 (1978) 700-704.
75 I. Prigogine and I. Stengers, Order out of Chaos, Bantam Books, Toronto, 1984, pp. 1-349.
76 A.L. Robinson, Physicists try to find order in chaos, Science, 218 (1982) 554-556.
77 S. Frautschi, Entropy in an expanding universe, Science, 217 (1982) 593-599.
78 S.J. Gould and N. Eldredge, Punctuated equilibria: The tempo and mode of evolution reconsidered, Paleobiology, 3 (1977) 115-151.
79 S.I. Gould, Darwinism and the expansion of evolutionary theory, Science, 216 (1982) 380-387.
80 E.S. Vrba, What is species selection?, Syst. Zool., 33 (1984) 318-328.
81 N. Eldredge and S.N. Salthe. Hierarchy and Evolution. Oxford Surveys, Evol. Biol., 1 (1984) 182-206.
82 P.H. Gouyon, C.J. Gliddon and D. Couvet, The evolution of reproductive systems: A hierarchy of causes, in Plant Population Ecology, Symposium of the British Ecological Society, 1988, in press.
83 N. Eldredge and J. Cracraft, Phylogenetic patterns and the evolutionary process, Columbia Univ. Press, New York, 1980, pp. 1-350.
84 D.B. Wake, A View of Evolution, Science, 210 (1980) 1239-1240.
85 G.J. Vermeij, Evolution and Escalation, Princeton Univ. Press, Princeton, New Jersey, 1987, pp. 1-527.
86 K.J. Hsü, Environmental changes in times of biotic crisis, in D.M. Raup and D. Jablonski (Editors), Patterns and processes in the history of life, Springer-Verlag, Berlin, 1986, pp. 297-312.
87 D. Jablonski, Evolutionary Consequences of Mass Extinctions, in D.M. Raup and D. Jablonski (Editors), Patterns and processes in the history of life, Springer-Verlag, Berlin, 1986, pp. 313-329.
88 T.F. Malone and J.G. Roederer (Editors), Global Change, Cambridge Univ. Press, Cambridge, 1985, pp. 1-510.
89 R.P. Detwiler and C.A.S. Hall, Tropical forests and the global carbon cycle, Science, 239 (1988) 42-47.
90 M.I. Dyer, F. di Castri and A.J. Hansen (Editors), Geosphere-Biosphere Observatories. Their definition and design for studying global change, Biology International, Special Issue 16 (1988) 1-40.
91 F.E. Smith, Ecosystems and Evolution, Bulletin Ecological Society of America, 56 (1975) 2-6.
92 R.M. May, The evolution of ecological systems, Scientific American, September Issue (1978) 118-133.
93 P. Blandin, Evolution des écosystèmes et spéciation: le rôle des cycles climatiques, Bull. Ecol., 18 (1987) 59-61.
94 P. Blandin and M. Lamotte, Ecosystèmes et Evolution, Le Courrier du CNRS N° 59 (1985) 25-33.
95 R.P. McIntosh, Ecosystems, evolution and relational patterns of living organisms, American Scientist, 51 (1963) 246-267.
96 R.M. Darnell, Evolution and the Ecosystem, Am. Zoologist, 10 (1970) 9-15.
97 F. di Castri, L'organisation du vivant et l'environnement, Chapitre III, in Sourcebook pour une éducation relative à l'environnement dans l'enseignement universitaire général, Unesco, Paris, 1988, in preparation.

98   F. di Castri, History of biological invasions, with special emphasis on the Old World, in J.A. Drake, F. di Castri, R.H. Groves, F.J. Kruger, H.A. Mooney, M. Rejmánek and M.H. Williamson (Editors), Biological Invasions: A global perspective, Wiley, Chichester, 1988, in press.
99   F. di Castri, Mediterranean-type shrublands of the World, in F. di Castri, D.W. Goodall and R.L. Specht (Editors), Mediterranean-type Shrublands, Ecosystems of the World 11, Elsevier, Amsterdam, 1981, pp. 1-52.
100 G.S. Puri, The forest/savanna boundary changes interpreted as ecosystem evolution, Tropical Ecology, 6 (1965) 88-97.
101 R. Lewin, Paleoclimates in Southern Africa, Science, 227 (1985) 1325-1327.
102 P.W. Rundel, The matorral zone of Central Chile, in F. di Castri, D.W. Goodall and R.L. Specht (Editors), Mediterranean-type Shrublands, Ecosystems of the World 11, Elsevier, Amsterdam, 1981, pp. 175-201.
103 B.B. Simpson and J. Haffer, Speciation patterns in the Amazonian forest biota, Ann. Rev. Ecol. Syst., 9 (1978) 497-518.
104 J. Salo, R. Kalliola, I. Häkkinen, Y. Mäkinen, P. Niemelä, M. Puhakka and P.D. Coley, River dynamics and the diversity of Amazon lowland forest, Nature, 322 (1986) 254-258.
105 M.J. Dunbar, The evolution of stability in marine environments. Natural selection at the level of the ecosystem, Am. Nat., 94 (1960) 129-136.
106 J.H. Steele, A comparison of terrestrial and marine ecological systems, Nature, 313 (1985) 355-358.
107 P. Blandin, Evolution des écosystèmes et stratégies cénotiques, in R. Barbault, P. Blandin and J.A. Meyer (Editors), Recherches d'écologie théorique: Les stratégies adaptatives, Maloine, Paris, 1980, pp. 221-235.
108 P. Blandin, R. Barbault and C. Lecordier, Réflexions sur la notion d'écosystème: le concept de stratégie cénotique, Bull. Ecol., 7 (1976) 391-410.
109 R. Lewin, Plant communities resist climatic change, Science, 228 (1985) 165-166.
110 R.L. Specht, Structure and functional response of ecosystems in the mediterranean climate of Australia, in F. di Castri and H.A. Mooney (Editors), Mediterranean Type Ecosystems. Origin and Structure. Ecological Studies 7, Springer-Verlag, Berlin, 1973, pp. 113-120.
111 F. di Castri, Evolution of hierarchically nested ecological systems, Workshop on self-organization processes, emerging properties and biological complexity, Paris, France, April 7-9, 1988, IUBS-ICSU, in preparation.
112 G. Rougerie, Géographie de la biosphère, Armand Colin, Paris, 1988, pp. 1-288.
113 G.C. Briggs, Biogeography and plate tectonics, Elsevier, Amsterdam, 1987, pp. 1-204.
114 F. di Castri and H.A. Mooney (Editors), Mediterranean Type Ecosystems. Origin and Structure. Ecological Studies 7, Springer-Verlag, Berlin, 1973, pp. 1-405.
115 H.A. Mooney (Editor), Convergent Evolution in Chile and California. Mediterranean Climate Ecosystems, Dowden, Hutchinson and Ross, Stroudsburg, Pennsylvania, 1977, pp. 1-224.
116 A.V. Milewski, A comparison of ecosystems in Mediterranean Australia and Southern Africa: Nutrient-poor sites at the Barrens and the Caledon Coast, Ann. Rev. Ecol. Syst., 14 (1983) 57-76.
117 E. Pignatti and S. Pignatti, Mediterranean type vegetation of SW Australia, Chile, and the Mediterranean basin, a comparison, Ann. Bot. (Roma), 43 (1985) 227-343.
118 F. di Castri and V. Vitali-di Castri, Soil fauna of mediterranean climate regions, in F. di Castri, D.W. Goodall and R.L. Specht (Editors), Mediterranean-type Shrublands, Ecosystems of the World 11, Elsevier, Amsterdam, 1981, pp. 445-478.

119 J. Blondel, F. Vuilleumier, L.F. Marcus and E. Terouanne, Is there ecomorphological convergence among Mediterranean bird communities of Chile, California, and France?, Evol. Biol., 18 (1984) 141-213.
120 M.L. Harmelin-Vivien, Etude comparative de l'ichtyofaune des herbiers de phanerogames marines en milieux tropical et tempéré, Rev. Ecol. (Terre Vie), 38 (1983) 179-210.
121 R. Galzin, Structure of fish communities of French Polynesian coral reefs. I. Spatial scales, Mar. Ecol. Prog. Ser., 41 (1987) 129-136.
122 R. Galzin, Structure of fish communities of French Polynesian coral reefs, II. Temporal scales, Mar. Ecol. Prog. Ser., 41 (1987) 137-145.
123 M.L. Cody, Optimization in ecology, Science, 183 (1974) 1156-1164.
124 W.B. Watt, Power and efficiency as indexes of fitness in metabolic organization, Am. Nat., 127 (1986) 629-653.
125 F.I. Woodward, Plant Ecology - Trends and Traits, Tree, 2 (1987) 252-254.
126 P. Jacquard and K. Urbanska, Population genetics, whole-plant physiology, and population biology: interactions with ecology, Acta Oecologica, Oecol. Plant., 9 (1988) 3-10.
127 R.V. O'Neill, D.L. DeAngelis, J.B. Waide and T.F.H. Allen, A Hierarchical Concept of Ecosystems, Princeton Univ. Press, Princeton, New Jersey, 1986, pp. 1-253.
128 J.S. Nicolis, Dynamics of Hierarchical Systems. An Evolutionary Approach, Springer-Verlag, Berlin, 1986, pp. 1-397.
129 J. Lawton, Problems of scale in ecology, Nature, 325 (1987) 206.
130 F. di Castri and M. Hadley, Enhancing the credibility of ecology: Interacting along and across hierarchical scales, GeoJournal, 1988, in press.

Ecological Assessment of Environmental Degradation, Pollution and Recovery,
Lectures of a course held at the Joint Research Centre, Ispra, Italy, 12–16 October
1987, O. Ravera (Ed.), pp. 31–53
© Elsevier Science Publishers B.V., Amsterdam — Printed in The Netherlands

ECOLOGICAL SUCCESSION AND EVOLUTION OF FRESHWATER ECOSYSTEMS

I. Ferrari and V.U. Ceccherelli
Istituto di Zoologia, Università di Ferrara
Ferrara 44100 - Italy

1. INTRODUCTION

Several qualified reviews of the major research lines regarding ecological succession and evolution of aquatic ecosystems have been produced. In particular, we refer to the fundamental ecology texts by Odum (ref.1) and Margalef (ref.2), as well as to the more recent text by Colinvaux (ref.3), and to the excellent books on limnology by Wetzel (ref.4) and Goldman and Horne (ref.5). Many of the statements of these authors have been emphasized in this article, with the aim of retracing the most significant contributions to modern ecology. Our greatest concern has been to associate an analysis of the problems inherent to succession and ecosystem evolution to a more general evaluation of the most successful approaches in ecological research. The recent book by McIntosh (ref.6) has been particularly stimulating for this purpose.

We have avoided a detailed presentation of results of the most significant research, which the reader can easily find in any textbook. Instead, we have preferred to utilize examples drawn from our own work, although certainly of more modest entity, carried out in different aquatic environments, particularly on seasonal succession of zooplankton. This choice has allowed us to link the discussion on general subjects to personal experience. We have grouped in an appendix six "cases" which are all relevant, in some aspects, to the matter of this article.

2. ECOLOGICAL SUCCESSION: THE CLEMENTS' VIEW

The first general formulation of the concept of succession is due to Clements (refs.7,8), an American botanist, whose thought has had a profound impact not only on the plant ecologists of his own generation and the following ones, but upon the entire ecological culture up to the present time. Clements was influenced by the original work of Cowles (ref.9) on the succession of plants in the sand dunes of Lake Michigan. His scheme of interpretation is rather simple: in the process of primary succession, the first colonizers, that form the pioneer stages, modify the habitat, making it unfavourable to themselves, and favourable to the other invading species, which will characterize, with their presence, the subsequent developmental stages (seres). Clements' views are effectively summarized by McIntosh (ref.6) as follows: "the universal tendency of the vegetation in a uniform climatic area was to converge from diverse bare areas to a climax stage which was controlled by the climate; the climax association was, barring external disturbances, a stable and self-reproducing collection of populations which was the culmination of the developmental sequence and formed a superorganism inclusive of the seres, from which it came".

The core of Clements' theory was that the community is a complex organism, and that it changes, like an individual organism, not in a random way but by progressive development. Thus, succession is viewed as a progressive, linear and sequential process through which vegetation reaches a stable state.

Some of Clements' contemporary ecologists did not appreciate totally his doctrine as it is documented by McIntosh (ref.6). According to Cowles, the succession is not a linear process; its stages may often be retrogressive. Cooper (ref.10) did not accept the idea of a linear succession and of a homogeneous climax community; he described the forest as a mosaic of patches of different ages, in a state of continuous change. Many ecologists of that generation did not share the idea of a community barring disturbances; instead, they believed that "disturbance was omnipresent, while equilibrium is an unlikely and ephemeral event" (ref.6).

Summing up, the basic themes concerning the exact succession mechanisms (that are still far from being settled) were already present in the debate among plant ecologists in the first decades of this century. In a sense, they were forerunners of ecosystems ecology, inasmuch as they were concerned with the larger systems transcending plants; their attention was focused on abiotic factors (water, nutrients, organic matter, light) and on the role of animals in initiating or inhibiting plant succession (ref.6).

Clements' emphasis on succession and on the community as a superorganism, in the context of a holistic vision of the universe, is not shared by the majority of present-day ecologists, but it permeated for a long time the major general ecology textbooks. Odum himself (ref.1), after defining succession as "an orderly, directional and predictable process", added: "the development of ecosystems has many parallels in the development biology of organisms...".

3. CHANGES IN ENERGY FLOW PARAMETERS DURING SUCCESSION

Lindeman is famous as the leader of the ecological energetics school. He has proposed the well-known model of energy flow through the trophic levels of ecosystems. In his fundamental essay of 1942 (ref.11), Lindeman's main concern was to describe the process of ecological succession in terms of community energetics. He assumed that, during the process of succession, productivity increases gradually, until it reaches a maximum in the complex vegetation of the climax. Measures of production made over the last forty years are adequate to refute Lindeman's hypothesis (ref.3).

Ecosystems ecologists who have most intensely strived towards a definition of rules of succession have been Odum (ref.1) and Margalef (refs.2,12-14). Firstly, in order to give a global characterization of the functioning of ecosystems, they have singled out some macroscopic energy parameters: gross primary production (P), average biomass (B) and total respiration of the community (R).

Odum re-proposed the classical distinction, due to Tansley (ref.15), between autogenic succession (influenced by processes "within the system") and allogenic succession (influenced by forces "outside the system"). In an autogenic succession and, in particular, in an autotrophic succession, the pioneer stages are characterized by a ratio $P/R > 1$ and by high P/B ratio. As the process evolves towards more mature stages, the biomass tends to increase, the P/B ratio tends to decrease, and P/R tends to 1. These trends may be observed both in a forest and in a microcosm succession, the only important difference lying in the time-scale: in a laboratory microecosystem the conditions of steady-state which a forest takes many years to achieve, are reached in only few weeks.

The evolution of energy flow parameters during succession is accompanied, according to Odum, by other significant changes: in pioneer stages food chains are linear and grazing predominates; in mature stages, they become longer (i.e. they comprise a greater number of trophic levels) and weblike, and detritus predominates. In the developmental stages, inorganic nutrients are generally extrabiotic, mineral cycles are open, and nutrient exchange rates between organisms and environment are high. In mature stages, nutrients become intrabiotic, their cycles tend to become closed and their exchanges between organisms and environment become slow.

Also Margalef singles out some regularities in the succession process, in particular an increase in biomass, especially of its less active portion, a decrease in concentrations of photosynthetic pigments and a greater determinism in the transfer systems of nutrients and energy as a consequence of increased environmental stability. Furthermore, both Odum and Margalef stress the development of homeostatic mechanisms, and the increase in biotic diversity as general trends of the succession.

4. SUCCESSION, DIVERSITY AND STABILITY

In the conceptual elaboration of Odum and Margalef, a basic point regards the analysis of changes, in the course of ecological succession, of diversity, assumed as a measure of the organization and complexity of a community.

Both authors point out that in the pioneer stages with high productivity, diversity - measured according to information theory - is low. On the other hand diversity is high in mature stages with high biomass values and a greater efficiency in energy utilization. This suggests the hypothesis that succession proceeds to maximize information, or order (ref.3).

In a sense, succession is a process of accumulation of information which, Margalef says, takes place through a transfer in time of the energy surplus available in the less mature stages. This energy is used to build up the organization, to increase biotic and biochemical diversity, to develop effective homeostasis mechanisms, and to increase the degree of auto-organization in ecosystems. An essential condition for succession to take place is that information may be transferred through time: in other words, the environment must be stable, the information transmitted must not be subjected to interferences, and noise must be negligible.

In aquatic environments, for instance, the plankton subsystem is intrinsically unstable, owing to water turbulence, its predictability is low, the random component prevails over the deterministic one. Diversity, that may be viewed as a measure of the information channel width, is not efficiently used by pelagic communities. Consequently, plankton is unable to increase its degree of maturity and to make its organization more complex and stable. Therefore, the plankton subsystem is permanently immature. The excess of planktonic production is transferred by sedimentation to the benthos, and goes to enrich the organization and increase the stability of the latter subsystem (refs.2,12).

Margalef (ref.14) proposed a general scheme of the relationships between diversity, productivity and stability in the most important aquatic and terrestrial ecosystems. According to his scheme, the human impact on the ecosystems determines regressive trends in the succession process. Agricultural systems and hydrophyte cultures are examples of regressed, unstable, highly productive and poorly diversified systems. The eutrophication of lakes is interpreted as a regression process, while natural evolution would lead towards a reduced production in bodies of water.

Conceptual models of the succession by ecosystems ecologists were fascinating but they have not gained from the popularity enjoyed over a number of years. An uncritical reading and a dogmatic reduction of the thinking of these authors have led to articles of faith and ill-founded assumptions taking root. For a long time it has been believed that an increase in stability is causally linked to an increased diversity, and it has been thought that a high diversity is, in all cases, a desirable property of ecosystems.

On the occasion of the first International Congress of Ecology (ref.16), the need was felt to revise critically and to define strictly concepts of extreme ambiguity, such as organization, maturity, diversity and stability.

May (ref.17) refused the view that complexity begets stability; the mathematical models devised by this author suggested that as a system becomes more complex it becomes more unstable. Jacobs (ref.18) argued that both components of diversity, defined by the Shannon-Wiener formula, richness and evenness, increase in initial stages of succession for mere statistical reasons. He went on to say that in later stages of succession the fate of diversity is not predictable at all. Jacobs gave also interesting examples that show that there is no obligatory causal connection between productivity (yield) and diversity.

Another separate line of thought about diversity-stability relationships arose from the research of evolutionary ecologists. The questions put forth by Hutchinson (ref.19) in his famous paper "Homage to Santa Rosalia" still remain without satisfactory answers (ref.20). However, some significant advances have been made towards a general theory of diversity. From a methodological point of view, a highly stimulating contribution came from the equilibrium theory of Island Biogeography by MacArthur and Wilson (ref.21). Further, Brooks and Dodson (ref.22) approach to size selective predation exerted by planktophagous organisms as a control factor of diversity in zooplankton communities has been very successful: adding predators to an ecosystem can increase diversity by reducing competition in the prey populations. Colinvaux (ref.3) names this phenomenon the "cropping principle".

More recently, Connell (ref.23) and Hubbel (ref.24) have developed some generalizations that have led to a new hypothesis on the mechanisms governing gradients, or clines, of diversity: the intermediate disturbance hypothesis. Connell, in particular, regarded the commonly observed high diversity of trees in tropical rain forests and of corals on tropical reefs as a non-equilibrium state which, if not disturbed further, will progress towards a low diversity equilibrium community. However, tropical forests and reefs are subjected to disturbances favouring different species which, otherwise, would be outcompeted by the dominant ones. From a more general point of view it has been suggested that disturbance may allow the coexistence of species with different life history strategies. At intermediate rates of disturbance, the colonizing species may be eliminated locally by the succession within patches of competitively superior species, but they may persist globally through the colonization of newly available resources in recently disturbed areas.

Huston (ref.25) proposed the general theory of the "dynamic equilibrium of diversity maintenance". He related maintenance of diversity to the interaction of two population parameters: the potential rate of population increase and the frequency of population reduction (by predators, adverse environmental conditions, or any other factor causing mortality). According to Huston's theory, diversity will be the highest when disturbance is just frequent enough to prevent competitive exclusion (but not to destroy completely species populations, of course).

Colinvaux (ref.3) devotes the last chapter of his ecology textbook to an analysis of geographical and historical patterns of species diversity, and to an outline of the main processes leading to diversity clines. The author states that any general connection between diversity and population stability is negative, and concludes that ecosystem stability depends to a large extent on environmental stability: the physical environment controls the processes setting species number.

5. SUCCESSION AS A CONSEQUENCE OF SPECIES STRATEGIES

According to the point of view of evolutionary ecologists, i.e. those concentrated on ecological and evolutionary interactions between species, the ordered development and the community organization revealed by succession can be explained as a result of coevolution and coexistence in a given area of species with different strategies and life histories. This is the "r and K hypothesis of succession" (refs.3,26). Essentially, succession is viewed as a function of invasion and extinction rates of species with different life histories: the species of pioneer communities are r-strategists; they will be displaced by the progressive invasion of K-strategy species. The final number of species at climax will depend upon the rates of continued invasion and local extinction. Evidently, this approach is an extension of MacArthur and Wilson model of Island Biogeography.

Colinvaux (ref.3) supports openly this hypothesis; he rejects Clements' point of view and stresses the soundness of the "individualistic hypothesis of succession" by Gleason (ref.27). He refuses also Margalef's models based on physical concepts as enthalpy and information, and about the Odum's model of a mature ecosystem, where gross primary production equals community respiration, he says: "To some ecologists this may seem a definition that lakcs interest, for there is no obvious reason why species replacements... should be influenced by the attainment of equilibrium between production and decomposition".

6. SUCCESSION IN AQUATIC ENVIRONMENTS

According to Margalef (ref.2), plankton seasonal succession due to life duration of the organisms involved can be considered homologous to the terrestrial succession taking place over tens of years. In particular, algal succession proceeds from the dominance of small sized forms during full circulation periods to prevalence of larger-sized forms, more adapted to the low nutrient concentrations, in stabilized waters during thermal stratification. The seasonal dynamics of zooplankton roughly follows the same pattern. Several evidences from research on seasonal change in biocoenotic indexes both in sea and in freshwaters are in accordance with this scheme.

However, in the past recent years, research on plankton succession shifted from a mere description of seasonal trends of communities to the analysis of factors and mechanisms which play a determinant role in structuring community, affecting biotic diversity and timing succession.

In the case of zooplankton, for example, the emphasis of research is upon the environmental determinants of the main life history traits (population dynamics, diapause, polymorphism ...) and upon biological interactions connecting species or trophic compartments. Structure and evolution of non-predator zooplankton are envisaged by some authors (ref.28) as largely influenced by the composition of food (live cells, detritus and bacteria), while the role of predation by fish and invertebrates as determinant factor in structuring zooplankton is stressed by some others (refs.4,29,30), resuming Brooks and Dodson model. These approaches stimulated investigations on behavioural aspects like feeding habits, prey defenses and so on and, in particular, on the adaptations to life in low Reynold's numbers conditions (volume edited by Kerfoot, ref.31). The development of this research may lead to devise suitable tests for the alternative hypotheses put forward to explain the paradox of the plankton (refs. 3,32,33).

Margalef (ref.2) dealt also with the subject of succession in running waters. River plankton is composed, generally, of taxa with high turnover rates (diatoms, ciliates, rotifers), typical of the pioneer stages of succession. Along the river course, the unidirectional flux of matter and energy (and of genotypes) corresponds to the temporal fluxes observed during succession. These ideas anticipate some of the guidelines of the more recent theory of the River Continuum (ref.34).

We shall omit to discuss the very important subject of succession in benthos communities. We shall only recall that a large number of studies have suggested the major effects of fish predation on population dynamics, productivity and seasonal succession of benthic invertebrates in lakes (ref.4).

In the appendix we have reported some results of our own research on zooplankton succession. The seasonal succession in a ricefield (Case 1) has been studied using methods derived from the Island Biogeography Theory. It suggests the classical trend from pioneer stages where primary production is principally sustained by phytoplankton and the main food chain is that of grazing, to a final phase where the detritus chain prevails.

Seasonal succession has been investigated also in a small eutrophic lake (Case 2). During summer we observed a progressive enhancement of the detritus component and a contemporary change in zooplankton structure: microfiltrators tend to replace macrofiltrators.

Cases 3 and 4 refer to time changes of rotifer taxocoenosis structure in small water bodies (a mountain lake, an experimental canal). The seasonal pattern appears strongly influenced by predation of Asplanchna upon microfiltrators.

Plankton succession in time and space in the River Po is analyzed in Case 5. In particular, the disturbance effect induced by a slight increase of the river flow upon taxocoenosis diversity during summer months has been examined.

Finally, in Case 6 we have studied the process of recolonization and accelerated eutrophication in a lagoon of the River Po Delta. The lagoon, in the previous period, had been affected for a long time by an inflow of fresh waters which prevented the development of an autochthonous strong zooplankton component.

7. LAKE EVOLUTION AND EUTROPHICATION

The long-term successional development of lake ecosystems is a process controlled by a number of complex factors affecting water production. The general successional trend is from lower to higher production (ref.4); long-term lake evolution results in a decrease of mean depth owing to sedimentation of both allochthonous and autochthonous materials. As the lake depth diminishes so does the volume of the hypolimnion: it follows that there will be a greater involvement of sediment-stored nutrients in the overall productive cycle of the lake and a decrease in dissolved oxygen content in the deeper layers of waters.

Both filling up and eutrophication (i.e. the process of increased productivity resulting from greater availability of essential plant nutrients such as nitrogen and phosphorus) would thus occur at the same time. Succession proceeds until lakes are obliterated and incorporated into the terrestrial landscape; in the late succession stages autochthonous primary production shifts from phytoplankton dominance to total dominance by littoral communities (ref.4).

The concept that most temperate lakes progress from oligotrophy to eutrophy is drawn from the typologistic limnology approach (according to Rigler, ref.35): yet it is not universally accepted. Margalef (ref.2) - as noted above - believes that the natural succession of lake over centuries proceeds from eutrophy to oligotrophy with a tendency towards increased structural complexity of the biocoenoses. Goldman and Horne (ref.5), on the contrary, state that "the main vegetation in lakes, the phytoplankton, is not necessarily specialized for a particular lake type; most species are quite generalized and are found in lakes with a variety of trophic conditions, although their relative abundance often reflects their trophic status".

This dispute caused lively debate among limnologists in the past years but now seems to be outdated. A thorough definition of conceptual terms such as eutrophic and oligotrophic, which are not easy to quantify, is certainly a more important task. Goldman and Horne suggest the use of those terms only in relation to other lakes within a single lake district.

Indeed, succession from oligotrophy to eutrophy and then to extinction is only one of the several possible trends of lake evolution; on the other hand, the generalization of a successional process towards oligotrophy is not particularly convincing. According to Goldman and Horne, for many lakes lifetimes will be more likely determined by geologic or climatic events than by the gradual sedimentation or inflow of nutrients. Most large deep lakes such as the Great Lakes or Lake Baikal will more likely maintain their present trophic state until they are destroyed by some catastrophic event.

Furthermore, paleolimnological records from deep undisturbed lakes have indicated strong fluctuations from oligotrophy to eutrophy and back; these alternating trophic conditions were documented by Horie (refs.36,37) for Lake Biwa, Japan, over the last half-million years.

In recent decades limnologists have had to deal with processes of accelerated evolution of trophic state of aquatic environments (cultural eutrophication and oligotrophication, according to Goldman and Horne). These phenomena, as is well known, are widespread in all industrialized countries and in the great urban areas throughout the world. Yet they are not without historical precedent; the famous case of eutrophication of lake Monterosi, due to the deforestation of its drainage basin for the construction of Via Cassia (ref.38), comes to mind.

We would like to emphasize the great contribution that research on eutrophication problems has made towards the understanding of the functioning and evolution of aquatic ecosystems. Firstly, standardization problems of criteria and methods of trophic state assessment were tackled. Further, in order to face problems of safeguarding and restoring lakes, fairly effective predictive models have been devised (refs.35,39-42). Finally, research on the role of communities in water bodies subjected to rapid evolution of trophic conditions has been greatly stimulated (refs.30,43-45).

8. CONCLUSIONS

The separateness of the two major schools of ecology has delayed the foundation of a theory of species diversity and community organization. Ecosystem ecologists ignored the diversity of species and the role that each of them plays; evolutionary ecologists, on the contrary, neglected energetics in the study of interactions among species (ref.20). The lack of a unifying theory of diversity has probably made it difficult to identify regularities and rules of successional processes. Yet in the past decade some promising research lines on the mechanisms governing community diversity and its distribution patterns have been developed (ref.3).

There is an intrinsic difficulty in understanding the peculiarity of the successional processes in aquatic environments. Succession and climax in aquatic ecosystems are not conceptually as well-based as in terrestrial ones (ref.5). The debate on long-term evolution of lakes is an obvious reflex of this ambiguity.

There is no doubt about the historical importance of the role played by the holistic models of succession derived from ecological energetics and information theory. On the one hand, they have led to accumulation of basic knowledge and standardization of methods to describe community structure. On the other hand, they have stimulated useful interactions between ecology and other sciences. Yet their heuristic power seems to be now exhausted.

New fronteers are open to research and the results obtained up to day encourage the effort in understanding the fine mechanisms regulating community evolution and to find a theory of succession in aquatic ecosystems. Much is to be expected from paleolimnology, from experimental research in microcosms and mesocosms, from the new stream ecology, from the expansion of predictive models.

REFERENCES

1. E.P. Odum, Fundamental of Ecology, W.B. Saunders Co., Philadelphia, 1971.
2. R. Margalef, Ecologia, Ed. Omega, Barcelona, 1974.
3. P. Colinvaux, Ecology, J. Wiley & Sons, New York, 1986.
4. R.G. Wetzel, Limnology, Saunders Co. Publ., Philadelphia, 1983.
5. C.R. Goldman and A.J. Horne, Limnology, McGraw-Hill Book Co., New York, 1983.
6. R.P. McIntosh, The Background of Ecology, Concept and Theory, Cambridge University Press, 1985.
7. F.E. Clements, Research Methods in Ecology, Univ. Publ. Co., Lincoln, Nebraska, 1905.
8. F.E. Clements, Plant Succession: Analysis of the development of vegetation, Publ. Carnegie Inst., Washington, Publ. No.242, 1916.
9. H.C. Cowles, Bot. Gaz., 27 (1889) 95-117, 167-202, 281-308, 361-391.
10. W.S. Cooper, Bot. Gaz., 55 (1913) 1-44, 115-140, 189-235.
11. R.L. Lindeman, Ecology, 23 (1942) 399-418.
12. R. Margalef, Am. Nat., 97 (1963) 357-374.
13. R. Margalef, Perspectives in Ecological Theory, Chicago Press, 1968.
14. R. Margalef, Brookhaven Symp. Biol., 22 (1969) 25-37.
15. A.G. Tansley, J. Ecol., 8 (1920) 118-148.
16. W.H. van Dobben and R.H. Lowe-McConnell (Editors), Unifying Concepts in Ecology, Dr. W. Junk B.V. Publ., 1975.
17. R.H. May, in W.H. van Dobben and R.H. Lowe-McConnell (Editors), Unifying Concepts in Ecology, Dr. W. Junk B.V. Publ., 1975, pp.161-168.
18. J. Jacobs, in W.H. van Dobben and R.H. Lowe-McConnell (Editors), Unifying Concepts in Ecology, Dr. W. Junk B.V. Publ., 1975, pp.187-207.
19. G.E. Hutchinson, Am. Nat., 93 (1959) 145-159.
20. J.H. Brown, Am. Zool., 21 (1981) 877-888.
21. R.H. MacArthur and E.O. Wilson, The Theory of Island Biogeography, Princeton Univ. Press, 1967.
22. J.L. Brooks and S.I. Dodson, Science, 199 (1978) 1302-1310.
23. J.H. Connell, Science, 199 (1978) 1302-1310.
24. S.P. Hubbel, Science, 203 (1979) 1299-1308.
25. M. Huston, Am. Nat., 113 (1979) 81-101.
26. P. Colinvaux, Introduction to Ecology, J. Wiley & Sons, New York, 1973.
27. H.A. Gleason, Bull. Torrey Bot. Club, 43 (1926) 463-481.
28. A. Hillbricht-Ilkowska, Pol. Ecol. Stud., 3 (1977) 3-98.
29. C.E. Williamson and J.J. Gilbert, in W.D. Kerfoot (Editor), Evolution and Ecology of Zooplankton Communities, University Press of New England,

    Hanover, 1980, pp.509-517.
30 R. de Bernardi, Boll. Zool., 48 (1981) 353-371.
31 W.C. Kerfoot (Editor), Evolution and Ecology of Zooplankton Communities, University Press of New England, Hanover, 1980.
32 G.E. Hutchinson, Am. Nat., 95 (1961) 137-145.
33 G.E. Hutchinson, A Treatise on Limnology, Vol.II, Introduction to Lake Biology and the Limnoplankton, J. Wiley & Sons Inc., New York, 1967.
34 G.W. Minshall, K.W. Cummins, R.C. Petersen, C.E. Cushing, D.A. Bruns, J.R. Sedell and R.L. Vannote, Can. J. Fish. Aquat. Sci., 42 (1985) 1045-1055.
35 F.H. Rigler, Verh. Internat. Verein. Limnol., 19 (1975) 197-210.
36 S. Horie (Editor), Paleolimnology of Lake Biwa and the Japanese Pleistocene, Otsu Hydrobiol. Station, Kyoto University, 1974.
37 S. Horie, Verh. Internat. Verein. Limnol., 21 (1981) 13-44.
38 U.M. Cowgill and G.E. Hutchinson, Trans. Amer. Philos. Soc., 60 (1970) 37-101.
39 R.A. Vollenweider, Schweiz. Z. Hydrol., 38 (1975) 29-34.
40 W.T. Edmondson and J.T. Lehman, Limnol. Oceanogr., 26 (1981) 1-29.
41 R.H. Peters, Limnol. Oceanogr., 31 (1986) 1143-1159.
42 R.H. Peters, in Atti 3º Congresso S.IT.E., Siena, 21-24 Ottobre 1987 (in press).
43 J.P. Nilssen, Mem. Ist. Ital. Idrobiol., 36 (1978) 121-138.
44 J. Shapiro and D.I. Wright, Freshwater Biol., 14 (1984) 371-383.
45 D.I. Wright and J. Shapiro, Verh. Internat. Verein. Limnol., 22 (1984) 518-524.

CASE STUDIES

1. Heleoplankton succession in a ricefield

The seasonal succession of heleoplankton was studied from 1982 to 1986 in an experimental ricefield in the River Po plain at Santa Vittoria, Province of Reggio Emilia (refs.1,2). In the course of the first year (1982) of rice cultivation, three successional phases can be distinguished:

1) During the first weeks of submersion (May-June) the ricefield is characterized by high phytoplankton production and, therefore, by the prevalence of the grazing chain; zooplankton is mainly represented by euplanktonic rotifers (Brachionus, Keratella, Synchaeta, Polyarthra, Asplanchna, Filinia) brought into the ricefield by irrigation water.

2) Species of cyclopoids and cladocerans (particularly efficient filter-feeders such as Moina) appear later (June-July) with high abundances.

3) Finally, from mid-July to September, in relation to the development of an extensive macrophyte cover, detritivorous species become dominant; they include both phytophilous rotifers (mainly Lecane, Lepadella, Colurella) and cladocerans (mainly Simocephalus and Chydoridae); the detritus chain prevails over the grazing chain.

The evolution of the biocoenosis structure was studied according to the species-abundance relationship method. Samples from the beginning (May-June) and the end (August-September) of the rice-growing season were taken into account. Copepods and cladocerans, on the one hand, and rotifers, on the other, were considered separately. For copepods and cladocerans, seasonal succession occurred according to an increasing trend of diversity (Fig.1). The pattern for rotifers was totally different (Fig.2): a rather large number of species was already present in May-June and their abundances were fairly evenly distributed; these species, entering the ricefield from the irrigation canal, have short generation times, which allow them to establish a highly structured community in a few days. In brief, each taxocoenosis shows a different seasonal trend in relation to the different traits of organism body size, life history, etc.

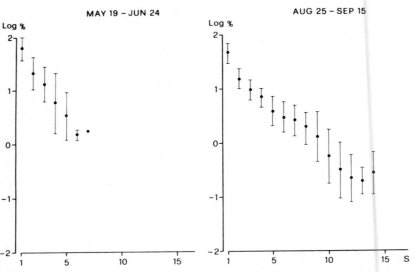

Fig. 1. Species-abundance relationships for cladocerans and copepods in the experimental ricefield (Case 1): log importance in percent ($\bar{x} \pm s.d.$) plotted against the species rank (S) from the commonest to the rarest.

Heleoplankton succession in the experimental ricefield was also studied by following the MacArthur and Wilson equilibrium model: the species turnover rate was analyzed by comparing immigration and extinction rates. The trends of immigration and extinction of systematic units are in accordance with those expected by the Island Biogeography Theory model (Fig.3).

Research in the same ricefield were carried out in years following 1982, taking cladocerans species only into account. An increase in the overall number of species, which rose from 13 in 1982 to 16 in 1984 to 20 in 1986, was noted; this increase is mainly due to detritivorous species of Chydoridae. In 1982 Chydoridae appeared only in August-September, while in 1986 they were present with remarkable densities throughout the entire rice growing cycle, starting in June in coincidence with the early appearance of widespread beds of filamentous

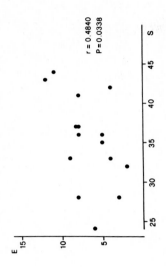

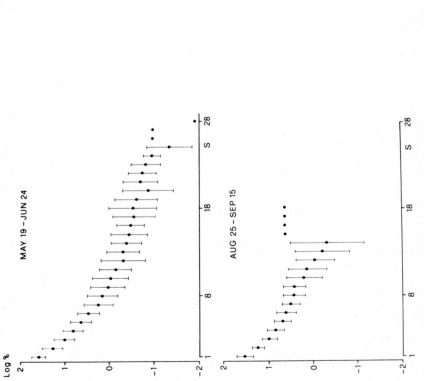

Fig. 2. Species abundance relationships for rotifers in the experimental ricefield (Case 1, see Fig.1).

Fig. 3. Number of immigrant (I) and extinct (E) species versus number of rotifer, cladoceran and copepod species (S) present in the experimental ricefield (Case 1).

algae. In 1986 the rising trend of diversity during the rice cultivation season was not particularly evident.

2. Zooplankton composition and succession in small eutrophic lakes

Research was carried out in two small reservoirs sited in the "Boschi di Carrega" (Val Taro, Province of Parma): Lake Ponte Verde (171 m a.s.l., maximum depth 6 m) and Lake Svizzera (136 m a.s.l., maximum depth 3 m). The lakes, each one about 1000 m^2, are located in two distinct catchment basins.

Zooplankton samples were collected weekly from June to September 1980 with a 90 μm mesh net. Physical and chemical data were taken at the same time (ref.3). The different trophic state of the two bodies of water is evidenced by a comparison of transparency data as well as data regarding percent saturation of dissolved oxygen (Fig.4).

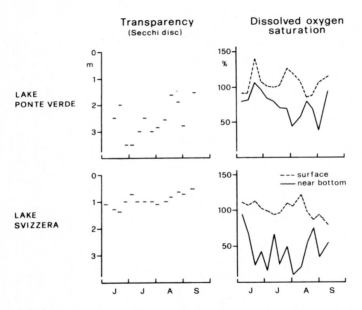

Fig. 4. Transparency and dissolved oxygen percent saturation in Lake Ponte Verde and Lake Svizzera (Case 2).

In Lake Svizzera recreational fishing is practiced; waters are turbid and invaded by macrophytes. Throughout the entire research period trophic conditions appear to be fairly constant. Zooplankton density is very high, despite the reduced number of macrofilter-feeders (calanoids are totally absent); rotifers and small cladocerans (Bosmina, Ceriodaphnia, Moina) predominate (Figs.5 and 6). Biocoenosis structure seems to be determined by two concomitant factors: prevalence of small bacterial-detritus aggregates in the food suspension available

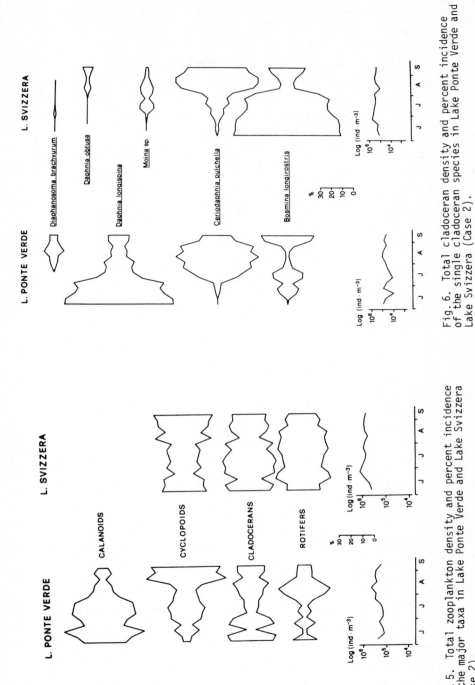

Fig. 5. Total zooplankton density and percent incidence of the major taxa in Lake Ponte Verde and Lake Svizzera (Case 2).

Fig. 6. Total cladoceran density and percent incidence of the single cladoceran species in Lake Ponte Verde and Lake Svizzera (Case 2).

for zooplankton and size selective predation of planktophagous fish.

In Lake Ponte Verde macrofilter-feeders (calanoids and large cladocerans, Eudiaptomus and Daphnia, respectively) are dominant in June-July. From July onwards massive blooms of large net algae (Ceratium and Volvox), unavailable as food for zooplankton, appear; consequently, microfilter-feeders (in particular small cladocerans) tend to become the prevailing component of zooplankton (Fig.6). After all, seasonal succession in Lake Ponte Verde is characterized by a shift of trophy conditions from springtime-early summer, when grazing food chain is predominant, to subsequent months when energy flow follows principally detritus pathways.

3. Zooplankton succession in an oligotrophic mountain lake

Lago Scuro Parmense is a lake of glacial origin located in the Upper Parma Valley at an altitude of 1527 m a.s.l. It is ice-covered from November to May. Its area is 1.2 ha, average depth 3.8 m, maximum depth 10.4 m. Physical and chemical data were taken and zooplankton samples collected (using a 50 μm mesh net) at weekly intervals from May to November 1986 (refs.4,5). Absolute values and space-time trends of nitrogen and phosphorus concentrations are typical of oligotrophic mountain lakes. The water oligotrophy is also evidenced by both high transparency values and low phytoplankton chlorophyll-a content.

The zooplankton include, above all, copepods and cladocerans, mainly represented by macrofilter-feeders species (Eudiaptomus intermedius, Daphnia longispina and Diaphanosoma brachyurum) and by one carnivorous species (Mesocyclops leuckarti).

Rotifers are also important in terms of both density and production: 19 systematic units were identified: Keratella cochlearis is the dominant species; Polyarthra dolichoptera-vulgaris, Ascomorpha spp. and Collotheca mutabilis have high densities as well. There are also two predator species: Asplanchna priodonta (present from late May to early September with significant density peaks in June and July) and Ploesoma hudsoni (with generally very low densities).

The succession pattern of rotifers is to a great extent determined by predation of A.priodonta which feeds on various microfilter-feeder species, in particular K.cochlearis. The disappearance of A.priodonta (most probably due to predation by M.leuckarti) is followed by explosive growth of K.cochlearis. Diversity and evenness indexes calculated for the entire rotifer taxocoenosis fall sharply from August onwards in concomitance with the decline of A.priodonta (Fig.7).

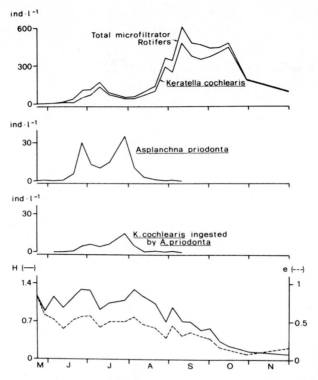

Fig. 7. Parameters of rotifer taxocoenosis structure in Lake Scuro Parmense (Case 3): density of dominant species, diversity (H) and evenness (e) indexes according to Shannon-Wiener formula; numbers of Keratella cochlearis ingested by Asplanchna priodonta are also reported.

4. Rotifer colonization in an experimental canal

Research was carried out on populations of planktonic rotifers in an experimental canal at the Isola Serafini station of National Electric Power Board (ENEL) on the River Po. Water from the river flowed into the canal, which was 1 m deep and had a constant volume of 12.6 m^3. The canal had been emptied before the research started; during the research period, water flow was kept at very low values so that a hydraulic residence time of about 4 days could be obtained.

Quantitative zooplankton samples were collected from August 27 to September 12, 1980 at 12-hour intervals using a 90 μm mesh net (ref.6).

Zooplankton density was very low during the first days of sampling, but then increased rapidly. Rotifers were the most important group; few cladocerans were found, while copepods (particularly cyclopoid nauplii and copepodites) were fairly abundant after September 8 (Fig.8). Two microfiltrator species were dominant among the rotifers: at the beginning Brachionus calyciflorus and

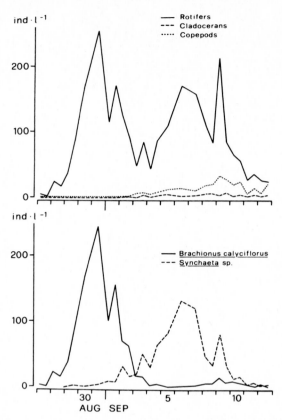

Fig. 8. Changes in density of the major zooplankton taxa and dominant species of rotifers in the experimental canal (Case 4).

subsequently Synchaeta sp. (Fig.8). A carnivorous species, Asplanchna girodibrightwelli, was also numerically important.

Asplanchna plays a basic role in the density control of the dominant rotifer species, which are its preferential prey (Fig.9). The increasing trend in diversity of rotifer taxocoenosis during the sampling period was related to the flattening of the successive dominance peaks determined by Asplanchna predation (Fig.10).

5. River Po plankton during the summer: flow rate and biotic diversity

Zooplankton samples were collected daily or every other day from July 23 to August 30, 1985 from a sampling station fixed in the middle reach of the River Po near Viadana. Sampling was effected using a 50 μm mesh net (refs.7,8). Changes in biocoenosis structure were analyzed in relation to variations in hydrometric and hydrological parameters (river flow rate, pH, dissolved oxygen).

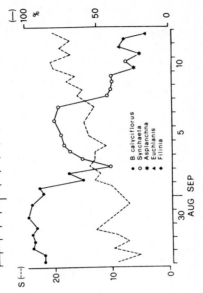

Fig. 10. Density of Asplanchna girodi-brightwelli and parameters of rotifer taxocoenosis structure in the experimental canal (Case 4): diversity according to Shannon-Wiener formula (H), number of species (S) and percent incidence of the dominant species at different dates.

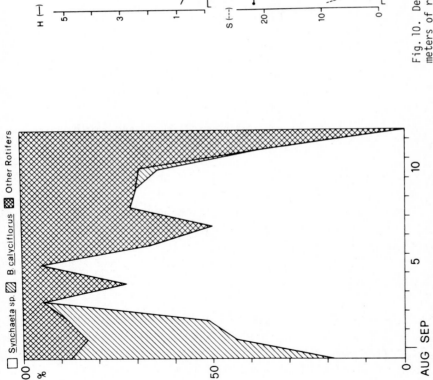

Fig. 9. Percent composition of microfiltrator rotifers in the diet of Asplanchna girodi-brightwelli (Case 4).

During reduced flow phases, the river waters showed a high trophic degree comparable to that occurring in shallow, highly productive water bodies and zooplankton was mainly represented by rotifers which attained very high densities (up to 6600 ind·l^{-1}); Brachionus calyciflorus was the dominant species (Fig.11).

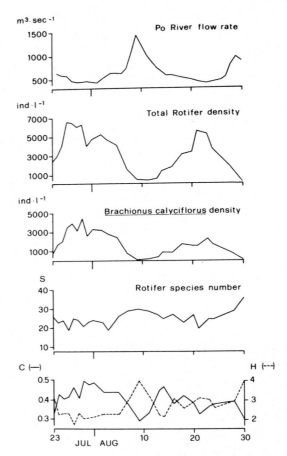

Fig. 11. River Po flow rate at Viadana (Case 5) and parameters of rotifer taxocoenosis structure: total density, density of the dominant species, number of species (S), diversity index according to Shannon-Wiener formula (H), dominance index according to Simpson formula (C).

Under these conditions the biocoenosis was characterized by a marked persistence in time and space of composition, abundance and diversity. During the summer low water periods, River Po plankton structure seems typical, according to Margalef, of a successional pioneer stage characterized by high primary production and relatively low diversity of the most important taxocoenoses (diatoms and rotifers). These features do not undergo substantial changes throughout the course of the river to the sea.

A modest flow rate increase destabilizes the river zooplankton structure leading to an increased biotic diversity as a consequence of an increased number of species (owing to washing out of benthic and phytophilous forms which are dislodged by the current from the bottom and side arms), but above all of a greater evenness of euplanktonic species abundances. Spate normally reduces drastically both density and diversity of river plankton. But a weak increase of flow during the summer promotes diversity reducing dominance of species, such as Brachionus calyciflorus, which are ill-adapted to high current velocity and turbidity and strengthening the relative importance of more typically reophilous species, i.e. those of the Keratella genus which have a thicker lorica and a more hydrodynamic shape. The intermediate disturbance hypothesis, according to which a moderate physical disturbance keeps up diversity, may be evoked in this case.

6. The accelerated trophic evolution of a deltaic lagoon

Zooplankton and zoobenthos distribution in relation to trophic state gradients was studied in the lagoons of the Northern Adriatic Sea, in particular by Ceccherelli and Ferrari (ref.9) in the Valli di Comacchio, and by Ferrari et al. (ref.10) and Ceccherelli et al. (ref.11) in the Sacca del Canarin and Sacca di Scardovari, two lagoons located in the River Po Delta.

We report here data concerning trophic state evolution of the Sacca del Canarin, a small lagoon with an area of 6 km^2 and an average depth no greater than 1 m. Maximum water level variation during a tidal cycle is about 1 m at spring tide. Up to 1979 the lagoon received freshwater from two canals, Busa Bonifazi and Busa Bastimento, connected to two delta branches, the Po di Pila and the Po di Tolle, respectively. In autumn 1979, owing to the construction of a power station, Busa Bonifazi (which had previously conveyed about 6% of the total river flow) was closed.

Before canal closure the strong inflow of freshwater into the lagoon prevented the development of a stable autochthonous zooplankton component, which was at that time limited to cirriped larvae. After closure of Busa Bonifazi, the lagoon underwent important hydrological changes and was recolonized by several zoobenthos species (chiefly bivalves). Significant changes also took place in zooplankton composition and production. In particular we observed a strong density increase of typical neritic forms and, above all, an increase in both density and variety of the native components (not only cirriped larvae but also rotifers, mollusc larvae and most probably juvenile stages of meiobenthic copepods)(ref.12). This is clearly evidenced by the average density of zooplankton sampled at the sea mouth of the lagoon during two 24-hour cycles in July 1979 and July 1980, respectively.

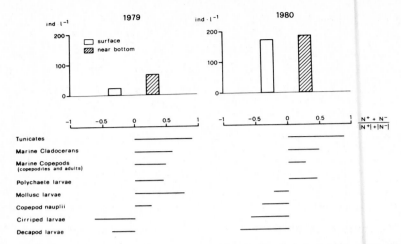

Fig. 12. Zooplankton density at the seamouth of the Sacca del Canarin (Case 6) over two 24-hour sampling cycles in July 1979 and July 1980; balance of organisms entering (N^+) and leaving (N^-) the lagoon during the 24 hours (see text) is shown.

The diel sampling cycles were effected under similar conditions of tidal range and River Po flow. Fig.12 reports, for the major taxa, the values of an index $(N^++N^-)/(|N^+|+|N^-|)$ which provides synthetic information on the balance of organisms entering (N^+) and leaving (N^-) the lagoon during the 24 hours. The index ranges from -1 to +1 and tends to -1 for taxa prevalently going out of the lagoon and to +1 for those prevalently coming from the sea. In comparison to the previous year, in 1980 a more intense export of lagoon zooplankton towards the sea was observed.

Research carried out more recently (from 1985 to 1987) evidenced further trophic evolution of the lagoon (ref.13): first of all, a strong increase in primary production (both of phytoplankton and macrophytes) was noticed; secondly, there was a further increase in density of small sized autochthonous zooplankton forms (rotifers and meroplankton), which showed large abundance fluctuations in time and space. It must be added that in spring-summer 1987 large populations of Acartia margalefi, a copepod species usually confined to the inner, more productive zone of coastal lagoons (ref.14), appear for the first time in the Sacca del Canarin.

The recent trophic state evolution seems likely to lead to summer dystrophy phenomena well known in the largest lagoons (Sacca di Scardovari, Sacca di Goro) of the Po Delta.

REFERENCES Case Studies

1. I. Ferrari, A. De Marchi, M.G. Mazzocchi, P. Menozzi, F. Minzoni, F. Piccoli and A. Moroni, Atti Convegni Lincei, 62 (1984) 175-196.
2. I. Ferrari, A. De Marchi, P. Menozzi, F. Minzoni and F. Piccoli, Verh. Internat. Verein. Limnol., 22 (1984) 1711-1716.
3. B. Malcevschi, Dinamica dello zooplancton in due laghetti eutrofici dei Boschi di Carrega, Tesi di Laurea, Università di Parma, 1980.
4. R. Antonietti, I. Ferrari, G. Rossetti, L. Tarozzi and P. Viaroli, Verh. Internat. Verein. Limnol., 23 (1988) 545-552.
5. G. Rossetti, P. Viaroli and I. Ferrari, in Atti 3º Congresso S.IT.E, Siena, 21-24 Ottobre 1987 (in press).
6. I. Ferrari, A. Caleo, M.G. Mazzocchi and G. Matteucci, in Atti IV Simposio Dinamica Popolazioni, Parma, 1984, pp.51-57.
7. I. Ferrari, R. Mazzoni and A. Solazzi, Atti 7º Congr. AIOL (1987) 261-266.
8. I. Ferrari and R. Mazzoni, Chemosphere (1988) (in press).
9. V.U. Ceccherelli and I. Ferrari, Estuar. Coast. Shelf Sci., 14 (1982) 337-350.
10. I. Ferrari, V.U. Ceccherelli and M.G. Mazzocchi, Oceanologica Acta, N SP (1982) 293-302.
11. V.U. Ceccherelli, V. Gaiani and I. Ferrari, Chemosphere, 16 (1987) 571-580.
12. I. Ferrari, V.U. Ceccherelli, M.G. Mazzocchi and M.T. Cantarelli, Neth. J. Sea Res., 16 (1982) 333-344.
13. R. Ambrogi, I. Ferrari and S. Geraci, in Proc. 22nd EMBS, Barcelona (1988) (in press).
14. I. Ferrari, A. Carrieri and R. Coen, Oebalia, 11 (1985) 187-201.

CONCEPT OF ECOSYSTEM STRESS AND ITS ASSESSMENT

R.D. GULATI

Limnological Institute, Rijksstraatweg 6, 3631 AC Nieuwersluis (The Netherlands)

1. INTRODUCTION

The increasing pollution and eutrophication of lakes connected with disposal of wastes originating in industrial emissions and in industrial, agricultural and house-hold uses of water are well known. Experts estimate that waste water fed directly into waterbodies may introduce as many as a million different pollutants into our aquatic ecosystems (1).

Management problems are related mainly to these as well as to energy sources and options and impacts of different uses on the environment. Environmentalist oppose large scale power plants because high temperatures increase mortalities of certain organisms and promote growth of others, the more undesirable ones. We know that acid mine drainage and acid rainfall can result from conversion of coal or from effects of thermal effluents from fossil fuelled power plants.

Major factors for sensitivity of aquatic ecosystems to pollution, besides role of these systems as receiving bodies for effluents, may be related with structure of food chain. Subject of heavy metals pollution of different aquatic environments has received the much needed attention recently (1). Also, recent studies have shifted emphasis towards impacts on metabolism of aquatic organisms and their ability to accumulate both essential and non-essential metals. Impacts caused in an aquatic system may depend on the toxicant distribution within the three compartments: water, suspended matter and sediment.

In view of the aforementioned, the science of ecology is increasingly being applied to different environmental problems. Among these, environmental management is a subject of great concern with a rapid increase in the nature of problems and their complexities. Ecology policy designs and their legal and financial implications involved vary greatly depending on the situation to be corrected and its geographical boundries and national and international importance.

Auerbach (2) cites the work of Clark and his coworkers who point out that 'the aim of sound ecological policy is not to predict and eliminate future surprises but rather to design resilient systems which can absorb, survive and capitalize on unexpected events when they occur'. For such a policy one needs to have a clear understanding of the ecosystem, especially with regard to the stress-related characteristics like resilience and stability. Stress assessment is of paramount importance in developing a good basis for studies relating to environmental or ecosystem management.

The practice of environment impact assessment is difficult and frustrating (3). The approach to stress-related environmental problems ought to be a triangular one, linking observational, theoretical (modelling) and experimental approaches (Fig. 1). The interest in research programmes involving experimental perturbations in laboratory microcosms and field microcosms, the so called enclosures or limnocorrals, and in isolated ponds, has attracted lot of attention. Most of such works are aimed at studying nutrient enrichment effects on biota, or at a better understanding of food-web relations under extreme stress, or relieving it altogether.

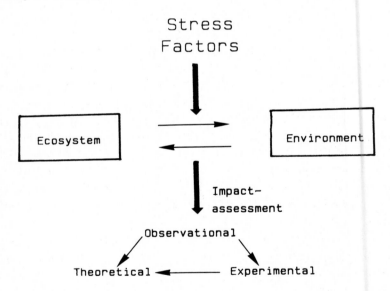

Fig. 1. Stress-related interactions of ecosystem and environment in general and scientific strategy of impact assessment.

In the last one decade or so the subject of ecosystem behaviour and response to stress has received increasing attention. This is evident from number of papers dealing with different aspects of stress ecology appearing in recent years, see e.g. several papers in Barrett and Rosenberg (4) and in Thorpe and Gibbons (5) and a comprehensive review by Rapport et al. (6). Many of these studies emphasize on integrative approach in ecology and environmental sciences, but at the same time recognize the importance of these two disciplines independent of one another.

Main object of this paper is to present to the reader a brief but overall view of stress concept and its assessment in the light of experience with aquatic ecosystems. Stress definitions and terminology precede a description of factors causing it and of means to quantify it. Although specific case studies are avoided, an attempt is made to analyse both endogenous and extraneous factors that cause stress in ecosystems as well as symptoms and syndromes. Modelling stress, though a challenge now may be a mean to predict stress and help prevent it. Besides, it may facilitate making decisions which are not only sound but acceptable both scientifically and socially.

2. DEFINITION OF STRESS

There are at least three types of stress definitions: biological, psychological and physical. A biological definition, "the sum of all non-specific biologic phenomena" (Selye and Horava, 1952), appears less applicable to a model of ecosystem stress than does the psychological definition. This latter defines stress as "any influence whether it arises from the internal environment, or external environment, which interferes with the satisfaction of basic needs or which threatens to disturb the stable equilibrium" (8). In physical context stress is defined as the force which produces or tends to produce deformation in a body measured by the force applied per unit area" (9). The actual deformation in physical systems is called strain, which is measured by the ratio of amount of change produced to the total value of the dimension in which the change took place. According to Hooke's Law we know that stress is proportionel to strain. Stress can be internal or external to a system, i.e. can arise because of changes within the system, or caused by factors outside the system.

The three definitions as such cannot be directly applied to ecological systems. However, combining the conceptual relationship of stress to the physical system with 'disturbance of equilibrium', as contained in the stress definition for psychological system, a definition applicable to ecology may be formulated. In ecosystem context, stress would tend to produce deviations from maximum system order (10). It is difficult to imagine that all variables in an ecosystem are simultaneously optimized. Therefore, all systems should be to some degree under stress and exhibit some degree of strain, even under conditions approximating steady state. According to Perkins (11) a linear relationship between stress and strain is quite justifiable, being more manageable mathematically and having proved effective in demonstrating more general stochastic situations, including those with large scale perturbations.

Lugo (12), who recently reviewed concepts of stress in relation to ecosystem, points out that response of biological material to stress has a common pattern irrespective of whether the biotic materials are cells, organisms, populations or ecosystems.

There have been numerous other definitions and concepts of stress offered by research workers during the past two or three decades. According to Auerbach (2), irrespective of the definition of stress and the stressor involved, the stress concept usually implies an interference with the normal functions of a system; its effects are most dramatically observed after certain thresholds of tolerance are exceeded and beyond the thresholds any recovery is problematical, or at least difficult. Bayne (13) used the term 'steady state values' for organisms under normal conditions. He defined stress as a measurable alteration of a physiological steady state (including the behavioural, biochemical and cytological aspects) which is induced by an environmental change, and which renders the individual more vulnerable to further environmental change.

Despite broader application of stress to ecosystem level and its generalization, the definitions are derived mainly from studies on a few taxonomic groups of organisms, namely mammals, fish and mussels.

Barrett *et al.* (14) defined stress as a perturbation (stressor) applied to a system: a) which is foreign to that system; and b) which is natural to that system, but applied at an excessive level. Application of nitrogen or phosphorus or temperature in excess, all three of which are natural, will be examples of the second category. The definition covers both man-made and natural stresses, but does not refer to negative deflections which are included in the derivation of Odum *et al.* (3). Lugo (12) emphasized that energy expenditure is increased during stress but the potential energy decreases. From the stress definitions of both Odum *et al.* and Lugo it is clear that stress puts the organisms and the system in which they live at a disadvantage since excessive energy expenditure is incompatible with survival (15).

Some earlier workers, e.g. Bayne (13) and Brett (16), point out the need for incorporating in the stress definition the altered function, in order to be able to assess the effects. It will enable to detect sub-lethal stress effects consequent upon low stress level but long-term exposure to it. The stress indicators which pinpoint such long-term alterations need to respond rapidly as well as reliably across a variety of organisms.

2.1. Alternative stress terminologies

Odum *et al.* (3) attempted to redefine stress in order to remove some of the existing confusion. They have illustrated this with the help of a hypothetical performance curve for a perturbed ecosystem (Fig. 2). Perturbation is one such term which they commonly defined as an alteration or deviation from what is usual or expected. In ecological usage, perturbation is any deviation, or displacement from the 'nomimal state' in structure or function at any level of organization. The nominal state is the normal operating range, including expected variance. In practise an ecosystem would be considered to be perturbed if a significant deviation from the nominal (generally some designated control level) has occurred. The terms stress and perturbation will be essentially synonymous if a deviation is caused by external forces.

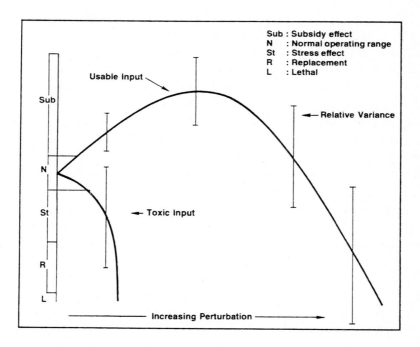

Fig. 2. Hypothetical performance curves for a perturbed ecosystem subjected to usable input and toxic input, simulating the output response measured, e.g. as rates of function, to increasing intensity of input perturbation (from Odum et al. (3)).

The term subsidy according to Odum et al. (3) is an antonym to stress in that it leads to favourable deflections leading to an improvement in the performance of the ecosystem. An example of subsidy effect would be enrichment of an oligotrophic lake which does not lead to algal bloom because zooplankton and fish respond promptly to consume the increased primary production. One such case has been reported by McAllister et al. (17) in an oligotrophic Canadian lake. They added nutrients in solution at the rate of 5 tons per week for 18 weeks to the lake, increasing the normal availability of nitrogen and phosphorus by factor 2 and 5, respectively. These workers succeeded in achieving their objective, which was to increase flora and fauna of the lake but to avoid conditions of undesirable eutrophication or a change in diversity of food organisms. The rate of primary production was at least doubled and biomass of zooplankton increased eight times and of fish by about 40%. However, such cases of apparent community level compensation to nutrient perturbation need to be monitored for longer periods of time before concluding if this is, indeed, all subsidy and no stress (3).

There may be cases in which the positive effects and negative effects interact, i.e. exhibit subsidy-stress interactions, also called 'push-pull' effects by H.T. Odum (18). For example, thermal plumes which stress some organisms can have positive effects on the whole community if the thermal energy is coupled through acclimated organisms, with the higher quality of energy of sunlight. Therefore, coupling and compensating capacities of organisms will determine the net effect on different levels due to combination of inputs.

Usage of the terms Stimulus and response is widespread in physiology and psychology. Thus, addition of phosphate to a system will be stimulus to which the algal community may respond as increased growth rate. The terms clearly distinguish between cause and effect. However, one has to carefully assess the response which may not necessarily be visible at the same trophic level but at an other one. Therefore, system approach is needed to clearly assign or attribute the effect or effects produced to the factor that caused it.

3. STRESS: CAUSES, QUANTIFICATION AND DYNAMICS

The agents which cause stress are: 1) thermal extremes; 2) toxic pollutants; and 3) radiation. Also within the ecosystem, both predation and competition and diseases may cause stress that may be reflected in changes in structure and in energy flow rates.

Many studies restrict measurement or quantification of stress to parameters causing stress, rather than also measure parameters with which synergistic interactions are expected. Thus, in such cases it is the cause (stress) rather than the response (strain) which is being measured or monitored. Strain caused by stress in an ecosystem, is easier to express as precise environmental index than quantifying the factors causing stress. In other words, response, which is proportional to the cause, provides a better means to study the environment impacts.

Quantification is needed of both stress factors (biocides, heavy metals, pollutants, nutrients and temperature) and strain they produce, e.g. changes in species diversity, density, biomass, production and energyflow rates and in equilibrium.

An imbalance between stress and the ability of system to compensate for that stress results in non-steady state conditions (11). Increased compensation for this imbalance is seen as succession or increased ordering of a system structure and function (Fig. 3). In systems which have been stressed beyond their normal tolerance process it is called "repair". As seen in figure 3 collective disordering functions (stress) generally fluctuate over a limited range with time and the system becomes capable of compensating for a greater proportion of stress as succession proceeds. It is not the mean value of stress but deformation (or formation) of the system (strain) which undergoes changes in succession. This change in system order is manifest in characteristics relating to changes in species number, organism size, gross community metabolism, efficiency of mineral cycles and changes in energy flow rates.

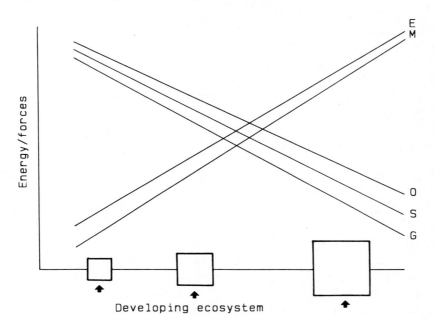

Fig. 3. Schematic representation of stress in a developing ecosystem (the 3 squares) and its relationship to energy and forces as the variables; codes: E, energy assimilated; M, maintenance energy; O, ordering forces exerted by the ecosystem; S, strain or resultant force; and G, energy required by develop order. The arrows indicate stress and its fluctuations; Modified from Fig. 1 in Perkins (11).

Quantifying stress in ecosystem in non-steady state is a formidable task, with regard to theoretical and procedural problems. As climax or steady state conditions are approached, all energy assimilated by the ecosystem is used in maintenance (19, 20).

External stress consists of any disordering function impinging on the ecosystem from outside, both biotic and abiotic. Internal stress originates within the system, causing inefficient geochemical cycling, unbalanced predator-prey relations and pollution from underground seepages containing pollutants (heavy metals, nutrients).

It is efficient to apply these stress and strain estimates simultaneously, although they do not identify total ecosystem response nor do they distinguish between internal and external stress. Because in a steady-state situation maintenance costs reflect energy assimilated any deviation from it reflects the effects of total stress.

3.1. Derivation of External Stress Index

Irrespective of whether the populations are stressed, it is of great interest to ecologist and environmentalist to measure total effects on the ecosystem of external stress, i.e. stress due to factors other than basal maintenance. In this regard Perkins (11) has formulated an index which demonstrates the feasibility of quantifying environmental stress. This index is based on energy output in ecosystems. Modelling work shows that consumer respiration is very sensitive to external stress and is a more sensitive indicator of function than biomass. The ratio of consumer respiration to total ecosystem respiration was, therefore, used as component in the index to identify degree of external stress in steady-state ecosystems. The index called the External Stress Index (ESI) is formulated as:

$$ESI = 1 - 0.5 \log(t+1) \left[\frac{0.5(F_i - F_e)}{F_i + SC} - \frac{R_c}{R_t} + 1.5 \right]$$

where t = annual turnover rate
F_i and F_e are energy input and output (Kcal.m^{-2}.year^{-1}), respectively
SC = standing crop biomass (Kcal.m^{-2})
R_c = consumer respiration (Kcal.m^{-2}.year^{-1})
and R_t = total ecosystem respiration (Kcal.m^{-2}.year^{-1})

The log-transformation reduces the effects of great variation found in the turnover rate t, which may fluctuate 50-fold among the ecosystems. The conceivable range of ES1 based on the above formulation is from zero to 2, its magnitude being inversely proportional to external stress. By comparing relatively stressed systems with those not stressed the index can be tested.

3.1.2. Applications of the stress index

Application of the stress index in eight ecosystems by Perkins (11) showed that there was apparently -perhaps superficially- a direct relationship among the environments to support populations of organism and the amount of external stress. It appeared that aquatic systems generally function under more stress than terrestrial systems, although more data are needed to confirm this. The index has served to compare the degree of environmental stress and validity and feasibility of quantifying ecosystem stress in terms of ecosystem parameters.

The main problem with the index, as Perkins (11) himself also points out, is that it presupposes steady state conditions and, therefore, its reliability decreases with increasing rate of change of ecosystems due to stress or succession. Also, the relationships presented need to be refined especially by using data on energy flow within ecosystems, both before and after stress applications. By having improved measurements of stress parameters the index will not only help us in having better insight into the functioning and structure of ecosystems but will also help us evaluate, manage and restore perturbed and polluted systems better. Lastly, as an essential component in the environmental models the strain index may improve the predictability value for application of these models in resource management practices.

3.2. Biochemical indicators of stress

Ivanovici and Wiebe (15) emphasize the need of studying biochemical variables for application to environmental problems. They examine the biochemical index AEC (Adenylate Energy Charge) of Atkinson (21) as a potential indicator of stress. The index can be calculated from the measured concentrations of adenine nucleotides: Adenosine triphosphate (ATP), Adenosine diphosphate (ADP) and Adenosine monophosphate (AMP). The AEC may be thus derived as:

$$AEC = (ATP + \tfrac{1}{2}ATP)/(ATP + ADP + AMP)$$

The index is a measure of potential energy available to organisms at the time of sampling. The values of the index vary from zero to one, increasing value indicates an increased metabolic potential, and *vice versa*. Ivanovici and Wiebe (15) believe that such an index can be helpful in quantifying stress which is otherwise, despite its common use, an obscure term.

3.2.1. Applications of the biochemical index

Literature on the applications of the biochemical index (15) shows that in the index range 0.8-0.9, encountered typically in cultures, the environmental conditions are optimal and growth and reproductive rates high. The index values between 0.5 and 0.7, on the other hand, frequently occur if organisms are growing and reproducing at reduced rates, being limited due to suboptimal conditions. AEC values of about 0.5 are accompanied by loss of viability.

At ecosystem level, a decrease in temperature from 39°C to 23°C resulted in decrease in AEC from 0.46 to 0.19; similarly, change from summer to winter conditions reduced the AEC from 0.68 to 0.32 for the *Spartina alterniflora* salt marsh sediments (see Table III in Ref. 15).

Compared with other methods of assessing sublethal stress AEC index has several main advantages: 1) it detects the presence of stress in organisms as well as indicates the severity; 2) it correlates with growth state; 3) it can be measured in a wide biological spectrum from bacteria to mammals, e.g. Chapman *et al.* (22); 4) individual AEC contents of the organisms have low variations so that even a few replicates may provide high reliability; and 5) its response time is fast, varying from minutes in microorganisms to 24 hours in higher organisms.

One major drawback of the AEC appears to be lack of specificity of response to different stressors, so that it does not provide information on the type of stressor. One more disadvantage is the lack of linearity of AEC response, which appears to be a 'step' function.

Applying AEC as a sole method in stress assessment is not without problems and, therefore, not recommendable until these uncertainties are resolved. Also methodological problems, because of lack of standarized techniques with regard to nucleotides, may be another serious bottleneck which may prevent a popular application of AEC as stress indicator. Summarizing, it may be remarked that despite the problems discussed, applying AEC may be useful in situations in which the accompanying problems are recognized and proper precautions taken. The applicability has perhaps more chances of success at individual or population level (e.g. in aqua culture) rather than at the community or ecosystem level.

4. SYSTEM APPROACH IN STRESS STUDIES

4.1. General remarks on stressor types, their effects

For studying the effects of stress, one needs to have systems which are initially similar with regard to their input and output, but one of them is stressed by changing the input and the other serves as a control. The change in input will produce changes in the system which will finally culminate in output differences. The output effects which, depending on their nature, may be considered as stress, subsidy, or both. For instance, addition or input of a nutrient (P, N) limiting production in a lake is a type of perturbation which may produce increase in phytoplankton primary production and biomass as an output but decrease in diversity of phytoplankton; this is observed commonly in eutrophic lakes. In such cases enhancement of biomass and productivity are examples of subsidization, while reduction in species diversity is a stress effect.

These perturbation and subsidy effects, if further extended to system level may show a series of changes at a trophic level other than that at which perturbation took place. A good example of this is an increase in densities in late spring of filter-feeding zooplankton in response to increase in early spring of algal biomass. This will be for zooplankton a case of subsidy rather than stress within a system. Similar examples can be cited for changes in fish in response to increased productivity and biomass of zooplankton and zoobenthos. However, these are immediate responses to favourable food situations in an ecosystem and do not include the long-term feed-back effects before the system reaches a steady state, or a new equilibrium. The subsidy or stress effect in this new situation may considerably differ from those observed as short-term effects. If the system is being continually stressed, e.g. if there is a continuous input of nutrients successively for several seasons, the effects may quite differ. Several examples of such effects can be described from studies on ecosystems which have become eutrophic in the last three or four decades because of continuous enrichment over the years. Further, removal, reduction or alienation of such a stress effect over a period of time may lead to a new balance due to so-called recovery or restoration of the ecosystem (23). Many of the biomanipulation techniques in use during the last two decades (23) or so have stress removal as the basic underlying principle. Only the means and approaches used to remove stress depend on a variety of factors varying from type of stressors, lake usage, lake morphometry, nature of sediments, etc.

Stresses may be a) a part of an ecosystem's normal every day environment or b) added by man. The concept of stress, with regard to ecosystems, generally refers to agents or factors that cause it. One may identify them broadly into two types:
1. those originating outside the ecosystem, e.g. thermal extremes, toxic pollutants, or radiation; and
2. those involving the ecosystem structure, e.g. predation and competition; see also Perkins (11).

In dealing with concept and problem of stress effects one must distinguish between 'point' and 'non-point' sources (24). Much of the current activity in environmental assessment in the industrialized world, both in the United States and in West Europe, deals with 'point sources' (e.g. city sewage, power station discharges, and refinery and other industrial wastes). Apparently, the cumulative effects of some of these point sources, particularly industry, may have non-point source of stress. However, available techniques for assessing these cumulative effects appear to be still in infancy.

Non-point sources such as industrial emissions, acid rains and chemical pollutants in dispersed forms and certain airborne diseases, if present, may inflict vast geographical areas, including ecosystems and their catchment areas. The run-offs from the catchment may in such situations have additive effect in accentuating stress effects on ecosystems. The stresses exerted may effect individual species, thus the species composition and ecosytem structure. Such changes have, interestingly, become evident mainly from studies on point-sources with a distance-dependent decrease in the severity of effects of the pollutant. Acid-rain pollution of aquatic ecosystems is a good example of this non-point-source type stress in which the magnitude of stress increases with decreasing distance from the emission source. The non-point sources are potentially capable of simultaneously stressing many components of these systems (2). For assessing the non-point sources one needs to have empirical knowledge as well as theoretical and mathematical basis for inferring the effects at the ecosystem level. This will be the right step in the direction of forecasting the consequences, and eventually taking preventive measures.

4.2. Ecosystem stress analysis

Analyzing the ecological impacts is a difficult task. Winkle et al. (25) have attempted to present a graphical summary of the difficulties involved which is reproduced here with some modification (Fig. 4). The panels in the

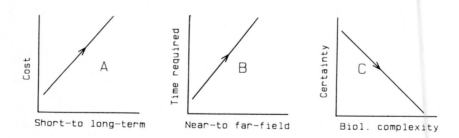

Fig. 4. Gradients of time (A), space (B) and biological complexity (C) associated with studying and analyzing the ecosystems impacts; difficulties associated with the three gradients, are increasing cost involved (A), time required (B) and decrease in level of certainty (C), respectively; derived from graphical representation of Winkle et al. (25).

figure depict that the magnitude of difficulties increases as one attempts to analyze the impacts ranging from:

1) short-term to long-term;
2) near-field to far-field; and
3) individual to ecosystem.

The factors which play important role in the 3 cases are the increasing cost, time and uncertainty, respectively, if one moves from the panel on the left side to that on the right in figure 4. For analyzing an impact it is, therefore, logical to begin on left-hand side of the gradient in each of the three depicted panels (Fig. 4), and at a later stage as money, time and know-how, respectively, allow move to the right-hand sides of the 3 panels. For this latter, we have for the last one decade or so simulation modelling as an important available tool to forecast the environmental impacts.

Although ecologist is obliged to predict effects quantitatively, with the help of simulation models, his intuition and judgement as professional ecologist may be equally important in decision making.

Attempts to quantify impacts at the ecosystem level show that one needs to choose parameters that provide information about the total ecosystem functioning. In study of the stress impacts such a parameter would be obviously the one which has exceeded a certain threshold level causing a series of changes both in structure and functioning of the ecosystem.

A functional harmony of autotrophs and heterotrophs and mineralization of sedimenting or sedimented sestonic material means a balance and synchrony between the nutrient release and nutrient and energy utilization processes. However, external stresses in the form of nutrient input, or other anthropogenic interferences, which tend to internal desynchronization in the functioning of the ecosystem, or both, will lead to unfavourable responses at population and community levels. These may be in the form of changes in the rates of energy flow and metabolic rates, or in the inter-trophic feed-backs. The disturbance in the so-called 'bottom-up' and 'top-down' trophic interrelations will be manifest as ecosystem deterioration and break-down, so that the recreative, social, economic and aesthetic values of such an ecosystem may diminish. By focussing on mechanisms which deregulate the system functioning it may be now possible to identify aspects that need to be monitored before a concerted remedial action can be taken. Such an approach will allow one to consider the ecosystem as an integrated system.

4.2.1. Endogenous and natural stress: some examples

Environmental fluctuations are stressful to ecosystems (26). At the ecosystem level an event in a sector of the system may cause mortality but the system as a whole may adapt to it. Several examples may be cited of stress events within the aquatic ecosystems which may be related to lake's location, morphometry, nature of bottom sediments, intensity of exposure to wind, temperature cycle, light climate, etc. These parameters may work in conjunction or in isolation to produce acute, short-term responses, i.e. alternating periods of recovery and recurring stress effects.

A breakdown of thermal stratification and events preceding it, namely the downward migration of thermocline, are essential part of temperature cycle of the temperate lakes, both monomictic and dimictic, but it creates problems for certain populations of bacteria, phytoplankton and zooplankton. However, for the lake's economy these phenomena are indispensable for heat redistribution and nutrient supplemenation and for removal of oxygen deficit that often occurs. Similarly, in shallower lakes, exposure to wind and stormy conditions lead to stirring up of the sedimented sestonic material that cause momentary sharp increases in turbidity which may lead to deteriorating light climate, thus adversely affecting growth of phytoplankton and of macrophytes in littoral region; strong water agitation may also cause mass mortality of some cladocerans because they trap airbubbles between their valves. However, these shallow ecosystems may easily adapt to these acute recurring stress phenomena. The main positive effects of such a stress would be: 1) to evenly distribute sestonic material which serves as food for suspension grazers; and 2) bring nutrients released at the sediments into circulation and most likely prevent the effects of nutrient and algal patchiness caused by e.g. differences in zooplankton distribution and grazing pressure, especially during calm periods.

In contrast with stress caused by human interference or allogenic forces, these natural stresses have a recognizable pattern and are better predictable because the ecosystems have tendency to return to a steady state. In such a case, therefore, harshness of the environment should not be confused with unpredictability. Moreover, such a natural stress may be considered as a normal condition of any environment.

4.2.2. Deviations from normal operating range

Studies dealing with ecosystem response to stress must include three important ecosytem parameters: community metabolism, nutrient cycling mechanisms and structural properties of the system. Any deviations of these parameters may be used to identify stress and stress intensity.

Of course, deviation on applying perturbation implies departure from the normal operating range as shown by Leffler (28) (Fig. 5). The initial ability of the ecosystem to resist deflections from the normal operating range (NOP) would vary greatly and this resistance (B in Fig. 5) is a difficult property to quantify. The next step the perturbation response is clearly a deflection from

NOP (Fig. 5). This stress trajectory can be defined as relative stability (27) as indicated by shaded area E in the figure. The extent to which this relative stability may be directly related with diversity. Also, nutrient reserves and availability have been proposed to have a major influence on response to stress. Some workers (e.g. see ref. 28) believe that nutrient availability may increase ecosystems resistance to stress but decrease its resilience (D Fig. 5). The parameters resistance and resilience are apparently opposed to one another. However, in less variable environments, i.e. systems with high stability, both the resistance and resiliece may be poorly manifest in case of severe stress (29); this, nonetheless, does not hold good for estuaries which despite their highly variable environment exhibit high levels of resistance and resilience. Leffler (28) in his work on aquatic microcosm experiments concluded that mechanisms by which increasing nutrient and energy subsidies affect the ecosystem stability are not known.

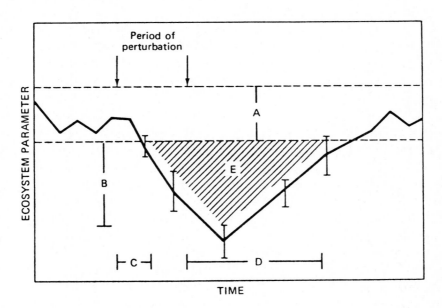

Fig. 5. Measures of relatively stability in a perturbed ecosystem: A, constancy; B, resistance; C, response time; D, resilience; and E, total relatively stability; dashed horizontal lines indicate the normal operating range (figure taken from Leffler (28)).

Golley (30) and Webster (27) proposed that high standing crop biomass in a system would reduce fluctuations caused by exogenous stress and thus import greater resistance against perturbations. Since there appears to be an inverse relationship between resistance and resilience, an increased resistance would

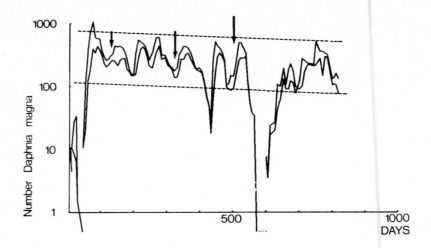

Fig. 6. Changes in the *Daphnia* population within normal operating range (dashed, horizontal lines); arrows A, B and C indicate the days on which stressor Dichlobenil was applied; for details see text (from Kersting (31)).

lead to slow recovery thus leading to a decreasing resilience. Nevertheless, Leffler (28) could found no support for these hypotheses from statistical treatment of his microcosm data.

Kersting (31) applied the concept of relative stability, resilience and resistance to a laboratory microcosm. He studied the effect of a herbicide (Dichlobenil) on *Daphnia magna* in an experiment lasting more than 800 days (Fig. 6). Dichlobenil was added to the *Daphnia* culture in concentrations of 1 ppm on day 136 and 2 ppm each on day ± 320 and day 434. On the first addition the population fluctuated within the NOP, i.e. there was no marked deflection. However, the subsequent two additions had response time of 88 days and 53 days, respectively. Response time is the duration between the herbicide addition and perturbation leading to a departure of the population from the NOP. It took 109 days and 144 days, respectively, for the population to recover and return to lie within the NOP.

4.2.3. Symptoms, syndromes and survival

As mentioned earlier, studying response of ecosystems to stress involves comparison with systems that are not distressed. In choosing objective scientific criteria judgement plays an important role (6). O'Neill and Reichle (32) consider nutrient recycling and rates of primary production to be of critical importance in trophodynamics studies. This, because despite changes in species composition in a system undergoing perturbation, the system may remain unaffected if nutrient and production dynamics do not change.

TABLE 1.

Symptoms of ecosystem distress; data from Table 1 of Rapport et al. (6), but restricting to aquatic systems. Codes: N, not measured; + or -, direction of change comparing with unstressed situation; and asterisk, occurrence of retrogression.

Type of stress	Response Variables				
	Nutrient pool	Primary production	Size distribution	Species diversity	System retrogression
Opportunistic fishing	N	N	-	-	*
Commercial fishing	N	N	-	-	N
Nutrient (lakes)	+	+	-	N	*
Toxins (rivers)	N	N	-	-	*
Shoreworks (Italian lakes)	N	N	-	N	*
Introduction of exotic species	N	N	-	N	*
Multiple causes	+	+	-	N	*

The symptoms of ecosystem change would, to a greater or lesser degree, depend on one's perspective. Rapport et al. (6) give several examples of manifestation of responses to stress in different ecosystems. Their data on aquatic ecosystems (Table 1) show that there may be several possible primary factors causing retrogression. Fishing and all the other four stress types summarized in Table 1 are attributable to human activity. First, harvesting in the form of a) selective and b) overfishing can affect the top-down relations in food chain so that the type or form of harvesting stress may be reflected in the altered structure of zooplankton and phytoplankton. The stress effects may be further accentuated if there is a simultaneous bottom-up effect via nutrient loading. Relieving stress, or even its symptoms, may be in such cases difficult, as demonstrated in some lake rehabilitation studies (23). Second, we all know well that additions of a) nutrients, both through sewage discharge and agricultural wastes, b) industrial pollutants, including PCBs, SO_2, pesticides, heavy metals and c) oil spills, irrespective if they enter the ecosystem via air or land, will cause drastic changes in the symptoms. Third, physical restructuring of the lakes either by changes in its shoreline, sand winning and deepening or by similar changes in lake's immediate catchment area, dumping of solid wastes, rubble, etc., or by other alterations which make the lake more vulnerable to wind action and erosion, may produce stress symptoms via changes in light climate, temperature cycle, extent of turbulence and nature of lake bottom. Clearing cutting of forested areas (33) may lead to nutrient accumulation in neighbouring ecosystems.

Introduction of exotic species of both plants and animals, either intentionally or carelessly, may disturb the existing equilibrium. Extreme natural events, though rare, may manifest in an unacceptable response; good examples are prolonged droughts, floods, volcanic eruptions and climatic shifts. Such effects will be both direct as well as via changes in the ecosystem catchment.

Nutrient increases resulting in a tripling of the average algal biomass and 20-fold increase in productivity in Lake Erie in the United States is a good example of the effect of land-base activities (34). Even though this means a nearly 7-fold increase in P/B ratio, this may not be true if the eutrophication stress leads to dominance of blue-green algae which are capable of building up huge biomass, being poorly edible, and growing better than other algae under poor light conditions. Therefore, the stress systems cannot always be generalized in terms of increase in P/B ratios alone.

Also reduction in species diversity has often been used as indicator of stress, but evidence of this is equivocal (28). Internal stress may even lead to increase in diversity (6), so that ecosystems may adapt to such stress without exhibiting any stress symptoms. In short, the value of diversity as a stress parameter may be limited and even superficially misleading. Intensive size-selective predation by fish on larger crustacean zooplankton may stimulate diversity and increase among small-sized zooplankton, including rotifers (35) which normally cannot compete with the larger crustacean forms for food. Accompanying these internal stresses -predation and competition- external factors causing eutrophication stress, may lead to changes in size. Regier (36) cites an interesting case of decrease in size of fish in relation to eutrophication of Laurentian Lower Great Lakes; in these lakes fish larger than 5 kg contributed 50% of total fish biomass before pollution problems, compared with fish < 1 kg that contributed to half the total biomass thereafter. The changes in species dominance, i.e. from white fish, lake trout, walleye and pike before eutrophication to mainly alewife, rainbow and smelt now are equally remarkable. Some of these shifts may be, as pointed out by Rapport *et al.* (6), related to increased susceptibility to infections and diseases caused by use of insecticides, and increased population density and amplitude of fluctuations.

In aquatic ecosystems, the effects produced by the stressor may be similar, irrespective of the type of stress (37). This may be specially true in the replacement of longer-lived, larger species by short-lived, opportunistic species. About the other changes caused by stress, namely reduction in stability and diversity, there is less unanimity. There may be a progression in stress symptoms, rather than that the symptoms appear immediately. For example, in response to acidification stress there may be a reduction in the number of zooplankton species, especially those of crustacea, and of benthic molluscs followed by lack of young fish because of reproductive failure. Phytoplankton composition may change, e.g. blue greens may be drastically reduced because of sensitivity to pH in acid range. These changes may culminate in predominance of benthic algae, sphagnum and other mosses.

Besides adapting to stress, other mechanism involved in coping with stress is capacity to resist; however, initially resilience may play an important role. One mechanism by which ecosystem can resist stress is replacement of the stress sensitive species. According to Margalef (38) aquatic ecosystems may develop 'external loops' that extend outside the system into the atmosphere for oxygen and nitrogen, and into the sediment for carbon and phosphorus. This 'overfeeding' may accelerate areal primary production and energy flow and thus counteract the stress. Lugo (12) who reviewed metabolism approach to stress, remarks that during the resistance phase community respiration rate may increase. Such changes, i.e. drains in energy flow, provide information on stress (39). Respiration to biomass indices have their broad applicability as stress indicators.

In aquatic ecosystems both synergistic and antagonistic interactions among the stresses have been described (40). Several heavy metals (e.g. Cu and Ni) are known to act synergistically on algal species, whereas increasing calcium levels reduce the toxic effects. Tabata's (41) work on *Daphnia* clearly demonstrates that concentration of heavy metals necessary to reach the level of lethal dosage may be, with the exception of mercury, up to an order of magnitude greater (e.g. Cd, Cr, Ni) if water hardness is high (400 ppm), than if the hardness is low. By differentiating between calcium and magnesium hardness Tabata established that for *Daphnia* the influence of calcium hardness was dominant. Thus the stress symptoms are a resultant of antagonistic effects that can be explained as chemical reactions between metal ions and inorganic compounds (carbonates of calcium). Similarly, organic ligands entering via sewage may complex copper and reduce its toxicity to phytoplankton. Tabata (41) likewise determined considerable reduction of heavy metal stress by organic substances, namely E.D.T.A., and sodium citrate. On the other hand, these organic metal-complexing agents can produce negative effects by first complexing metals in the sediments which may be then released because of rapid degradability of complexing agents.

5. STRESSORS AND STRESS MODELS

Lugo (12) developed a conceptual model of an idealized, unstressed system in order to analyze the possible ways in which stressors affect individual eco-

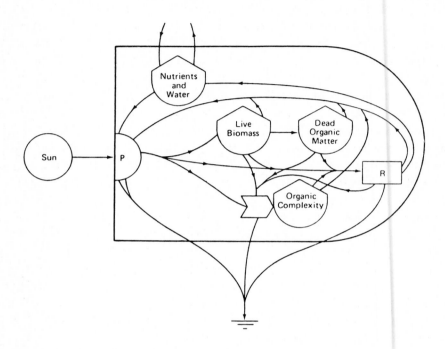

Fig. 7. Simplified model of an ecosystem without stress (Fig. 4 in Lugo (12)). The large bullet-shaped outline is the boundary of the system; P is photosynthesis; and R is respiration. Arrows show energy flows; tanks are state variables; organic complexity includes diversity which is subject to impact of stressors reaching via the big arrow-head. Both P and R and the many cycles and feed-backs contribute to the system homeostasis. For effects of stressors on the model see figures 5-8 in Lugo (12).

systems (Fig. 7). He used nutrient concentration, live biomass, dead organic matter and organic complexity (as a measure of species diversity) as state variables and photosynthesis, respiration and nutrient cycles and feed-backs as the functional ecosystem processes.

He applied five stressors to the model. These relate to: 1) the primary (solar) energy; 2) diversion of primary energy before being transformation by plants; 3) diversion of primary energy after it is transformed by plants, e.g. by created nutrient limitation in case of aquatic ecosystems; 4) removal of storages in the four state variables by predation or harvesting; and 5) acceleration of respiration rates, e.g. by temperature increase.

The model suggests that different stressors affect different parts of the ecosystem and the response perhaps also depends on the stressor type. The overall effect of the stressors is to reduce the efficiency in the ecosystem of energy flow, i.e. to prevent energy upgrading to higher trophic levels. Although overall organization and homeostasis of the system are altered, not all the stressors handicap the system to the same magnitude. For example, if the species diversity is reduced, but primary production rate continues to be high, system continues to photosynthesize at high rate and is apparently not stressed. However, it needs to be realized that reduction in diversity of phytoplankton, e.g., may be concomitant with deterioration of food quality for zooplankton grazing so that this would reduce the rate of upgrading of energy to higher trophic levels. This is often the case in shallow, hypertrophic Dutch lakes in which higher nutrient input (P, N), primarily an allogenic stressor in these lakes, leads initially to increased primary production especially of green algae that need relatively higher nutrient level. Depleted nutrient conditions but particularly self-shading which follows lead to growth limitation of green algae, thus stimulating dominance of blue-green algae, with constantly high production rates; these algae being inedible build up high standing crop but have low diversity. In short, poor light climate which is the main consequence (42) of the eutrophication stress in these shallow Dutch lakes acts itself as a stressor for the green algae and macrophytes both of which appear to be first reduced and then disappear altogether.

Similarly, in the Loosdrecht lakes in The Netherlands stressors that lead to a drastic decrease in the density of piscivorous fish (stressor type 4: e.g. by harvesting and intensive fishery) has immediate consequences in the form of feed-backs that cause an increase in planktivorous fish (Pers. Comm. van Densen), especially bream, which in turn predate heavily on the large grazer species of zooplankton (*Daphnia* spp., e.g.) relieving zooplankton grazing pressure on phytoplankton, thus leading to an increase in latter's biomass. Therefore, such endogenous stress problems have important implications in lake management studies dealing with removal of exogenous stress. That the zooplankton composition is unchanged despite the recovery measures (23) may indicate that external measures are ineffective in removing internal stresses.

5.1. Modelling stress

Most of the ecosystem models that exist relate to response of community to stress. Among these some appear to predict a non-linear response to exogenous stress and may be quite realistic (see e.g. refs. 43, 44, 45). These models are capable of predicting significant changes in response to external stresses caused by temperature, nutrients and light. However, there is no explicit mention in these simulations about critical deviation or a change in function. These models can be considered as low stress models because they do not tell us if the system would change character under increased stress.

The deficiencies of models as cited above are the cause of suspicion among the ecological managers about the modelling process (see e.g. ref. 46). What the managers would ideally like is to understand the 'collapsing' ecosystem syndrome. One example of such a collapse comes from recent field study of a Danish lake Søbygård (47). In this lake even after a 50-fold reduction in external phosphorus loading, algae and zooplankton crash every year 1-3 times. The uncoupling of the grazing food chain appears to be the work of high pH (> 10.5), an internal source of stress rather than stress caused by high predation of zooplankton by planktivorous fish. Therefore, modelling such an unstable system with recurring high stress phenomena may need modelling the combined effects of internal and external stresses that lead to such drastic deregulation in the ecosystem function.

6. STRESS PREVENTION

First, ecologist needs to provide data which have been collected systematically in order to give an insight into the ecosystem level of organisation. Such data, relating to both structural and functional system level parameters, can help to delineate the stress effects at different trophic levels in order to make generalization at the ecosystem level feasible. Secondly, with the availability now of sophisticated computer facilities and simulation models, the scientific input can facilitate valid predictions about the stress effects.

Based on these, the ecologist is required to take decisions which are acceptable both scientifically and socially. In case of non-point source stressers, e.g. acid rains, or other forms of air pollution, the decision making may involve not only scientific knowledge and social aspects but also a concerted international approach and political will of the countries involved. The acid rain problems in the Scandinavian and other West-European countries can only to a small part be solved by these countries themselves because major part of this pollution originates elsewhere, but carried with wind it deposits the emissions hundreds of kilometres away from place of origin. It is in this field that not only an international monitoring and solutions are required but professional ecologists and environmentalists -both from the terrestrial ecosystems and aquatic ecosytems- need to make their observations available to one another in order to develop a joint strategy of prevention and dealing with the cause of this non-point source of stress.

7. CONCLUDING REMARKS

Barrett (41) rightly remarks that 'ecology should remain the scientific discipline that deals with the structure, functioning, and behaviour of ecological systems'. The field of stress ecology has a unifying character in that it brings the ecologist and environmental scientist to a common platform for a proper understanding of the ecosystem problems and their solutions.

The research in stress ecology has not yet evolved into quantitative science. This is related to a lack of a quantitative definition of stress so that the stress in different systems can be compared. Also, on ecosystem basis we are confronted with the problem of trophic level differences with regard to stress definition (subsidy-stress gradient), i.e. non-linear response of the system components to the stress inputs as noted in figure 2.

The capacities of the ecosystems to resist outside stimuli (stress) differ; also the response time before the stress systems become manifest is variable. Therefore, a comparison of responses of the systems to stress is almost impossible, let alone deriving generalisations from it. The variations in resilience of the system and our inability to predict this only adds to the existing difficulties and dilemmas. Also, changes in diversity and diversity indices in response to stress have their inherent drawn-backs.

Lastly, energy budget of the ecosystem, based on the trophic level energy-flow, has often a negative balance, e.g. if energy consumed exceeds the energy produced within the system, it does not necessarily indicate a stress response. However, in case the system collapses to a point of no return stress may invariably be the cause, though it is often too late to quantify the cause and correct the damage.

8. ACKNOWLEDGEMENTS

It was a pleasant experience to be associated with ISPRA Course which brought 'land-based' ecologists, environmentalists and aquatic ecologist together for a week of deliberations. I am highly grateful to Prof. O. Ravera for inviting me to give two lectures, one of which formed the subject matter of this 'stress' paper. I also owe my special thanks to Miss Cecilia Kroon for her painstaking typing work and for looking after the lay-out of the typescript. Messers K. Siewertsen and E. Mariën were very helpful in preparing the illustrations and photographs.

9. REFERENCES

1. U. Förstner, G.T.W. Wittmann (Eds.), Metal pollution in the aquatic environment, Spring-Verlag, Berlin, 1979, 486 p.
2. S.I. Auerbach, Ecosystem Response to Stress: A Review of Concepts and Approaches, in G.W. Barret and R. Rosenburg (Eds.), Stress Effects on Natural Ecosystems, John Wiley, 1981, pp. 29-41.
3. E.P. Odum, J.T. Finn and E.H. Franz, Perturbation theory and subsidy-stress gradient. Bioscience, 27 (1979), 349-352.
4. G.W. Barrett and R. Rosenberg (Eds.), Stress Effects on Natural Ecosystems, John Wiley, 1981.
5. J.H. Thorp and J.W. Gibbons, (Eds), Energy and Environmental stress in aquatic systems. US Department of Energy Symposium, Series 48, CONF 77114, 1978, 854 pages.
6. D.J. Rapport, H.A. Regier and T.C. Hutchinson, Ecosystem behaviour under stress. The American Naturalist, 125 (1985), 617-640.
7. H. Selya and A. Horava, Second Annual Report on Stress. Act. Inc. Montreal., 1952.
8. G. Engle, Mid-Century Physchiatry. Charles C. Thomas, Springfield, Illinois, 1953, 51 pages.
9. R.C. Weast, Handbook of Chemistry and Physics. Chemical Rubber Co. Cleveland, 1969.
10. B.C. Patten, An introduction to cybernetics of the ecosystem: The trophic-dynamic aspect. Ecology, 40 (1959), 221-231.
11. R.J. Perkins, Stress and its relation to ecosystem. Int. J. Ecol. Environm. Sci., 1 (1974), 119-127.
12. A.E. Lugo, Stress and ecosystem (pp 61-101). In: Energy and Environmental Stress in Aquatic Ecosystems (Eds. J.H Thorp and J.W. Gibbons) DOE Symposium Series (Conf. 771114), Oak Ridge, Tenn., USA, 1978, 854 pp.
13. B.L. Bayne, Aspects of physiological condition in Mytilus edulis I., with respect to the effects of oxygen tension and salinity, Proc. Ninth Europ. Mar. Biol. Symp., 1975, 213-38.
14. G.W. Barrett, G.M. Van Dyne, and E.P. Odum, Stress ecology. Bio Science, 26 (1976), 192-194.
15. A.M. Ivanovici and W.J. Wiebe, Towards a working 'Definition' of 'Stress': A review and critique, 1981 (pp. 13-27), in Stress Effects on Natural Ecosystems (Ed. G.W. Barrett and R. Rosenberg), John Wiley.
16. J.R. Brett, Implications and assessments of environmental stress, pp. 69-83 in The Investigation of Fish - Power Problems, H.R. MacMillan Lectures in Fisheries (Ed. P.A. Larkin), University of British Columbia Press, Vancouver Canada, 1958.
17. C.D. McAllister, R.J. Le Brassem and T.R. Parson, Stability of enriched aquatic ecosystems. Science, 175 (1972), 562-564.
18. H.T. Odum, Energy cost-benefit models for evaluating thermal plumes, in J.W. Gibbons and R.R. Sharitz (Eds.), Thermal ecology, Conf. 730505 Natt. Tech. Inf. Serv., US Department of Commerce, Springfield, V.A., 1974, pp. 628-649.
19. E.P. Odum, The strategy of ecosystem development. Science, 164 (1969), 242-270.
20. R. Margalef, Stress in Ecosystems: A Future Approach. In Stress Effects in Natural Ecosystems (Eds. G.W. Barrett and R. Rosenburg) John Wiley, 1969, 281-289.
21. D.E. Atkinson, Citrate and the citrate cycle in the regulation of energy metabolism, Biochem. Soc. Symp., 27 (1968), 23-40.
22. A.G. Chapman, L. Fall and D.E. Atkinson, Adenylate energy charge in Escherichia coli during growth and starvation, J. Bacteriol., 108 (1971), 1072-1086.
23. R.D. Gulati, Concept of stress and recovery in aquatic ecosystem. ISPRA Course on "Ecological assessment of environmental degradation, pollution

and recovery, Elsevier, Amsterdam, in press.
24. S.I. Auerbach, Current perceptions and applicability of ecosystem analysis to impact assessment, Ohio J. Sci., 78 (1978), 163-174.
25. W. Winkle, S. van Christensen and J.S. Mattice, Two roles of ecologist in defining and determining the acceptability of environmental impacts. Int. J. Environm. Stud., 9 (1976), 247-254.
26. M.J. Dunbar, The evolution and stability environments, natural selection at the level of ecosystem, Am. Nat. 94 (1960), 129-136.
27. J.R. Webster, An analysis of Potassium, Calcium Dynamics in Stream Ecosystems on three southern appalachian watersheds of contrasting vegetation, Ph.D. Thesis, University of Georgia, Athens, 1975.
28. J.W. Leffler, Ecosystem response to aquatic microcosm, (pp. 102-119). In: Energy and Environmental Stress in Aquatic Ecosystems (Eds. J.H Thorp and J.W. Gibbons) DOE Symposium Series (Conf. 771114), Oak Ridge, Tenn., USA, 854 pp. 1978.
29. A. Jernelov and R. Rosenberg, Stress Tolerance of Ecosystems. Environm. Conserv., 3 (1976), 43-46.
30. F.B. Golley, Structural and functional properties as they influence ecosystem stability, in Proceedings of the First International Congress of Ecology, The Hague, 1974, pp. 97-102, Center for Agricultural and Documentation, Wageningen.
31. K. Kersting, Development and use of an Aquatic Micro Ecosystem as a Test System for Toxic Substances, Properties of an Aquatic Micro-Ecosystem IV. int. Rev. Ges. Hydrobiol. 69 (1984), 567-607.
32. R.V. O'Neill and D.E. Reichle, Dimensions of ecosystem theory, p. 11-26 in R.W. Waring, Ed., Forests: fresh perspectives from ecosystem analysis. Oregon State University Press, Corvallis, 1980.
33. G.E. Likens, F.H. Bormann, R.S. Pierce and W.A. Reiners, Recovery of a deforested ecosystem, Science, 199 (1978), 492-496.
34. H.A. Regier and W.L. Hartman, Lake Erie's fish community: 150 years of cultural stresses, Science 180 (1973), 1248-1255.
35. R.D. Gulati, Zooplankton structure in Loosdrecht lakes in relation to the trophic status and the recent restoration measures, Developments in Hydrobiology, Junk, The Haque, etc., in press.
36. H.A. Regier, Changes in species composition of Great Lakes fish communities caused by man. Trans. N. Am. Wildl. Nat. Resour. Conf. 44 (1979), 558-566.
37. H.A. Regier and E.B. Cowell, Application of ecosystem theory: succession, diversity, stability, stress and conservation, Biol. Conserv., 4 (1972), 83-88.
38. R. Margalef, Stress in ecosystems: a future approach, pages 281-289 in G.W. Barrett and R. Rosenberg (Eds.), Stress effects on natural ecosystems, Wiley, London, 1981.
39. H.T. Odum, Work, circuits and systems stress, pages 81-138, in H.E. Young, Ed., Symposium on primary productivity and mineral cycling in natural ecosystems, University of Maine Press, Orono, 1967.
40. T.C. Hutchinson, Comparative studies of the toxicity of heavy metals to phytoplankton and their synergistic interactions, Water Pollut. Res. Can., 8 (1973), 68-90.
41. K. Tabata, Studies on the toxicity of heavy metals to aquatic animals and the factors to decrease the toxicity, Bull. Tokai Reg. Fish Res. 58 (1969), 203-261.
42. R.D. Gulati, K. Siewertsen and G. Postema, Zooplankton structure and grazing activities in relation to food quality and concentration in Dutch lakes. Arch. Hydrobiol. Beih. Ergebn. Limnol., 21 (1985), 91-102.
43. S.W. Nixon and J.N. Kremer, Narragansett Bay. The Development of a Composition Simulation Model for a New England Estuary. In: Ecosytem Modelling in Theory and Practice: An Introduction with case histories (Eds. A.S. Holland and J.W. Day), John Wiley, 1977, 621-673.
44. D.M. Di Toro, D.J. O'Connor, R.V. Thomann and J.L. Mancini, Phytoplankton

zooplankton Nutrient Interaction Model for Western Lake Erie, in Systems Analysis and Simulation Ecology, Ed. B.C. Patten, Academic Press, Inc. New York, 1976, 3, 423-473.
45. R.A. Park, D. Scavia and N.L. Clesceri, (LEANER: The Lake George Model in Ecological Modelling in a Resource Management Framework) (Ed. C.S. Russel), 49-82. Resource for the future, Inc. John Hopkins, University Press, Baltimore, 1975.
46. R.E. Ulanowicz, Modelling Environmental Stress (pp 1-18) In: Energy and Environmental Stress in Aquatic Ecosystems (Eds. J.H Thorp and J.W. Gibbons) DOE Symposium Series (Conf. 771114), Oak Ridge, Tenn., USA, 1978, 854 pp.
47. E. Jeppesen, M. Sondergard, O. Sortikjaer and P. Kristensen, Interactions between phytoplankton, zooplankton and bacteria in a shallow hypotrophic lake: a study on short term changes in a trophic relationship in lake Søbygård, Denmark. Developments in Hydrobiology Junk, The Hague, Boston, in press.
48. G.W. Barrett, Stress Ecology: An Integrative Approach, in G.W. Barrett and R. Rosenberg (Eds.), Stress Effects on Natural Ecosystems, John Wiley, 1981, 3-12.

CONCEPT OF STRESS AND RECOVERY IN AQUATIC ECOSYSTEM

R.D. GULATI

Limnological Institute, Rijksstraatweg 6, 3631 AC Nieuwersluis (The Netherlands)

1. INTRODUCTION

Accelerated eutrophication and acidification of aquatic environments have become regional, even global problems (1) in the last 3 or 4 decades. Threat caused by these and localized cases of both inorganic and organic pollution sources is aggravated by our poor knowledge of the environmental pathways, metabolism and toxic effects of pollutants discharged, often indiscriminately into our aquatic systems.

The recovery potential of a damaged ecosystem depends on the extent of damage caused, type of perturbation and mode of operation of the stressor (2). The major types of perturbation include nutrient enrichment, among which phosphorus increase is the cause of chief concern in most cases. Acidification and addition of toxic substances are other causes of ecosystem damage, at almost all trophic levels. Quite a few other types of stresses which have called attention for corrective measures are thermal discharges from power plants as well as the related impingement and entrainment of many planktonic organisms. In many cases introduction of alien species to a lake may have serious 'side effects' making recovery difficult.

The extent to which a system may respond to anthropogenic stresses and the degree to which the system can recover after the stresses are removed are subjects of fundamental importance in resource management studies, see e.g. Schindler *et al.* (3). However, only a few recovery studies deal the effects of stresses on whole ecosystems. Several factors make an ecosystem approach difficult in recovery studies. Among these, lack of basic background data for the period before the lake got polluted is of great importance. This may especially be important in understanding the 'chronology' of deterioration processes and mechanisms causing them. Also, it may provide insight into mechanisms behind reversal processes in the course of rehabilitation. It is difficult to generalize mechanisms and sequence of steps in order to initiate recovery studies. Invariably, the location, depth and area of lake to be restored determine the decision making; other important factors in this regards are the costs involved and economic and recreative utility of the water-body, etc. Also, nature of measures planned for recovering a lake depends to an important extent on if the lake is deep and stratifying type, or shallow and continually mixed; the nature of hypolimnetic oxygen conditions, bottom sediments and extent of aquatic macrovegetation are some of the factors which need to be considered in the planning of recovery studies.

The choice of restoration measures and procedures for management of lake eutrophication depend on the nature of waterbody and its function. Lakes and reservoirs, even if they are generally similar fundamentally in their physico-chemical and biological conditions may differ essentially in their drainage areas and their hydrological and morphometrical features. The reservoirs have usually greater surface area and greater mean and maximum depths but shorter hydraulic residence time (4). These characteristics, together with relatively higher municipal and agricultural discharges into the reservoirs, lead to much higher nutrient loading rates in reservoirs than in lakes. In short, these differences emphasize the need to recognize that in lake management studies, dealing with eutrophication processes in these systems, strategies and approach may differ as well.

The recovery and restoration of lakes is generally a reversal of man-made pollution. In most cases recovery should lead to reduction in eutrophication levels acceptable to man.

More than 2 decades ago, i.e. in mid-sixties, certain Swedish lakes (5) and Lake Washington (6) in the United States were among the first lakes in which restoration work was succesfully attempted. The rehabilitation measures in most such studies have been aimed at reducing drastically both the internal and external nutrient loading rates. To achieve these purposes, several direct options now in use relate to diversion of sewage input and its treatment, inactivation of nutrients in the input water, hypolimnetic aeration, artificial bubbling and regulation of pH, etc. Recently, liming of acidified waters, and biological regulation, the so-called biomanipulation, have been practised with success in many waters, simultaneous with the other restoration measures. The latter technique offers same promising field of biological approach using the feed-back regulation of the food chain.

This paper discusses briefly some of the major techniques that have been more universally employed in the last 2 decades or so, both in Europe and North America, even though with limited success. Some case studies are described with remarks on the successes and failures. The choice of the examples cited is rather arbitrary but covers a wide range of geographical conditions and differences in approach and techniques used.

2. LAKE RECOVERY: CONCEPTS, FEASIBILITY AND TECHNIQUES

2.1. Concepts and principles

The concepts and principles of current lake restoration do not appear to essentially differ from those in the days when approach to limnology was a more classical one. However perspectives and approaches appear to have changed with the availability of new techniques for applying their concepts and principles to lake management (1). Vollenweider (1) provides a scheme of strategies pertinent to lake management. First, the 'object of and for restoration' need

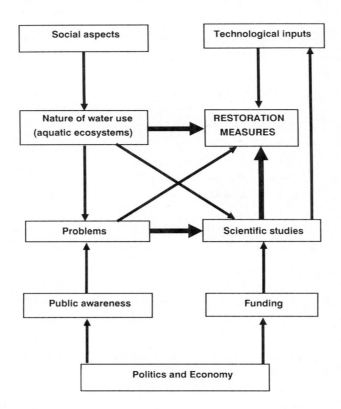

Fig. 1. Strategic principles of lake restoration based on modification of figure 1 in Vollenweider (1).

to be defined clearly. An important factor in this respect is a proper understanding of the nature of problems relative to specific water use before corrective measures are possible. The nature of these measures will depend on a thorough scientific knowledge of the structure and functioning of the ecosystem involved. Public awareness of the problem area will generally initiate a series of steps. First among these is basic research to determine the nature of problems. This, together with extent of funding, will form the basis of management measures to be taken applying the technology available so that users objectives are met with.

The scheme in figure 1 depicts strategy of lake restoration in relation to social aspects and public awareness on one side and funding and technological inputs on the other. Nature of water use and problems will determine the course of scientific studies and choice of restoration measures. Last, but not least, role of prevailing political climate and economy may not be underestimated, because of its ultimate role in decision making and its execution.

2.2. Diagnosis and feasibility studies

Although several restoration methods are now available (6), they may not be universally applicable to all lakes, nor they may be equally successful in all lakes. Before implementing a recovery technique the uses of the lake and causes and problems must be critically evaluated.

The diagnosis and feasibility studies involve insight into inlake limnological conditions and watershed characteristics. Lake area and depth contour studies are essential to get information on mean depth, water volume, etc. Measurements of physico-chemical factors include water temperature, light climate and water chemistry (dissolved oxygen, pH, alkalinity, nutrient concentration, i.e. nitrogen and phosphorus, both in dissolved and particulate forms) and analysis of lake sediment cores to determine fractions of phosphorus, both organic and refractory. Biological examination will include study of: 1) composition and biomass of phytoplankton, as chlorophyll or as biovolume, as well as of seston, expressed as carbon and dry weight, light and nutrient limitation; 2) zooplankton species abundance, relative dominance of the different taxonomic groups in the biomass, extent of dominance of large-sized herbivores, and a broad knowledge of food and feeding conditions of the zooplankton; 3) areal coverages of the different macrophyte types; 4) fish, including knowledge of the important planktivores and piscivores and their major food sources; and 5) benthic organisms. The biologial aspects mentioned, together with the physico-chemical conditions, will give a global insight into the structure and functioning of the lake ecosystem being diagnosed.

Major points that need special attention in such diagnostic and feasibility work are: 1) the existence of blooms among phytoplankton, especially the ones dominated by blue-green algae (cyanobacteria); 2) absence or low densities of large-bodied cladocerans among the crustacean zooplankton with the relative increased dominance, instead, of small bodied cladocerans and rotifers; and 3) if the condition and density of piscivore fish are strikingly poor, or low and if these predatory fish are in a state of equilibrium with their food resource, i.e. their major prey the planktivore fish. The existence of conditions summarized under these three points would generally indicate: a) deterioration of light limitation and poor utilization of the primary producers in food chain; and b) high size-selective predatory pressure of planktivores on zooplankton caused generally by poor growth conditions of piscivore fish. Briefly, in such cases response of the corrective or restoratory measurements, aimed at external nutrient load, may be slow or inadequate. Thus, there may be indispensible need to combine these steps with inlake measures especially that lead to a more stable biological equilibrium accompanying a decreased nutrient loading.

From the data collected, namely those pertaining to Secchi-disc depth (for light climate), chlorophyll, phosphorus and nitrogen, Carlson's index (7) of eutrophication may be determined. Such an index can be computed from one of the three equations depending on availability of data on the first three factors. This index may provide useful comparative information if several lakes in the same vicinity are involved in restoration study. However, the value of this index may be limited if the waters are coloured or turbid, as is often the case in shallow, well mixed lakes with rich peaty bottom and high content of dissolved humid substances. Also, some lakes undergoing restoration may exhibit nitrogen deficiency during the recovery phase and thus deviate in several ways from Carlson's index (8). This will be true particularly in reservoirs. Also, according to this index, macrophyte-dominated lakes, having relatively high clarity, because of low chlorophyll concentrations, may be misinterpreted. Nevertheless, if such situations, as mentioned, do not exist in lakes under restoration a decrease in Carlson's index would generally indicate the effectiveness of recovery measures.

Collection of information on watershed will comprise deliniating the drainage basin of the lake under study, identifying the point-source discharges (including storm and sanitary drainage) as well as non-point sources of discharge. A clear distinction needs to be made of nature of these discharges and contributions herein of urban, rural and agricultural areas.

Determinations in the inlet water of nutrients (N & P) loads, suspended solids and of dissolved materials, though difficult and time consuming may provide useful information on nutrient budgets and the role of external loading. Similar data on water exported from the lakes is also needed to put up loading models. Combined with inlake limnological data and sediment-nutrient dynamics, input-output model of phosphorus may be constructed for use in lake management. A lot of basis work has been done in the last 15 years or so (see e.g. 9, 10, 11, 12). The major difference among the models lies in phosphorus sedimentation estimates, as well as in retention and release coefficients within the lake.

2.3. The available techniques

Lake recovery generally involves rehabilitating a lake by manipulating with lake ecosystem to remove the undesirable condition. Ryding (13) and Uhlmann (14) have briefly discussed the widely used lake rehabilitation techniques. Uhlmann gives an excellent evaluation of the strategies involved in controlling eutrophication in lakes and reservoirs. Ryding has given an extensive list of references available up to the late seventies.

There are two types of techniques now available for rehabilitating lakes. First, the methods involved limit fertility by reducing nutrient loading or sedimentation rates, or both, thus dealing directly with cause of lake problems. Second, there are now procedures available which, without reducing the fertility, involve management of lakes by improving the oxygen regime by aeration, destratification or oxidation of sediments.

Techniques which have been successfully employed in lake rehabilitation may be divided into two groups: 1) in which emphasis is on nutrient control and 2) that emphasize on direct control or removal of biomass. This division is essentially the same as given in some recent other literature on the subject (6). Vollenweider (1) divides these techniques on the basis of their nature in physical, chemical, or biological. Here, they are discussed shortly without clearly catagorizing them, but based more on their importance, applicability and historical development. Many of these techniques may not work in isolation but in conjunction. Moreover, choice of a techique may often depend on specific situation rather than on general preference for a certain technique.

2.3.1. Nutrient diversion or removal

Diversion of nutrient input is widely recognized as essential first measure to obtain improvement in lake trophic state (6). Nutrients enter water body from different external sources, both point and diffuse. The diversion involved re-routing the nutrient-rich water from lake's drainage basin and nutrient removal and sediment reduction is achieved by retaining them by chemical inactivation and setting the treated water before it is let into the lake.

Despite highly efficient treatment plants to remove phosphorus from inlet water, phosphorus inputs via point-sources may continue to be high. Also such treated waters may often be significant sources of specific ions used in flocculation processes, such as aluminium, iron, calcium (13). Also, even if P-concentrations may have significantly declined, improvement to a lower trophic state may occur in a limited number of cases (15). This may generally happen because either nutrient diversion is insufficient or volume of water let in annually may exceed a certain level that will produce higher than permissible P-load, thus producing no effective response to diversion.

Also, a continued nutrient release from the lake sediment into water column will retard or even prevent a response of lake under restoration to nutrient diversion and removal. However, depending on lake mean depth and water volume and flushing rate, data on inlake nutrient dynamics as well as on the loading from external sources, will provide information on response time to reach a desired low nutrient concentration. The techniques of nutrient diversion if combined with sediment removal and dilution may accelerate recovery.

Data in examples of lakes restored (section 3 of this paper) will show how very important the nutrient diversion and nutrient removal are among the techniques in use to rehabilitate lakes.

2.3.2. Sediment removal

Because phosphorus loading from sediment forms a significant proportion of total phosphorus loading in many lakes, sediment removal is likely to eliminate the internal supply considerably so that overall water quality may improve. Using this technique the lake may become deeper depending on the thickness of 'mobile' sediment. An other advantage of removing sediment is that toxic materials may also be got rid off. Also macrophytes may be removed. All these manipulations would generally lead to uncovering a sediment stratum that is poor in nutrients so that internal loading is reduced drastically.

Peterson (16) has discussed several problems related to sediment characterization and dredging and their potential adverse effects on the lake's environment. First, dredging and sediment removal may result in resuspension of the sediment in the water column, leading to nutrient release. Wind action will stimulate mixing of nutrients within the euphotic zone, thus increasing the potential for algal blooms. However, this is not always the case because high turbidity due to sediment resuspension may limit algal increase. There is evidence of this from sediment removal by cutter-head hydraulic dredging in Lake Herman, South Dakota, USA (17). Problems of increased algal production, even if they occur, are likely to be of short-term and negligible compared with the long-term benefits (6).

Secondly, because toxic substances are generally attached to small-sized sestonic particles (diameter < 74 um) as observed by Murakami & Takeishi (18) for polychlorinated biphenyls (PCB), low settling rates of the resuspended sediments may be potentially a serious problem.

Also, benthic organisms which form a major food source for fish, may be removed with the sediment so that both commercial and game fishery will be adversely affected. This problem can be, however, greatly overcome by leaving a portion of the lake bottolm undredged for the benthic food organisms to recolonize the dredged area; the re-establishment of these organism in the dredged areas is apparently quite fast, as demonstrated by the work of Andersson *et al.* (19). The problem of disposing the dredged sediments, although a non-lake problem, may not work stimulatory in choosing sediment removal as lake rehabilitation-technique. This will be especially true if there is evidence of sediment contamination with toxic materials, including heavy metals. The promulgation of Public Law may prohibit without a permit the use of wetland areas as disposal sites if these wetlands exceed certain areal limits. In the United States the area limit above which disposal is not allowed is 4 ha. Also, flooding the wooded areas with the dredged sediment slurry may kill vegetation, including trees. Diking in upland areas with the sediment is also not without problems - groundwater contamination if underlying bottom is sandy, or contains gravel.

In short, the problem of sediment disposal has to be included in the preliminary planning and programme of restoring a lake by sediment removal. This will also include a thorough examination of the sediment chemistry with special reference to the contaminants and their toxicity potential for lake biota.

Among the lakes restored using sediment removal, Lake Trummen in Sweden is perhaps the most well known example (13, 19, 20, 21, see also section 3 in this paper).

There have been several attempts in the United States (6) to first predict via model the depth to which a lake must be dredged to control adverse nutrient exchange from the sediment in summer; also dredging depth needed to control the regrowth of macrophytes has been predicted in several Wisconsin lakes. This latter was done by regressing the maximum depth of plant growth (Y) on average summer water transparency depth (X) as measured by a Secchi disc. The relationship is described by the equation $Y = 0.83 + 1.22X$, in which both X and Y are in metres.

However, Canfield *et al.* (22) have re-examined the relationship in more than 100 lakes ($LogY = 0.26 + LogX$), more than half from Wisconsin (N=55) and rest from Finland (N=27) and Florida (N=26). Based on an emperical model they demonstrated that factors other than Secchi-disc depth (SD) may also contribute to the maximum depth of colonization (MDC) of macrophytes, such that for the same SD the MDC can widely vary. Differences in nutrient availability, nature of substrate and geography may all contribute to the variations.

A complete macrophyte removal may, however, be undesirable considering the needs for fish spawning, bird feeding, wild-life habitat, etc. In some recent lake management studies, macrophytes were introduced after lake-level drawdown and refilling in order to create fish refuges (23).

Concluding, a number of variables may determine the suitability of a lake for dredging, but the shallow ones are more suitable. Also, low sedimentation rates, richness of organic material and long hydraulic residence time will contribute positively to suitability for sediment removal by dredging. Cost of dredging are apparently prohibitive. Avoiding extreme cases cited by Cooke *et al.* (6), sediment removal method is about 30 times more expensive than, e.g., nutrient inactivation method (mean costs of the methods: $ 12691 ha^{-1} *versus* $ 400 ha^{-1}). However, the costs need to be evaluated on the basis of long-term benefits, i.e. longevity of the desired effects. If the costs for the 2 techniques are amortized over the post-treatment years, in which period benefits of restoration still exist, the cost differences between the 2 techniques may not diverge that much. One such comparison made by Cooke *et al.* (6) shows that costs in Lake Trummen nine years after the treatment (dredging) translate to ca. $ 159 ha^{-1} as against $ 85 ha^{-1} for alum-treated lake West Twin (Ohio, USA), 5 years after treatment. Lastly, sediment removal by dredging is undoubtedly an effective lake restoration technique, provided there is a thorough preimplementation evaluation regarding the choice of equipment, sediment disposal areas and potential lake uses after lake restoration.

2.3.3. Nutrient inactivation

The nutrients (P) may be inactivated or precipitated in order to make them unavailable to plankton and waterplants. The technique has been used for removing phosphorus from the photic zone and for preventing its release from the sediments and recycling in the lakes. Phosphorus which is sorbed on to salts of aluminium (hydroxide, sulphate), the most commonly used P inactivants, is not released even after the onset of anoxia.

Phosphorus will be generally removed as $Al_2(SO_4)_3$ precipitate by sorption of P on the surface of $Al(OH)_3$ polymer or on floc and by entrapment of P containing particulate matter in the floc (6). This latter is most efective in the pH range of 6-8. Removal of inorganic P is directly dependent on Al:P ratio. The per cent removal of organic P is apparently less than other P forms and may be of critical importance if P-limiting coditions prevail. Aluminium may be added to lake surface as in several earlier studies, or to hypolimnion, thus avoiding the biota of littoral zone to exposure of aluminium dose.

The dose of aluminium sulphate to be added must be determined before hand. For this, information on pH, alkalinity and possible toxic effects of metal in dissolved form (Al III) is needed. Initially on adding aluminium sulphate the pH decreases in the range 6-8 will be accompanied by formation of $Al(OH)_3$, with formation of very little dissolved Al. However, if the pH decreases below 5.5 concentration of dissolved aluminium increases rapidly. Cooke *et al.* (24) adopted 50 ug Al.l^{-1} in dissolved form as the safe upper level, based on long-term studies on rainbow trout. Generally, the concentration of dissolved Al will be lower than 50 ug.l^{-1} because pH values of most lakes are in the range 5.5 - 8.5. However, in lakes receiving acid wastes, via point sources or acid rains, the buffering capacity may be low so that Aluminium Al (III) may reach values that are toxic to biota, even though data in this respect are quite restrictive.

The problem of acidity (pH < 5.5) on adding aluminium sulphate may be overcome by using sodium aluminate, instead. From the studies of Kennedy and Cooke (25) it appears that in lakes with pH of about 7.4, a dose of about 15 mg Al.l^{-1}, added as aluminium sulphate, will reduce the pH to ca. 5.7 so that dissolved Al concentration will be within safe level.

The literature on nutrient inactivation of lakes shows that the Wisconsin Department of Natural Resources (USA) did pioneering work with respect to use of aluminium sulphate to rehabilitate lakes. In Horseshoe lake (Wisconsin), a eutrophic hard water lake, the P concentration on whole lake basis, was in 1982 about 17% of the concentration before the treatment in 1970. Apparently, the aluminium hydroxide layer was buried but in the twelve years after this has continued to control P release (26). The low concentrations in the water column of the lake (0.55 mg P.l^{-1}) twelve years after the treatment reflect at reduced loading from the sediments despite continuing agricultural run off. Also, light climate and plankton abundance indicate an improved situation. This study, together with one on Lake Snake, a small lake, demonstrates that aluminium treatment is effective on long-term basis and treated sediments continue to sorb P for 10-12 years after the application.

Cooke *et al.* (6) cite several other examples of lakes thus treated from other parts of the United States, including some shallow ones. In these latter type of lakes due to holomixes and stirring up of the sediments aluminium hydroxide floc may redistribute, thus exposing old untreated sediments which may freely release P into water column. Also, Welch *et al.* (27) pointed out that alum application may not be appropriate in shallow lakes with macrophyte problems. This is because alum treatment reduces P availability to phytoplankton, thereby increasing light penetration so that macrophyte growth may be enhanced. This may necessitate employing other rehabilitation techniques, namely dredging, harvesting or draw-down. It is likely that addition of alum annually, at a time when macrophytes die and their decomposition rate is at its maximum, may efficiently intercept P-release.

A more effective means of reducing phosphorus in the incoming water is by combining detention of the water in a pre-reservoir to eliminate first a major part of P by sedimentation before applying a treatment. The incoming water may also be treated with aluminium sulphate or iron, and diverted into a sedi-

level and trophic state. The results obtained would be usually temporary unless the source of loading is controlled. The extent of success, i.e.-long-term recovery achieved, may vary from lake to lake. To maintain the recovery or restoration level obtained, follow-up studies in the rehabilitated ecosystems should concentrate on an eventual reappearance of stress and strain and on methods of correcting them.

2.3.9. Liming

Because of acidification of thousands of lakes by the deposition of acid in precipitation and dryfall, both in Europe and North America in the last 2 decades or so, liming has been applied as a recovery technique, especially in Sweden, Norway, and in Eastern Canada and New York State in North America (45). In coloured lakes, e.g. in Wisconsin and Michigan in the United States, liming has been used to increase productivity. Most popular materials used for liming are pulverized limestone ($CaCO_3$), slaked or hydrated lime ($Ca(OH)_2$), and quicklime (CaO). The first-named form of lime is preferred because, though less reactive than hydrated lime, or quick lime, it is less caustic and less expensive. For a summary of techniques used for application of lime, the reader is referred to Fraser et al. (45). In relatively small lakes with short renewal times and high humate content in the sediments, soda ash (Na_2CO_3) diluted with lake water is injected into sediment as a restoration measure. This method utilizes ion exchange capacity of the humates in the neutralization process to exchange H^+ for Na^+ (46). In a small Swedish lake Lilla Galtsjön, average alkalinity and pH in the critical spring run off period increased from zero to more than 120 meq.l^{-1} and from < 5.0 to 6.7, respectively (47).

The increase in alkalinity on adding lime, e.g. dissolved limestone, is generally dose dependent in that for each mole of limestone addition alkalinity increase by two moles. The decrease in acidity manifests as a decrease of one mole with one mole lost as the evolved CO_2. Longevity of the chemical effects of liming has been a subject of discussion and disagreement. Whereas the Swedish scientist argue that $CaCO_3$ continues to dissolve for several years after liming -so that recovery effects may continue- the Canadian workers are of opinion that $CaCO_3$ becomes non-reactive a short period after application, either due to complexing with sediments or coating with metals or hydroxides. In short, there is a great uncertainty about long-term, or sustained recovery on lime treatment. The changes in lake chemistry and consequently in the lake biota and food chain because of liming depend greatly on the extent of acidity prevailing prior to treatment and on the nature and extent of treatment. The word 'recovery' in such cases is perhaps a misleading usage since the lake may not necessarily revert to old limnological conditions but a new equilibrium between the prevailing abiotic and biotic factors may occur. The dominance of dinophycean phytoplankton in the acid lakes and their slow replacement by, e.g. diatoms or chrysophycean phytoplankton, on liming is among the most common feature of lakes recovering from acidification (48). Similarly, *Sphagnum* dominating the macrophytes of acidified lakes may decline; also zooplankton will decrease.

Accompanying the recovery process in the lakes being neutralized there are potential adverse affects on biota, e.g. toxicity of aluminium, especially during its hydrolysis. Fish kills abserved on liming were attributed to aluminium toxicity (48) and are apparently the greatest in the pH range of 5.2 to 5.4, in which range $Al(OH)_3$ is insoluble and may, therefore, precipitate and accumulate on the fish gill surfaces (6). This type of aluminium toxicity is not essentially connected with the neutralizing affect of liming, but is also likely to occur during the acidification phase when the pH decreases to levels in the range 5 to 5.5, as discussed earlier.

Through the liming technique is an effective means of neutralizing acification, there are, nevertheless, potential adverse affects on biota due to aluminium toxicity. Also, it appears that liming is only a short-term cure, with long-term recovery and improvement expected from reducing industrial emissions, a global problem at the time being.

2.3.10. Biomanipulation

Most of the available lake restoration techniques discussed so far, and those on which the case studies are based (section 3), relate to manipulating with the abiotic factors, i.e. reducing or removing or diverting the external nutrient (N and P) inputs to the lakes and regulating the internal loading by inactivating nutrients in the sediments, or removal of sediments by dredging, etc. However, while monitoring lake recovery the biotic components (algal growth and composition, chlorophyll, zooplankton abundance and composition, fish, etc.) or factors controlling their effects (e.g. under water light climate, nutrient availability, zooplankton grazing pressure, and predation by fish, etc.) are used as restoration indicators.

Recently, however, the so-called 'biomanipulation technique' has opened a new area of studies in lake rehabilitation although the technique is complementary rather an alternative to the aforementioned restoration methods (2.3.1 to 2.3.9). The emphasis of these studies lies in manipulating with the food chain (49, 50): algae - herbivore zooplankton - planktivores (fish) - piscivores (fish). This biological approach is aimed mainly at altering the food chain to favour increased grazing on algae by zooplankton (by relieving zooplankton predation by the planktivorous fish). This is achieved by bringing more piscivorous fish in the system to reduce or eliminate the planktivores and fish that recycle nutrients. The biomanipulation technique is no alternative to the non-biological means of lake rehabilitation discussed in this chapter, but if applied in conjunction with the other methods rather than in isolation can hasten recovery process. For a detailed discussion on the applications and results achieved see useful reviews by Shapiro and co-workers (49-52), Benndorf and co-workers (53), Bernardi (54), and section 3.2.5.

3. RECOVERY OF LAKES UNDER STRESS: SOME CASE STUDIES

3.1. General considerations

Reducing pollution of aquatic ecosystems by remedial measures is removing stress. The response is reflected in recovery, the rate of which is a measure of effectiveness of the preventive steps taken. However, interaction among the abiotic and biotic variables makes predictions difficult. The choice of rehabilitation measures in lakes will vary depending upon the nature of stress or pollution effects which need to be removed. In most lake ecosystems integrated implementation of several techniques may be needed to restore lakes.

Ryding (13) presents a scheme of interacting factors which needs consideration for assessing the recovery of polluted ecosystems (Fig. 2). The

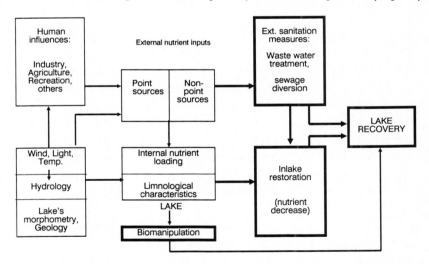

Fig. 2. Schematic representation of factors causing lake pollution and measures (thicker lines) leading to recovery; modified from Ryding (13).

scheme outlines interactions among the important characteristic -both inlake as well as in the water shed- which affect the recovery processes of lakes. The emphasis in the scheme lies on decreasing nutrient concentration, especially in the input sources, by both lake remedial measures, e.g. sewage treatment and nutrient diversion as discussed in the earlier section. Manipulations with the in-lake chemical and biological characteristic will also contribute to decreased nutrient supply.

3.2. A survey of case studies

Large number of lakes in Europe and North America have been restored in the last 15 years or so, or are in the process of restoration (55). The time taken for the recovery is apparently connected with the morphometry, size, depth and water volume, nature of bottom sediment and extent of their enrichment with nutrients. However, the acceleration of recovery process is also greatly dependent on the control measures to reduce external loadings, besides reducing the internal supply, especially from the sediments.

There have been also attempts to model restoration process although with partial success. One has to bear in mind that initial recovery is no guarantee that predictions will be *per se* valid in longer-term. Therefore, the lasting positive effects can only show us about the fruitfulness of measures taken and their general applicability on a wider scale, at least in lakes of similar type for which such a model was initially developed.

There have been some attempts to collate the existing information on restoration work and lake recovery. In a summary of restoration work presented by Forsberg in Sas and Vermij (56) five types of lake responses can be distinguished among the 46 lakes examined (Table 1). In four of the five types identified, there was a reduction of P but only in the first two response types also chlorophyll decreased. The change in trophic state was limited to the

TABLE 1

Summary of lake response types based on a world-wide survey by Forsberg of lakes in which P-reduction was attempted as a restoration measure; cross, no response; dash, response not always clear; and R, response to restoration measures; for details see in Sas and Vermij (56).

Parameter	Response type				
	1	2	3	4	5
Phosphorus reduction	R	R	R	R	x
Chlorophyll reduction	R	R	x	x	x
Restoration success	R	-	-	x	x
No. of lakes	15	12	7	8	4

response type 1 only (15 out of 46 lakes). It appears that at P concentration of 100 ug.l^{-1} (response types 3, 4 and 5) resilience frequently prevents trophic change. This resilience may be apparently connected with N/P ratios and light climate which favour only a few algal types. For example, species like *Oscillatoria agardhii* would tend to dominate despite initial P-reductions, provided the light-conditions are poor so that competition for nutrients from other algal groups is negligible.

3.2.1. Swedish lakes

Ryding (13) gives examples of a dozen Swedish lakes, mostly shallow (mean depth < 6.0 m in ten cases), in which external phosphate supply was reduced in

TABLE 2

Changes in three restoration parameters in 12 Swedish lakes after a reduction in phosphorus supply during 1972-74; data are means of values in June-September; values in parentheses are for 1977 i.e. 3-5 years after restoration (modified from Ryding; see text).

Lake	Year of restoration	Total-P $(mg.m^{-3})$	Chlorophyll *a* $(mg.m^{-3})$	Transparency (m)
Malmsjön	1973	820 (225)	84 (44)	0.4 (0.8)
Uttran	1973	69 (60)	20 (24)	1.7 (1.4)
Ekoln	1972	78 (44)	25 (10)	1.6 (1.5)
Sillen	1972	57 (64)	48 (56)	1.0 (0.9)
Boren	1974	40 (16)	9 (5)	1.4 (2.1)
Ryssbysjön	1973	460 (116)	39 (30)	0.8 (1.1)
S. Bergunda-sjön	1973	1160 (670)	150 180)	0.4 (0.5)
Molkomsjön	1972	16 (24)	12 (8)	1.9 (1.6)
Glaningen	1973	720 (240)	22 (13)	1.3 (-)
Gåran	1972	116 (63)	30 (35)	1.1 (-)
Ramsjön	1972	605 (334)	160 (89)	0.4 (-)
Näsbysjön	1972	87 (67)	48 (28)	0.7 (0.6)

1972-74 by between 30 and 95%. The effects of these reductions on chlorophyll *a* concentration, N:P ratios and transparency were followed up to 3 to 5 years later. The main conclusions from these studies were: 1) a reduction in P but an increase in N contributed to increased N:P ratios; and 2) in half of these lakes that were recovering, P became growth limiting instead of the N as before the recovery process. The trends of changes in the common eutrophication parameters, phosphorus, chlorophyll and transparency (Table 2), after reduction in supply of nutrients to these Swedish lakes are difficult to explain. Short residence times, varying from 7 days to one year except in 2 lakes, do not seem to prevent the lakes from accumulating large amounts of P in the sediments which P may easily recycle back into water, a process that will delay recovery. Water temperature and pH may also play an important role.

The schemes to reduce nitrogen may actually result in conditions that promote nitrogen-fixing algae, thereby frustrating the efforts to rehabilitate. If no signs of decrease in algal standing crop are observed, carbon:nitrogen: phosphorus ratio of the algae may be helpful in predicting the future recovery rates. The study by Ryding (13) ends in a none-too-optimistic note: "no approach seems to be available today for accurately describing the recovery of shallow, polluted lakes".

Lake Trummen Recovery Project: Lake Trummen, a small (1 km²), shallow (max. depth 2.2 m), situated in South Sweden in the proximity of the town of Växjo received up to late fifties (1958) grossly polluted sewage. The lake thus became highly polluted. The initial efforts to improve water quality by sewage diversion did not produce the desired results. Suction dredging of sediments and macrophyte elimination were carried out in 1970 and 1971 (20). The recovery of the lake was followed in the 3 subsequent years. Bengtsson et al. (21) compared data of the prerecovery and postrecovery periods. The decreases in concentrations of P in the sediments and in interstitial water are striking (Table 3). In the upper 10-cm sediment postrecovery P concentrations were only 0.2-4.0% of those before recovery. Also, P release rates of the anoxic sediment decreased considerably.

The gross composition of the new surface sediments became comparable with that of oligotrophic lakes in the region. The interstitial water from the new sediment, unlike from the old sediment, inhibited rather than stimulated algal growth in bioassay.

A four-fold decrease occurred in organic carbon which was 80 mg $C.l^{-1}$ before restoration. The particulate material comprised only a minor part of the

TABLE 3

Range of phosphorus concentrations (PO_4-P mg.l^{-1}) in interstitial water of sediment in lake Trummen (Sweden) before (2 Feb.-4 Sept., 1969) and after recovery (13 Feb.-5 Sept., 1973); data are from Table 1 in Bengtsson et al. (21).

Sediment stratum (cm)	Before restoration	After restoration
0-10	2.38 - 2.39	0.006 - 0.099
15-25	1.14 - 1.49	0.020 - 0.154
30-40	0.65 - 0.75	-

organic carbon concentration of 20 mg $C.l^{-1}$ after restoration so that there was sufficient dissolved organic material for bacterial growth and abundance, both of which do not seem to have been affected by the restoration measures. The number of heterotrophic bacteria that developed under anaerobic conditions increased by an order of magnitude (57) after restoration.

Before restoration the phytoplankton, which was comprised mostly of filamentous blue-green algae (cyanobacteria), as well as one colonial form among these (*Microcystis aeruginosa*) in summers, changed drastically (57). Chrysophycean forms (*Mallomonas* sp., *Synura* spp and *Dinobryon* spp.) became abundant after restoration in winter. In summers, diatoms, green algae and some cyanobacterial species occurred. The species diversity expressed as Shannon index increased from < 2 before restoration to 2-4 thereafter. The total phytoplankton biomass was also reduced considerably.

The phytoplankton primary production (A_{max}) decreased from about 10 mg $C.l^{-1}.d^{-1}$ before restoration to one-fifth of this rate after restoration. The areal production decreased less after restoration because of increase in the trophogenic depth in response to deeper light penetration. The nanoplankton (< 10 um) contributed to two-third of the community primary production after restoration; although comparative fractionation of primary production was not done before the restoration period, the phytoplankton fraction below 45 um constituted 20-40% of the total production.

Increased nanoplankton production appears to be important for zooplankton development (19). The initial response of the crustacean zooplankton as well as rotifers was a drastic decrease after restoration (except *Keratella cochlearis*, a rotifer, which increased). Such a decrease appears to be caused by a decrease in phytoplankton production rates and in phytoplankton biomass (57).

The macrobenthos was present in low abundance (1000-2000 ind.m^{-2}) before restoration perhaps because of harsh environmental condition prevailing in the sediments (19). The impact of dredging was to stimulate tubificids and chironomids, especially *Chironomus plumosus*. A study of effects on fish was complicated by fish-kill in severe winter of 1970 so that the catches in 1971 onwards do not give a true picture regarding the effects of restoration. Besides, the near-disappearance of crustacean zooplankton after restoration as well as removal of reed-beds (fish habitat) during the dredging process may produce both food shortage and stress; and may have contributed to shortage for spawning grounds for fish. The data, therefore, do not permit conclusions if the changes in fish structure and population are caused by restoration alone.

Other Swedish lakes: In many other Swedish lakes real improvements have been observed by reduction in phosphate levels existing 40-50 years ago (58). Delayed recoveries were usually found in lakes with high internal loading from the sediments in summer.

In Lake Vättern, a well-known oligotrophic lake, the second largest (1912 km², mean depth 39 m) among the Swedish lakes, the water quality deteriorated during the 1960's by municipal and industrial discharges. By 1967, the Secchi-disc depth had decreased to an annual minimum of 6 m from 15-17 m, 55 years earlier. Through sewage treatment and marked reduction in discharge of organic matter, the annual P-input has been reduced from 200 tons P.yr^{-1} in late 1960's to 80 tons P.yr^{-1} now. The lake has recovered from the pollution. This is noted by an increase in Secchi-disc values to ± 15 m in 1976-80, reduction in phytoplankton standing crop and decrease in total P-concentration (see Fig. 3 and Table 1 in Ref. 58), besides shifts in algal composition from greens and blue-greens to chrysophycean forms.

Lake Mälaren, another big lake (1140 km^2) was also restored in first half of 1970s, with significant reductions in concentrations of P and chlorophyll as first important indicators of recovery; also following the recovery chlorophyll concentrations decreased remarkably in Ekoln bay of the lake. The recovery of Lake Norrviken, a much smaller lake (area 2.7 km^2; mean depth 5.4 m), which was heavily polluted until 1969 has been well documented (59). This could be achieved by sewage diversion. However, remobilizations of N and P from the sediments have delayed further improvement of Lake Norrviken and the lake has stabilized at the 1976 level.

Thousands of lakes in S- and SW Sweden were acidified in the early eighties with serious biological damages. Coincidently, ecological studies in these lakes were started just before acidification was first observed. Sulphate reduction has resulted already in partial recovery of a number of these lakes.

3.2.2 North American Lakes

Lake Washington: The recovery studies on this lake, which received increasing amounts of secondary sewage effluents from 1941 to 1963, are one of the first and are among the longest monitored both before and after restoration measures. As early as 1955, a detailed study of the lake was initiated after the appearance of *Oscillatoria rubescens*. Although nutrient loading data are documented (60) from 1950 the sewage input which was the cause of deterioration was diverted in the 1963-1968, with almost a complete diversion of the effluents by early 1977.

The total phosphorus and total nitrogen inputs to the lake were reduced from 204 x 10^3 kg P.year^{-1} and 1419 x 10^3 kg N.year^{-1} in 1964 to about one-fifth and one-half of these rates, respectively, in 1976. This reduction was achieved by building secondary sewage treatment plants in the towns around the lake and by diversion of effluents from *ca.* 75000 m^3.d^{-1} discharged to lake in 1963 to zero in 1968. Subsequently, only fluvial and atmospheric nutrient inputs have contributed to external nutrient loadings. The nutrient outflow from the lake comprised from 10 to 75% of total P. Therefore, the net annual P-input, although significantly reduced, was quite variable despite sewage diversion, being between 13 x 10^3 kg P.year^{-1} in 1968 and 73 x 10^3 kg P.year^{-1} in 1972.

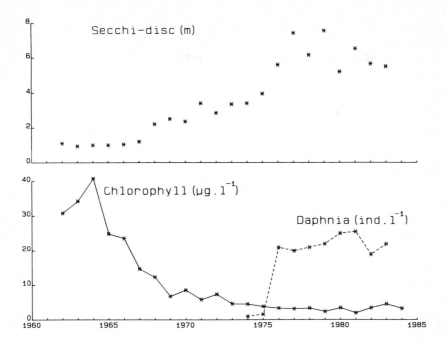

Fig. 3. Means of Secchi-disc transparency (July-August, chlorphyll in surface water and of all *Daphnia* species in upper 1 m (June-August) in Lake Washington (redrawn from Fig. 1 in Edmondson and Litt (62)).

The variations in net loading of N were less striking.

The effects of these nutrient reductions became apparent in the years 1964-1969 in which years major decreases in chlorophyll occurred (Fig. 3). The decreasing tendency continued in the subsequent one decade. The proportions in the phytoplankton biomass of blue-green algae, mainly *Oscillatoria rubescens* and *O. agardhii*, remained high (> 80%), exhibiting a rather delayed response. It was in 1973 that the annual mean share of these blue-greens in the phytoplankton biomass fell below 70%.

In 1976, the lake entered a new phase with major changes in the community structure of plankton (61). *Daphnia* which was consipicuous by its near-absence in the previous years became relatively abundant (Fig. 3) and dominated the zooplankton almost throughout the year (62). This affected phytoplankton and seston concentrations which declined.

The decreases in seston concentration which started first in 1967-'68 continued up to 1980. These changes were reflected both in mean and maximum transparency values. The annual minima and maxima for Secchi-disc depth in July and August, which were very similar (± 1m) in the prerestoration period (early sixties) -reflecting a constantly poor light climate- diverged and increased considerably (Fig. 3); their values in the late seventies (1978-1980) ranged from 7 to 10 m and 14 to 15 m, respectively. Thus, it became clear that nutrients, and not light, limited production in the upper layers. Apparently, biotic changes, e.g. appearance of *Daphnia*, have also contributed to seston declines and increased clarity.

Interestingly, the ratio of total phosphorus to chlorophyll which fluctuated between 2 and 3 in the prerestoration period, i.e. before 1968, increased to lie generally between 3 and 9 in the period after 1968. Even though both total P and chlorophyll decreased in the recovery period, 1968-'80, the latter decreased strikingly more because of drastic changes in the blue-green, which in the earlier period comprised from 80% or more of the phytoplankton chlorophyll.

Brooker Lake: Situated in Florida, Brooker Lake is among a few lakes in which effects of restoration measures on lake limnology were monitored in detail as well as systematically for two years and compared with data collected during the 15-month prior to the restoration measures (42). This small lake (area 10 ha; mean depth 4 m) became highly eutrophic in late sixties and early seventies due to influx of nutrients and organic matter from a dairy feed lot. By 1978, legal steps led to termination of the dairy form and its inflows into the lake, though run off from the former pastures still occured.

In mid 1981 multiple inversion aeration system was installed so that bubbles generated by means of diffusers could displace about 15% water per day so as to bring about a complete circulation in 6-7 days. In a biweekly sampling starting 15 months before this operation began and continuing for 2 years after it, 24 physico-chemical variables and 4 biological variables were measured.

The vertical differences in O_2 concentrations decreased markedly and Secchi-disc transparency increased on aeration P (< 0.05). The factors which decreased significantly ($P < 0.05$) are: turbidity, pH and concentrations of alkalinity, total nitrogen, hydrogen sulphide and iron. Major nutrients (total-P, NO_2 and NO_3) and Ca, Na, and Mg increased significantly only during the first year of aeration, returning to pre-aeration levels in the second year. Total-N to total-P ratio increased to 21 in the first year of aeration, a threefold increase caused mainly by total-P which decreased to one-fourth of the pre-aeration level of 240 ± 40 ug.l^{-1}.

Productivity of the lake was not affected, nor was chlorophyll concentration reduced. However, the total phytoplankton standing crop decreased to $< 50\%$ of the pre-aeration levels because of a shift from blue-green dominated (*ca.* 66%) phytoplankton to one dominated by green algae and diatoms (62%) even though these last two groups did not change significantly in their concentrations.

There was an increase in the occurrence frequency of taxa of green algae per sampling date (especially chlorococcals) from a mean of about 5 in the year before aeration to ca. 15 and 21 in the first and second year after, respectively. Also total number of chlorophycean (green algae) taxa increased from 38 to 65 following aeration.

Trophic state index calculated using the annual means of the relevant data showed no changes; values for the three study years were 64.2, 60.8 and 61.7, respectively.

The decrease in the mean densities of crustacean zooplankton, both cladocerans and copepods, to 8% in the first year and 9% in the second year, of the mean level in the pre-aeration period, was remarkable. Numerical dominance (> 99%) of *Daphnia ambigua* and *Ceriodaphnia lacustris* in the pre-aeration period was replaced by two smaller cladocerans, *Bosmina longirostris* and *Eubosmina tubicen*. The number of cladoceran species increased from 7 in the first year to 11 in the second and 19 in the third year, respectively.

Rotifer densities increased in the first year after aeration. However, a decrease subsequently and variability in species composition, including reduction and elimination of some species and appearance and increase of others, do not allow conclusions. Increase of some taxa (e.g. *Anuraeopsis fissa*) which are typical of highly eutrophic waters (63) is quite surprising but may be related to marked decrease in the size of food particles.

TABLE 4

Annual mean densities (no. per Ekman grab) of the predominant benthic invertebrates in Brooker Lake (Florida, USA); year 1 represents pre-aeration period and in years 2 and 3 the lake was aerated; except for oligochaetes data of the first year differed significantly from data of the aeration period; for details see Cowell et al. (42).

Group/taxon	Year		
	1	2	3
Oligochaetes	17	4	24
Chaoborus sp.	331	36	39
Chironomids	97	188	175
Others	2	28	23
Total	447	256	261

The total mean densities of predominantly benthic invertebrates taxa differed considerably on aeration (Table 4). *Chaoborus punctipennis*, the most dominant benthic form before aeration, decreased by an order of magnitude. But aeration caused an increase in chironomid total numbers particularly at shallow stations. This was caused by mainly one (*Glyptotendipes paripes*) of the seven chironomid species encountered. On the other hand, *Tanytarsus* mean densities were sharply reduced.

3.2.3. Dutch lakes

Most Dutch lakes are shallow, interconnected and, being under strong influence of water from River Rijn, highly hypertrophic. Restoration of Lake Veluwe (area 32.4 km², mean depth 1.3 m) with its water retention time of about 250 days was attempted by Hosper and Meijer (64). Phosphorus was removed at the point source since 1979 and the lake flushed with P-poor but Ca- and N-rich water from Flevopolder in winter when *Oscillatoria agardhii* still occurred in high densities. P-concentrations were reduced from 400-600 ug P.l^{-1} before 1980 to about 100-200 ug P.l^{-1} after 1980; simultaneously, the chlorophyll concentrations decreased from 200-400 ug.l^{-1} to 50-150 ug.l^{-1} (Fig. 4). In 1985 when the P-concentration was < 100 ug.l^{-1} *Oscillatoria* was replaced by diatoms and green algae, possibly caused by the improved light climate, as noted from increase in Secchi disc depths. Interestingly, share of biovolume of algae other than blue-green algae to total algal volume increased from *ca.* 20% before restoration measures in 1979-'80 to > 80% now.

Flushing in winter in Lake Veluwe appears to have 2 main advantages, it broke the *Oscillatoria* 'cycle' in the lake as well as the high alkalinity (bicarbonates) of water (200-300 mg.l^{-1}) probably led to increased precipitation of $Ca_3(PO_4)_2$ which is quite stable at pH values prevalent in the lake.

Loosdrecht lakes, a system of shallow lakes (area 12.5. km²; mean depth 1.4 - 1.8 m), situated about 25 km south of Amsterdam, have been the subject of a multidisciplinary study since 1982 (65). The external loading of these lakes was reduced from 1 mg P.l^{-1} up to 1983 to about 0.3 mg P.l^{-1} in the period thereafter. This was achieved by treating the incoming water with $FeCl_3$ to complex major part of P in solution and in particulates. The follow-up studies up to 1987 have not shown a recuperation of the system. Internal loading and resilience of the system, because of the dominance of blue-greens, seem to have retarded the recovery process in the four years since the restoration measures were taken. However, C:P and N:P ratios of seston (< 150 um) have shown some increase (Fig. 5) and there are some indications from work in Laboratory Scale Enclosures that P rather than light alone may limit production in the course of summer (Pers. comm. M. Rijkeboer). Also, for the same chlorophyll level as before restoration there is less total phosphorus (pers. comm. Van Liere). Summer means of total phosphates to chlorophyll ratios have decreased from 1.0

- 1.2 to 0.67 - 0.73 at the three stations in the lake in the four years since 1983. However, there is no change so far either in the chlorophyll or seston concentrations; also primary production rates and trophic state parameters do not indicate a response so far to the restoration measures.

The abundance of planktivorous bream (*Abramis bramis*), through its intensive predation on large-sized zooplankton, perhaps contributes to the low grazing pressure of zooplankton which are thus ineffective in reducing algal biomass. Also, nutrient (P and N) regeneration by both crustaceans (66) and the rotifers (Pers. comm. Dr. J. Ejsmont-Karabin) appear to contribute significantly to

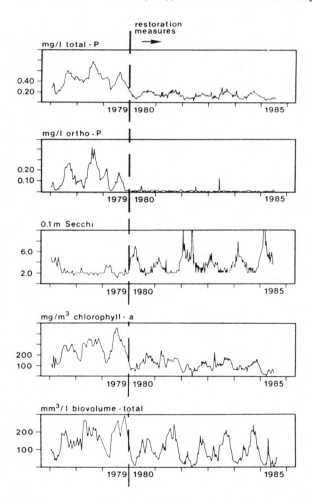

Fig. 4. Changes in different parameters on restoration measures in Lake Veluwe (The Netherlands); the different panels are from different figures (slightly modified) from Hosper and Meijer (64).

internal loading. This, as well as bioturbation by bream (Pers. comm. Dr. E. Lammens), may also delay the recovery effects (67) in the Loosdrecht lakes. This is also corroborated by lack of a major shift so far in composition or abundance of phytoplankton.

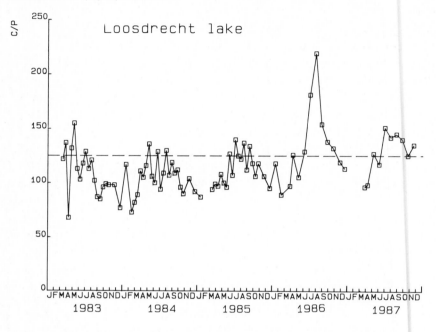

Fig. 5. Carbon:phosphorus ratios of seston (< 150 um) in Loosdrecht lake (The Netherlands) in the period before (1983), during (1984) and after (1985-1987) restoration measures; horizontal dashed line represents a C:P ratio of 125.

It is, therefore, suggested that in such cases a combination of P-reduction measures with biomanipulation at the fish level, i.e. introducing piscivore fish, may accelarate the recovery effects. However, sediment removal is likely to produce quicker and lasting effects.

3.2.4. Danish lakes

In a small Danish lake, Søbygård (area 0.5 km²; mean depth 1 m (56, 68) reduction in P-load from 27 $g.m^{-2}.yr^{-1}$ to 5 $g.m^{-2}.yr^{-1}$ led to an increase in chlorophyll from 300 $mg.m^{-3}$ to 800 $mg.m^{-3}$, with an accompanying decrease in transparency from 1 m to 0.4 m. Interestingly, a shift in summer phytoplankton composition, from a dominance of blue-greens (*Aphanozomenon*) to that of chlorococcales (*Scenedesmus* and *Chlorella*), was associated with a zooplankton decrease. High pH ($\pm$ 11) resulting from high photosynthetic activity because of high internal phosphorus loading perhaps contribute to low zooplankton levels.

In addition, fish predation which appears to have increased after the restoration measures may contribute to low zooplankton levels. Recovery which appears to depend upon P depletion may take 30 years according to the present theoretical estimates. Reducing fish population by biomanipulation may hasten recovery by allowing increased mortalitiy of algae by zooplankton grazing. Thus, a lowered photosynthetic activity and decrease in pH and in P release rate from the sediments may also contribute positively to the recovery process.

In lake Glumsø another small Danish lake (area, 0.3 km²; mean depth 1.8 m) P-load has been reduced by sewage diversion from 4.5-6.0 $g.m^{-2}.yr^{-1}$ in 1981 to one-fourth these levels now (56); Also chlorophyll level dropped simultaneously from 600-800 $ug.l^{-1}$ to *ca.* 200 $ug.l^{-1}$. Despite a low transparency level, 0.4 m, persistant dominance of *Scenedesmus*, a green alga, cannot be explained because cyanobacteria would be at an advantage under such a poor light climate. However, very high NO_3^- concentrations (10 mg $N.l^{-1}$) in lake Glumsø may favour the green algae because of their higher growth rates under these conditions (69).

The effects of sewage diversion in Lake Glumsø could be relatively successfully predicted because of the rather stable and simple biological communities of the lake (70). Apparently, intensive field and laboratory experiments facilitated a combination of simple empirical models and complex dynamic models.

3.2.5. East German Lakes

The restoration studies in German Democratic Republic (GDR) have concentrated mostly in drinking water reservoirs (71, 14). The country has a critical water balance with utilization (8 x 10^9 m³) exceeding resources up to 40%. This necessitates re-use of waste water; in other words, there is a constant need for water-quality managament. Among the 524 lakes (total area 1145 km²) classified by Klapper (71) on the basis trophic state, hydrographic factors and hygiene aspects, between 15 and 20 were oligotrophic and mesotrophic, the rest were either eutrophic (*ca.* 310) or highly eutrophic and hypertrophic (*ca.* 200). Besides, there are others in which silting was a main problem problem.

Among the problems affecting the lakes in East Germany mass algal growth, low oxygen content and increased amounts of ammonia are apparently the major ones. Klapper (71) listed 184 of the lakes in which hydrography was the cause of problems and considered chances of rehabilitating these lakes successfully to be better than the others. In five cases of lake restoration described by Klapper (71) Fe was used in different concentrations (0.32-12 g.m³), in three lakes in combination with Bentonite and polyacrylamide, with polyacrylamide and claypowder, or with Bentonite; in the other two lakes Aluminium sulphate ($Al_2(SO_4)_3$) was used in aqueous solutions in concentrations of 0.2 g Al and 8.1

g Al.m^{-3}. Orthophosphate elimination was quite variable, 30-90%, depending on the type of treatment, and removal of total P was less successful, even though in Lake Süsser See (dose 8.1 g Al.m^{-3}) algae and bacteria were co-precipitated leading to an increased transparancy to 3.4 m.

The pollution of lake Arendsee in the county of Magdeburg was tackled in two ways: 1) by diverting the sewage effluent into an other catchment area since 1970; and 2) by hypolimnetic depletion of nutrient-rich water (72), instead of a surface outflow. This latter resulted in a 6-7 times faster export of N and P than via surface outflows. Also, because O_2 deficient hypolimnetic water was removed lake's oxygen balance improved. By 1979, deep water depletion systems were completed in Bergsee and Montzener See and several other projects were in preparation.

In Weida reservoir the problem was high dissolved N (> 40 mg.l^{-1} NO_3) -higher than specified in the standard of drinking water- rather than high P such that only up to 20% of NO_3 could be eliminated after 113 days of retention. A steel cage (60 x 20 x1.5 m), with strawpackings in order to stimulate growth of microorganism, was installed at the lake bottom within the hypolimnetic water. Epilimnetic water rich in nitrate, together with a carbon source, was pumped into the cage. On denitrification by the facultative anaerobes, molecular nitrogen formed escaped from the lake and the nitrate concentrations was reduced to levels acceptable in drinking water.

In the framework of rehabilitation work, several lakes and ponds were desilted in late seventies. Also, hypolimnetic aerators requiring low energy and maintenance costs were developed for continuous use. Low-oxygen water collected a few metres above the lake bottom is pumped to surface, is enriched with air by hollow jet nozzle, transported to 10 m depth via a bubble-conveying pipe which ascends to surface carryig more hypolimnetic water due to admixed-air siphoning effect. The aerated water is finally returned to the upper hypolimnion and deflected horizontally.

An other good example of lake restoration-work in East Germany is that of Benndorf (73) on the hypertrophic Bautzen reservoir (area 533 ha; mean depth 7.4 m); this study involves manipulation in food chain. The native populations of predacious fish, pike perch (*Stizotedion lucioperca*) and pike (*Esox lucius*), were enhanced by stocking the reservoir pond-raised fish, as well as by restrictions on minimum catch size ($\geq$ 60 cm) for anglers and by prohibiting commercial fishery. This led to increased predation of the piscivores on zooplanktivorous fish, namely small perch (*Perca fluviatilis*), but allowing the piscivorous age classes to develop relatively better. This is apparently a key factor in regulating the young-of-the-year of all fish species, including the perch. In 1980, i.e. after three years of this 'field experiment', the changes

that reflect at improvement in reservoir became measurable and were followed in the subsequent years. Most notable among these are:
1) A marked increase in biomass of the most effective filterfeeding cladoceran, *Daphnia geleata*, and a three-fold increase in its share in total herbivorous zooplankton, namely from about 25% to 75%.
2) The mean individual body weight of the herbivore crustaceans more than doubled, *viz*. from < 10 ug wet weight.ind^{-1} in 1980 to about 20 ug.ind^{-1} in 1981-'83.
3) Strong initial increases in the invertebrate predators, *Chaoborus flavicans* and *Leptodora kindti*.
4) Appearance of extended clearwater phase in summer (maxima of Secchi-depth, 4-7 m) and a general increase in mean Secchi-disc transparency by > 0.5 m.
5) Increased proportion of inedible algae, but a variable response in the maximum phytoplankton biomass.
6) Total phosphorus increased by 150% average, even after correcting for change in external load.
7) Instability in the characteristics listed (1-6) was perhaps due to a) stochastic character of the meteorological conditions and b) time lag in various feedback mechanims within the ecosystem (73).

According to Benndorf (73) significant increase in P concentration on biomanipulation may be related to high P-import accompanied by reduced P-uptake rate by algae (because of increase in inedible forms) that is more pronounced than the increase in sedimentation loss rate. Thus, for a decrease in phosphorus concentration during a biomanipulated restoration of a lake, external P-loading needs to be reduced simultaneously to a level below the 'biomanipulation-efficiency threshold'. This threshold will be influenced also by several factors, other than external loading, e.g. mean depth, internal P-load, calcite precipitation and mixing process.

Concluding, top-down control of food chain must be supplemented by other control measures to reduce P-loading in order to control eutrophication effectively and to guarantee restoration success.

3.2.6. West German lakes

The general water pollution and lake restoration policy in Federal Republic of Germany, like in Switzerland, is that lakes should be mesotrophic (74). This can be achieved if total-P concentrations do not exceed 20-30 ug.l^{-1} during complete circulation.

The approach to lake rehabilitation in W. Germany is more based on lake sanitation, rather than on lake restoration. Therefore, the emphasis is generally on improving the quality of input water, both through waste water treatment and P removal and diversion of P-rich waters. Moreover, most work that has been done deals with drinking water basins.

One good example of restoration work is in the Wahnbach reservoir (74, 75), the source of drinking water for the city of Bonn and large areas of Rhine-Sieg district. This reservoir which was constructed in the fifties by building a dam in the Wahnbach valley so that the reservoir is long and narrow (area 2.1 km²; water volume 41.3 x 10^6 m³), with a catchment area of 70 km². The reservoir became eutrophic by 1964 when also blue-green alga, *Oscillatoria rubescens*, appeared. This restricted the recreation use and led to problems for drinking water. Among the main causes of problems were several diffuse sources of phosphorus in the lake catchment that is predominantly arable and highly fertilized for maize crop.

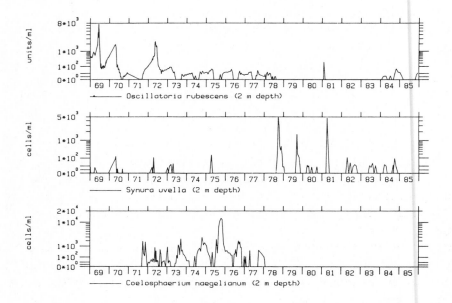

Fig. 6. Changes in the abundance of selected algal species in the Wahnbach reservoir (West Germany) before and after completion of phosphorus eleminiation plant in autumn 1977; figure from Clasen and Bernhardt (75).

In end of 1977 a P-elimination plant (PEP) became operational. The incoming water was treated with $FeCl_3$ in a dose ranging from 4 to 10 mg.l^{-1} ferric iron (Fe III). This led to sharp decreases in phosphorus, turbidity, seston and to some extent also in dissolved organic carbon (DOC). Within half a year of P-elimination programme total P-concentrations in the treated input water to reservoir dropped to 1-13 ug.l^{-1}, with soluble reactive phosphate being hardly detectable. *Oscillatoria rubescens* decreased dramatically and *Coelosphaerium* sp. disappeared (Fig. 6). In the first few years after restoration *Synura uvella*, having high chlorophyll content increased considerably, but thereafter decreased. The biomass production in the reservoir measured as chlorophyll ranges from 5 ug.l^{-1} which correponds to a mesotrophic state. The data collected in the period 1976-'81 explicitly answered several questions related with the success of the project. Application of the results to other reservoirs and lakes to control eutrophication (76) is apparently feasible if mean depth does not exceed 15-20 m and if shallow, branching bays are absent.

In Lake Ammer (area 46.6 km²; mean depth 37.5 m) sewage diversion measures and considerable decrease (40%) in detergent P-load in the river Ammer reduced also the annual P-total load, from 2.15 g.m^{-2} in 1973-'75 to 1.2 g.m^{-2} in 1984. Nevertheless, the measures were not sufficient to clearly improve the lake quality, except for a considerable reduction in phytoplankton production. Thus, even the more advanced waste treatment may be not effective if an important fraction of the phosphorus arises from diffuse sources (77).

3.2.7. Swiss lakes

Imboden (78) has summarized the restoration work on Lake Baldegg (area 5.2 km²; mean depth 34 m). External control measures, namely sewage treatment and control of fertilizing practice in agriculture, were taken first to stop pollution at P concentrations of 500 ug P.l^{-1}. These external measures were apparently not enough to reduce trophic state significantly. However, a combination of the external control with artificial mixing of lake water with air bubbles in winter and input of oxygen into hypolimnion in summer led to success. The annual mean P-content of the lake has halved in 10 years of restoration measures since 1975-76.

Availability of limnological data on Swiss lakes and those on sources of nutrients (79, 80) has enabled to estimate changes in P influx into lakes using different sanitation measures and legislative enforcements. These influx values are compared with tolerable and critical annual P mass loading of the individual lakes, derived with the help of Vollenweider model (81) and related model concepts for stagnant water (82, 83) to achieve mesotrophic conditions in Swiss lakes. In lake Geneva, e.g. waste-water treatment alone would not reduce phosphate below the critical loading to the lake of 640 tons.year^{-1}. But together with P-elimination via waste-water treatment, also careful agricultural practise is observed with the aim of reducing P-influx from non-point sources. In short, step-wise introduction of measures in several lakes e.g. Vierwaldstätter, Zuger, Brienzer, Thuner and Lake Geneva, has allowed the P-influx to be reduced so that P-values lower than critical loading and occasionally even lower than admissible loading have been achieved. Substantial decrease in P-concentration has also been achieved in Zürichsee/Walensee system (56) by sewage treatment plants; however, data on P-loading are not adequate.

3.2.8. English Broads

In a small waterbody (area 4.7 ha; mean depth 0.7 m), Alderfen Broad restoration was attempted by isolating it from outside input (84). In the first years after these steps algal biomass increased, P concentrations decreased and waterplants returned. The feedback mechanisms resulted in a series of changes: increased production of organic material, reduced circulation and increased P release at bottom followed by increase of algal biomass. *Aphanotheca*, a blue-green alga, dominant before isolation disappeared. The isolation technique in the absence of sediment removal appears risky but long-term effects will show its effectiveness.

In Cockshoot Broad, a relatively larger water body (area 3.3 km^2; mean depth 1 m), partial isolation and removal of P-rich sediment were the main restoration measures taken (84). The sequence of changes here was as follows: strong increase in zooplankton, fish being absent, led to a pronounced decrease in algal biomass in the first year. After that a fish population developed and reduced the zooplankton. Part of the broad has already been recolonized with several aquatic plants, perhaps because of improved light climate.

4. CONCLUDING REMARKS

The factors which threaten our lake ecosystems relate mainly to excessive nutrient inputs that invariably enhance production of bloom-forming cyanobacteria. These sully lake appearance, interfere with recreational activities and threaten food chain, particularly the survival of commercially important fish. Other adverse affects of nutrient addition and pollution of lake relate to deterioration of drinking water quality.

In many eutrophic and restoration studies phosphorus:chlorophyll relations have been used to describe the trophic state and changes herein. Nevertheless, resilience of the system under consideration may make such comparisons less fruitful. This is because even a three-fold change in P load, either increase, as during eutrophication, or decrease, as during restoration process, need not result in a corresponding change in algal chlorophyll or biomass. Thus, a reduction of P in the system may lead to a deceptive conclusion about success of the restoration effort. The change in species composition of phytoplankton upon decrease in nutrient loading rate may complicate the predictive capacity of loading models.

One would expect that response to restoration of shallow lakes may be generally more rapid, unless the sediments are rich in nutrients and retention time of water is low. In deep lakes, on the other hand, P reserves in the sediments may usually delay the response. Based on nutrient availability studies, one may expect that only in severely P-stressed situations the biomass of blue-green algae will be reduced, thus making a shift to chrysophycean and diatom flora possible.

From the literature on lake recovery studies it is quite clear that drastic reduction of external nutrient inputs, as well as in the internal inputs, deserves utmost attention. Moreover, the effects of nutrient reduction need to be constantly monitored for long periods, say for five years or more. Examples where integrated control measures have been applied are rather limited. This is true also for detailed baseline studies for prerestoration period, thus making interpretations difficult, especially in view of high resilience that the lake ecosystems often exhibit.

The multiple restoration techniques if applied may not be compatible, thus negating each other's effects. For example, flushing is incompatible with hypolimnetic withdrawals; likewise, nutrient (P) inactivation and sediment oxidation are not complementary with flushing. In lakes suffering from acid rain problem, aluminium used as inactivant may be released as dissolved Al and produce deleterious side effects for lake's organisms. Sediment removal may give disposal problems, especially if the sediments are rich in heavy metals or in other toxic substances. Besides, sediment removal may lead to increase in lake depth such that it may bring about a change in the subsoil water level, thus affecting agricultural needs of water in the cathment area.

The interest in biological control, the so called biomanipulation in recent years has also stimulated complementary approaches in lake restoration research, encouraged in many cases by the inability of abiotic control measures alone to improve trophic state. Nonetheless, long-term studies in which the abiotic and biotic measures were applied in conjunction are practically nonexistent.

Despite some promising studies on lake restoration in the last 2 decades, both in Europe and North America, we still greatly need to improve our understanding of the functioning freshwater ecosystems in order that we can rationally protect, restore and manage these systems. Last, but not least, the freshwater lakes and reservoirs need to be treated as integral part of man's total environment, rather than as 'dumping grounds' for our wastes.

5. ACKNOWLEDGEMENTS

I am thankful to Prof. O. Ravera for inviting me to participate in the ISPRA Course and to give two lectures, one of which related to "Lake Recovery", the theme of the present paper. I am indebted to Miss Cecilia Kroon for her skilful typing work and tremendous patience to accept and incorporate last-minute changes in the typescript. I owe my thanks to Mr. K. Siewertsen for assistance in preparing the illustrations and to Mr. E. Marien for the photography work.

6. REFERENCES

1. R.A. Vollenweider, Scientific concepts and methodologies pertinent to lake research and lake restoration, Schweiz. Z. Hydrol. 49 (1987) 129-147.
2. J. Cairns Jr., Restoration of damaged ecosystems, in: Research on Fish and Wildlife Habitat (Eds. W.T. Mason, Jr. and Sam Iker), Office of Research and Development, USEPA Washington DC, 1982, pp. 220-239.
3. D.W. Schindler, K.H. Mills, D.F. Malley, D.L. Findley, J.A. Shearer, I.J. Davies, M.A. Turner, G.A. Linsey and D.R. Cruikshank, Long-term ecosystem stress: The effects of years of experimental acidification on a small lake, Science, 228 (1985) 1395-1401.
4. K.W. Thornton, R.H. Kennedy, J.H. Carroll, W.W. Walker, R.C. Gunkel and S. Ashby, Reservoir sedimentation and water quality - a heuristic model. In Proceedings of Symposium on Surface Water Impoundments. Amer. Soc. Civil Engineers, Minneapolis, Minn. (1980), pp. 654-661.
5. S. Björk, Scandinavian lake restoration activities, in: Proceedings (pre prints) of International Congress of EWPCA (Rome, 1985), 1985, pp. 293-301.
6. G.D. Cooke, E.B. Welch, S.A. Peterson and P.R. Newroth (Eds.), Lake and reservoir restoration, Buttorworth (Ann Arbor Science) Boston etc., VIII, 329 pages, 1986.
7. R.E. Carlson, A trophic state index for lakes, Limnol. Oceanogr., 22 (1977) 363-369.
8. W.W. Walker Jr., Trophic state indices in reservoirs, In Lake And Reservoir Management, EPA 440/5-84-001, 1984, pp. 435-440.
9. P.J. Dillon and F.H. Rigler, A test of a simple nutrient budget model predicting the phosphorus concentration in lakewater, J. Fish. Res. Board Can., 31 (1974) 1771-1778.
10. R.A. Vollenweider, Input-output models with special reference to the phosphorus loading concept in limnology, Schweiz Z. Hydrol., 37 (1975) 53-84.
11. R.A. Vollenweider, Advances in defining critical loading levels for phosphorus in lake eutrophication, Mem. Ist. Ital. Idrobiol., 33 (1976) 53-83.
12. K.H. Reckhow and S.C. Chapra, Engineering Approaches for Lake Management: Vol. I. Data Analysis and Empirical Modelling, Butterworth, Boston etc., 1983.
13. S.O. Ryding, Reversibility of man-induced eutrophication. Experience of a lake recovery study in Sweden, Int. Rev. Ges. Hydrobiol., 64 (1981) 449-503.
14. D. Uhlmann, Evaluation of strategies for controlling eutrophication of lakes and reservoirs, Int. Revue Ges. Hydrobiol., 67 (1982) 821-835.
15. P.H. Uttormark and M.L. Hutchins, Input/output models as decision aids for lake restoration, Water Res. Bull., 16 (1980) 494-500.
16. S.A. Peterson, Lake restoration by sediment removal, Water Res. Bull., 18 (1982) 423-435.
17. C.L. Churchill, C.K. Brashier and D. Limmer, Evaluation of a Recreational Lake Rehabilitation Project, OWRR Comp. Report No. B-028-SDAK, Brookings, SO: Water Resources Inst., South Dakota State University, 1975.
18. K. Murakami and K. Takeishi, Behaviour of heavy metals and PCBs in dredging and treating of bottom deposits, in S.A. Peterson and K.K. Randolph (Eds.), Management of Bottom Sediments Containing Toxic Substances: Proceedings of the 2nd U.S./Japan Experts Meeting, EPA-600/3-77-083, Corvallis, OR: USEPA,

CERL, 1977.
19 G. Andersson, N. Berggren and S. Hamrin, Lake Trummen restoration project. III. Zooplankton, macrobenthos and fish, Verh. Internat. Verein. Limnol., 19 (1975) 1097-1106.
20 S. Björk, Swedish Lake Restoration Programme gets results, Ambio, 1, (1972), 153-165.
21 L. Bengtsson, S. Fleischer, G. Lindmark and W. Ripl, Lake Trummen restoration project, I. Water and sediment chemistry, Verh. Internat. Verein. Limnol., 19 (1975) 1080-1087.
22 D.E. Canfield Jr., K.A. Langeland, S.B. Linda and W.T. Haller, Relations between water transparency and maximum depth of macrophyte colonization in lakes, J. Aquatic Plant Manage., 23 (1985) 25-28.
23 E. van Donk, R.D. Gulati and M.P. Grimm, Restoration by biomanipulation in a small hypertrophic lake: Preliminary results, Hydrobiologia, in press.
24 G.D. Cooke, R.T. Heath, R.H. Kennedy and M.R. McComas, Effects of Diversion and Alum Application on Two Eutrophic Lakes, EPA-600/3-78-033, 1978.
25 R.H. Kennedy and G.D. Cooke, Control of lake phosphorus with aluminium sulfate, Dose determination and application techniques, Water Res. Bull., 18 (1982) 389-395.
26 P.J. Garrison and D.R. Knauer, Long term evaluation of three alum treated lakes, in Lake and Reservoir Management, EPA 440/5-84-001 1984, pp. 513-517.
27 E.B. Welch, J.P. Michaud and M.A. Perkins, Alum control of internal phosphorus loading in a shallow lake, Water Res. Bull., 18 (1982) 929-936.
28 H. Bernhardt, Recent developments in the field of eutrophication prevention, Z. Wasser Abwasser Forsh, 13 (1981) 14-26.
29 T.A. Haines, Acidic precipitation and its consequences for aquatic ecosystems: A review, Trans Am. Fish Soc., 110 (1981) 669-707.
30 J. Cairns, Jr., K.L. Dickson and J.S. Crossman, The response of aquatic communities to spills of hazardous materials, Proc. National Conference Hazardous Material Spills, 1972, pp. 179-197.
31 W. Ripl, Biochemical oxidation of polluted lake sediment with nitrate. A new lake restoration method, Ambio, 5 (1976) 132-135.
32 W. Ripl and G. Lindmark, Ecosystem control by nitrogen metabolism in sediment, Vatten 2 (1978) 135-144.
33 P. Stadelman, Der Zustand des Rotsees bei Luzern, Kantonales amt fur Gewasserschutz, Luzern, 1980.
34 E.B. Welch, The dilution/flushing technique in lake restoration, Water Res. Bull., 17 (1981) 558-564.
35 M.A. Perkins, Limnological Characteristics of Green Lake: Phase I Restoration Analysis, Dept. Civil Engr., Univ. WA, Seattle, 1983.
36 S. Björk, European Lake Rehabilitation Activities, Inst. Limnol. report, Univ. of Lund, 1974.
37 R. Gächter, Lake restoration by bottom siphoning, Schweiz Z. Hydrol., 38 (1976) 1-28.
38 J. Clasen, German experience in reservoir management and control, in: Restoration of Lakes and Inland Waters, EPA/440/5-81-010, U.S. Envir. Prot. Aq., Washington, DC, 1980, pp. 153-177.
39 R.A. Pastorak, T.C. Ginn and M.W. Lorenzen, Evaluation of Aeration/Circulation as a Lake Restoration technique, Report No. TC-3947, USEPA, 1980.
40 D.J. McQueen and D.R.S. Lean, Aeration of anoxic hypolimnetic water: Effects on nitrogen and phosphorus concentrations, Verh. Int. Ver. Limnol., 22 (1984) 267-276.
41 B.C. Cowell and C.J. Dawes, Algal studies of eutrophic Florida lakes: the influence of aeration on the limnology of a central Florida lake and its potential as lake restoration technique, Final Report, Fla. Dept. Nat. Resour. Tallahassee, 299 pages, 1984.
42 B.C. Cowell, C.J. Dawes, W.E. Gardiner and S.M. Scheda, The influence of

whole lake aeration on the limnology of a hypertrophic lake in central Florida, Hydrobiologia, 148 (1987) 3-24.
43 M.H. Garrell, J.C. Confer, D. Kirchner and A.W. Fast, Effects of hypolimnetic aeration on nitrogen and phosphorus in a eutrophic lake, Water Resour. Res., 13 (1977) 343-347.
44 C. Forsberg, B. Hultman, L. Söderberg and K. Wallen, Naturvårdsverkets RR-undersökning, 9. Restaurering au små näringsrika sjöar medelst reduktion av algmassa, Vatten, 32 (1976) 320-327.
45 J. Fraser, D., Hinckley, R. Burt, R.R. Severn and J. Wisniewski, A Feasibility Study to Utilize Liming as a Technique to Mitigate Surface Water Acidification, Rept. No. RP 1109-14, General Res. Corp. for Elect. Power Res. Inst., 1982.
46 G.K. Lindmark, Acidified lakes: Sediment treatment with sodium carbonate - a remedy? Hydrobiologia, 92 (1982) 537-547.
47 W. Ripl, Lake Restoration Methods Developed and Used in Sweden, Wayne, NJ: Atlas Copco Inc., 1981.
48 F. Eriksson, E. Hörnström, P. Mossberg and P. Nyberg, Ekologiska effekter av Kalkning i Försurade Sjöar och Vattendrag, Sötvattens - Laboratoriet, Drottningholm, Sweden, Nr. 6, 1982.
49 J. Shapiro, V. La Marra and M. Lynch, Biomanipulation: An ecosystem approach to lake restoration, in: P.L. Brezonik and J.L. Fox (Eds.), Water Quality Management through Biological Control, Gainesville, FL: Department of Environmental Engineering Sciences, Univ. Florida, 1975, pp. 85-96.
50 J. Shapiro, The need for more biology in lake restoration, in Lake Restoration, EPA-440/5-79-001, (1978) pp. 161-167.
51 J. Shapiro and D.I. Wright, Lake Restoration by biomanipulation: Round Lake, Minnesota, the first two years, Freshwater Biol., 14 (1984) 371-385.
52 J. Shapiro, B. Forsberg, V. La Marra, G. Lindmark, M. Lynch, E. Smeltzer and G. Zoto, Experiments and Experiences in Biomanipulation, Interim Rept. No. 19, Limnological Research Center, Univ. of Minnesota, Minneapolis, Minn., 1982.
53 J. Benndorf, H. Kneschke, K. Kossatz and E. Penz, Manipulation of the pelagic food web by stocking with predacious fishes, Int. Rev. Ges. Hydrobiol., 69 (1984) 407-428.
54 R. de Bernardi, Biomanipulation of aquatic food chains to improve environmental quality. ISPRA Course Proceedings "Ecological assessment of environmental degradation, pollution and recovery", Elsevier, Amsterdam, 1988, pp.....
55 International Congress Lake Pollution and Recovery, Rome (15-18 April, 1985) organized by European Water Pollution Control Association, Associazione Nazionale di Ingegnezia Sanitaria (Andis), 1987, 466 pp.
56 H. Sas and S. Vermij, Proceedings of the first workshop for 'Eutrophication management in international perspective', Watersupply Zürich, 1987.
57 G. Cronberg, C. Gelin and K. Larsson, Lake Trummen restoration project, II. Bacteria, phytoplankton and phytoplankton productivity, Verh. Internat. Verein. Limnol., 19 (1975) 1088-1096.
58 C. Forsberg, Lake recovery in Sweden, in: Proceedings (pre prints) of International Congress of EWPCA (Rome, 1985), 1985, pp. 272-281.
59 I. Ahlgren, Response of Lake Norrviken to reduced nutrient loading. Verh. Internat. Verein. Limnol., 20 (1978) 846-850.
60 W.T. Edmondson and J.T. Lehman, The effect of changes in nutrient income on the conditions of Lake Washington, Limnol. Oceanogr., 26(1) (1981) 1-29.
61 W.T. Edmondson, Recovery of Lake Washington from eutrophication, in: Proceedings (pre prints) of International Congress of EWPCA (Rome, 1985), 1985, pp. 228-234.
62 W.T. Edmondson and A.H. Litt, Daphnia in Lake Washington, Limnol. Oceanogr., 27 (1982) 272-293.
63 R.D. Gulati, Zooplankton and its grazing as indicators of trophic status in Dutch lakes, Environm. Monit. Assessm., 3 (1983) 343-354.

64 H. Hosper and M.L. Meijer, Control of phosphorus loading and flushing as restoration methods for lake Veluwe, The Netherlands, Hydrobiol. Bull., 20 (1986) 183-194.
65 L. van Liere, Loosdrecht lakes: origin, eutrophication, restoration and research programme, Hydrobiol. Bull., 20 (1986) 77-85.
66 P.J. den Oude and R.D. Gulati, Phosphorus and nitrogen excretion rates of zooplankton from the eutrophic Loosdrecht Lakes, with notes on other P sources for phytoplankton requirement, Developments in Hydrobiology, Junk, The Hague, Boston, in press.
67 L. van Liere, R.D. Gulati, F.G. Wortelboer and E.H.R.R. Lammens, Phosphorus dynamics following restoration measures, Developments in Hydrobiology, Junk, The Hague, Boston, in press.
68 E. Jeppesen, M. Sondergard, O. Sortikjae and P. Kristensen, Interactions between phytoplankton, zooplankton and bacteria in a shallow hypotrophic lake: a study on short term changes in a trophic relationship in lake Søbygård, Denmark. Developments in Hydrobiology, Junk, The Hague, Boston, in press.
69 L. Leonardson and W. Riple, Control of undesired algae and induction of algal succession in hypertrophic lake ecosystems, Developments in Hydrobiology, 2 (1980) 57-65.
70 L. Kamp-Nielsen, Modelling the recovery of hypertrophic Lake Glumsø (Denmark), Hydrobiol. Bull., 20 (1986) 245-255.
71 H. Klapper, Experience with lake and reservoir restoration techniques in the German Democratic Republic, Hydrobiologia, 72 (1980) 31-41.
72 H. Klapper, Oligotrophierung eines tiefen, geschichteten Sees in einem Erholungsgebiet durch Ableitung des Tiefenwassers, Limnologica (Berlin), 10 (1976) 587-593.
73 J. Benndorf, Food web manipulation without nutrient control: A useful strategy in lake restoration? Schweiz. Z. Hydrol., 49 (1987) 237-248.
74 H. Bernhardt, Strategies of lake sanitation, Schweiz. Z. Hydrol., 49 (1987) 202-219.
75 J. Clasen and H. Bernhardt, Chemical methods of P-elimination in the tributories of reservoirs and lakes, Schweiz. Z. Hydrol., 49 (1987) 249-259.
76 H. Bernhardt, J. Clasen, O. Hoyer and A. Wilhelms, Oligotrophierung stehender Gewässer durch chemische Nährstoff-Eliminierung aus den Zuflüssen am Beispiel der Wahnbachtalsperre (Gesamtbericht), Arch. Hydrobiol., Suppl. 70 (Monographische Beiträge), H. 4, (1985) 481-533.
77 B. Lehnhart and Ch. Steinberg, Auswirkungen der Phosphathöchstmengenverordnung am Ammersee, Vom Wasser, 67 (1986) 237-248.
78 D.M. Imboden, Restoration of a Swiss lake by internal measures. Can models explain reality? in: Proceedings (pre prints) of International Congress of EWPCA (Rome, 1985), 1985, pp. 29-40.
79 H. Ambühl, Phosphate und Gewässerschutz - Untersuchungen zur Eutrophierung der Alpenrandseen, Seifen-Öle-Fette-Wachse, 108 (1982) 453-459.
80 R. Gächter and O.J. Furrer, Der Beitrag der Landwirtschaft zur Eutrophierung der Gewässer in der Schweiz, Schweiz. Z. Hydrol., 34 (1972) 41-70.
81 R.A. Vollenweider and J. Kerekes, OECD Cooperative Programme for Monitoring of Inland Waters (Eutrophication Control), Synthesis Report, Paris, 1980.
82 D.M. Imboden, Modellvorstellungen über den Phosphorkreislauf in stehenden Gewässern, Z. Wasser-Abwasser-Forsch., 15 (1982) 89-95.
83 R. Gächter, D.M. Imboden, H. Bührer and P. Stadelmann, Mögliche Massnahmen zur Restaurierung des Sempachersees, Schweiz. Z. Hydrol., 45 (1983) 247-266.
84 B. Moss, H. Balls, K. Irvine and J. Stansfield, Restoration of two lowland lakes by isolation from nutrient-rich water sources with and without removal of sediment, J. App. Ecol., 23 (1986) 391-414.

CONCEPT OF STRESS, NATURAL AND ANTHROPOGENIC, IN TERRESTRIAL ECOSYSTEMS

Gundolf H. KOHLMAIER
Institut für Physikalische und Theoretische Chemie
Johann Wolfgang Goethe Universität
Niederurseler Hang, 6000 Frankfurt 50, Federal Republic of Germany

1 THE CONCEPT OF STRESS AND STRAIN AND HYSTERESIS WITHIN THE PHYSICS OF ELASTIC PROPERTIES OF SOLIDS

The term stress used in biological sciences is originally borrowed from the theory of elasticity of solids. In physics, stress is defined as a force F per area A applied either along one direction perpendicular to two opposing surfaces (longitudinal stress), tangential to one surface with the opposing surface fixed (shear stress), or perpendicular to all surfaces (volume stress) causing a deformation while strain is a measure of the degree of deformation (expressed as $\Delta l/l_0$, $\Delta l/h$ with h being the thickness of a slab, or $\Delta V/V_0$ correspondingly). The generalized Hooke's law states that strain and stress are proportional to each other within the limits where a linear relationship holds.

Physics distinguishes between outside forces acting on a system and internal restoring forces of the system itself, counteracting the outside force, observing Newton's 3rd law of motion: action force = reaction force. In Hooke's law of an extended spring, the reaction force is derived from the deviation of the spring from its equilibrium position:

$$F = -k \cdot (x-x_0) \tag{1a}$$

Similarly, although intuitively expected the opposite way, the stress is derived from the observed strain:

$$\text{stress} = E \cdot \text{strain} \quad \text{or} \quad \sigma = E \cdot \varepsilon \tag{1b}$$

where $\sigma = F/A$ and $\varepsilon = \Delta l/l_0$ (for the longitudinal stress), and where E is the elastic modulus (Young's modulus, shear modulus, or volume modulus) which measures the resistance to elongation, shearing or volume change. It should be noted that cause and effect are not stated in the common way, in which the effect variable appears on the left and the causing variable on the right hand side of the equation, and correspondingly in a graph with the strain variable on the abscissa and the stress variable on the ordinate (Fig.1). The advantage of this presentation is, however, given by the fact that in such a graph the elastic energy or the work to achieve the elongation can be interpreted geometrically as the area between the stress curve and the abscissa which is given for the longitudinal stress by:

$$w(0 \to \varepsilon) = \int_0^\varepsilon A \cdot \sigma \cdot l_0 \, d\bar{\varepsilon} = V \int_0^\varepsilon \sigma \, d\bar{\varepsilon} \qquad (2)$$

If the longitudinal stress is extended beyond the point A, the response becomes non-linear with the resistance $E(\varepsilon)$ becoming smaller than the linear portion. Relaxing the stress at a point B, the system now follows a path from B to C, such that for a stress zero a remaining stretch is observed (this phenomenon, by the way, can be similarly observed in the magnetisation of ferromagnetic substances, in which some magnetic domains still persist after the magnetising field is switched off). The curve integral for the elastic energy is larger than zero indicating that a portion of the elastic energy from O via B to C has been

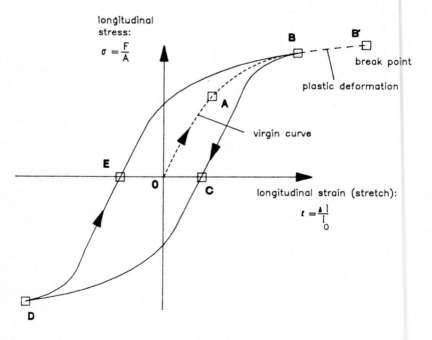

Fig. 1. The phenomenon of hysteresis in physics.
Longitudinal stress, applying Hooke's law $\sigma = E \cdot \varepsilon$ (valid for linear portion).

converted into heat. Further compression can take the system from point C to D, from where the system will return via E to B, where the applied forces are reversed. The described behaviour is called elastic hysteresis loop (hysteresis = lack behind). Extension of the system beyond B to B' is described as plastic or flow deformation; even after stress release the system shows persistent deformation. Beyond point B' the system is disrupted. Hysteresis and plasticity are often used terms to express certain phenomena of stress in ecosystems.

2 THE CONCEPT OF STRESS IN MAN AND HIGHER ANIMALS AS DEVELOPED BY SELYE

Man and mammals in general have a highly developed nervous system which interacts with the hormone system in response to any stimulus disturbing the physical or mental equilibrium. The hormone alarm system initiates a series of biochemical reactions to reduce, neutralize, or counteract the possible dangers or conflicts associated with the stress producing agents, called stressors. While in the physics of elasticity stress is the force which brings about a deformation, stress in the sense of Selye (ref.1) is the result of the action of stressors: activation of the nervous and hormone system, mobilization of energy and chemicals to counteract the stress situation, leading ultimately to serious injuries and malfunctions of various organs, if the stress persists for an extended period of time or if the stress reactions cannot be properly channeled due to the cultural environment. In Fig.2 it is pointed out that a specific stimulus (heat, frost, noise, hormones, drugs, but also emotional stresses like fear, joy, etc.) will according to Selye initiate two types of actions:

1) a stimulus specific action; and
2) a stimulus unspecific action, called stressor action, typical for the body's response to maintain homeostasis (Fig.2).

The same stressor can initiate different types of stress symptoms, many times depending on the internal disposition factors (genes, prior experience, prior exposure, ...) as well as on the external disposition factors (climate, nutrient status, chemicals, ...).

Following a single stress stimulus, three sequential phases can be distinguished:

1) the prephase, sometimes associated with a state of shock in which the resistance to the external stressor is reduced (leading in exceptional cases to death) and in which the energization of the body is prepared;
2) the acute alarm phase in which the stressor is counteracted; and
3) the rest phase in which the body functions normalize.

If the stressors act over a longer period of time, the rest phase is eliminated and substituted by a prolonged phase of resistance which, however, cannot be

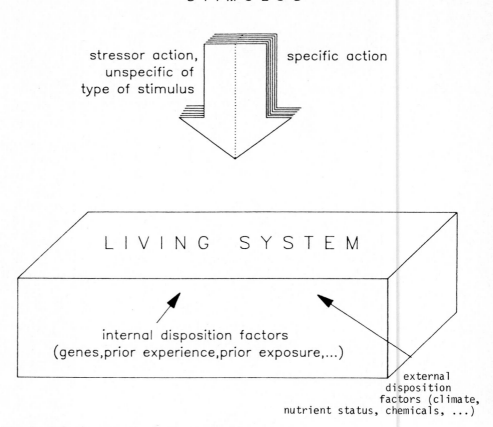

Fig. 2. Selye's concept of stressor action.

maintained indefinitely but will be succeeded by a phase of exhaustion which ultimately will lead to death.

These three phases, typical for the general adaption syndrome or perhaps more appropriately biological stress syndrome (Fig.3), are accompanied by a typical triade of the stress reaction: hypertrophy of the adrenal cortex, shrinkage of the thymus and lymph nodes and the appearance of gastrointestinal ulcers.

In summary we can say that Selye's stress concept describes the higher animals' response to stimuli of the environment with respect to the system's tendency to survive and to maintain homeostasis. The system's defense can break down both if the intensity or the duration of the stressors' action are exceeding a limit which is both dependent on internal and external disposition factors.

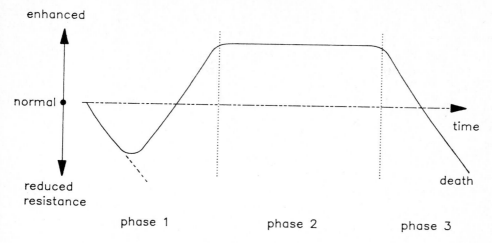

1. Stress is the unspecific response of the human body to any physical or mental strain with which it is confronted. Stressors of different type like heat, cold, light, noise, drugs, hormones, sorrow or joy, etc. apparently can cause an identical biochemical response.
2. The biochemical stress response is directed such as to reduce, neutralize or counteract the strain, in order to maintain homeostasis of the body. Selye distinguishes between the syntoxic and the katatoxic reaction type to maintain homeostasis.
3. The stress response is mediated by the nervous and hormone system of the human body.
4. Selye describes three phases of the general adaption syndrome in response to the continuous exposure to stressors: (1) alarm phase; (2) phase of resistance; (3) phase of exhaustion and death.

Fig. 3. Selye's concept of stress in man.

3 DIFFERENCES IN THE RESPONSE OF PLANT AND ANIMAL DEVELOPMENT TO ENVIRONMENTAL FACTORS

In the preceding section the response of an adult individual (man or higher animal in general) to stimuli of the environment was characterized by homeostasis. This principle of environmental homeostasis seems to be valid also for animal development in general, in as much it provides freedom from the influences of unprogrammed disturbances that might upset the delicate balance required to produce complex organized tissues. Plants, in contrast, are undergoing development often greatly influenced by external factors. The differentiation of specific tissues is carried out under the direction of hormones, the particular expression of which is mediated not only by genes, but also by external environmental factors. According to Raven and Johnson (ref.2), "both animals and plants develop according to a predetermined blueprint (genes), but in animals development reaches a conclusion - the adult individual - whereas in plants some cells undergo development continuously. In plants the course of the individual development is greatly influenced by signals from the external environment, mediated by hormones. In animals external influences exert a much weaker influence on the system. Plant development occurs as a result of the interactions between genetic factors and the external environment. The environment determines the ultimate expression of the genes in producing the phenotype". The major classes of phytohormones - auxins, cytokinins, ghibberillines, ethene, and abscisic acid - interact in a complex way to produce a mature, growing plant. Unlike the highly specific hormones of animals, phytohormones are not produced in definite organs nor do they have definite target areas, they function as rather generalized stimulators or inhibitors of growth. By integrating the complex signals that reach the plant from its environment they allow the plant to respond efficiently to the demands of the environment. Since plants, unlike animals, cannot move from place to place to seek more favourable circumstances, they try to adapt by the very slow process of growth.

4 CONCEPTS OF STRESS AT THE ECOSYSTEM LEVEL

Ecosystems are constituted by the living community or biocoenosis (including plants, animals and microorganisms) and the inorganic environment or biotope (surrounding air, water and soil space).

4.1 Stress in individuals and populations

As the individuals of one species make up a population and the populations interact in the community, one approach to determine the response to stressors in an ecosystem is to focus on the changes observed in the individual of a particular species. Depending on the mean life time of the reproductive cycle of the individuals (compare e.g. bacteria vs trees) and the intensity and duration of the stress (acute vs chronic exposure), the observations will include the generation changes of a particular population or will concentrate on the individual itself. Following this approach, Bayne (ref.3) defines stress as a measurable alteration of a behavioral, biochemical, physiological or cytological steady-state of an individual organism which is induced by an environmental change, and which renders the individual and as a result the population or community more vulnerable to further environmental change. Ivanovici and Wiebe (ref.4) have reviewed the stress effects in individuals and populations and came to the conclusion that most methods used to determine stress in individuals show high variability and extensive time needs (Table 1), except perhaps for the determination of the adenylate energy charge (AEC) which they recommend as a biochemical stress index:

$$AEC = \frac{ATP + 1/2ADP}{ATP + ADP + AMP} \qquad (3)$$

TABLE 1

Summary of the individual's response to the environmental sublethal stress (modified from Ivanovici and Wiebe (ref.4)

Examined variables	Limitations
Biochemical effects:	
Blood plasma: glucose	Highly variable, baseline data needed, interpretation difficult.
Enzyme function	Highly variable and contradictory.
Hormone changes	Limited application.
Free amino acids	Inadequate data; variable.
RNA:DNA	Inadequate data.
Adenylate charge	Variation in different organs, dependence on growth state, susceptable to environmental perturbations
0.8-0.9 optimal conditions	
0.5-0.7 non-optimal conditions	
$\leq$ 0.5 loss of viability	
Morphological effects:	
Various structures	Insufficient knowledge about normal structures, slow response time, too many structures and types for general application.
Physiological effects:	
Reproduction	Extensive time needs.
Scope for growth	None so far.
Respiration	Response inconsistent, interpretation difficult, facultative anaerobiosis.
Osmotic and ionic regulation	Response is variable.
Nitrogen excretion	Baseline data for more species needed.
Behavioral effects:	
Swimming and orientation	High individual variation.
Feeding and avoidance	Species-specific, objective assessment difficult.

with ATP, ADP and AMP being the corresponding concentrations of the tri-, di-, and monophosphates. AEC is known to be an important parameter in the cellular energy metabolism and its regulation (ref.5); it reflects the anabolic and katabolic activity of a particular organ or individual. The concept of energy drain following the stressor action is already contained in the original formulation of Selye and will be taken up later in the discussion of stress on the community or ecosystem level.

4.2 Towards a more general concept of environmental stressors: the time frame and intensity of environmental perturbations

The concept of stress in ecosystems is originally derived from Selye's concept of stress in man (and mammals). Both the intensity of the stressor and the repetition sequence of a series of stress stimuli or the continuation of one particular stress situation are of importance in the understanding of the stress syndrome. However, not all stimuli of the environment are considered, but only particularly those which are mediated by the highly developed neural-endocrine system. The question arises, whether for other vertebrates, invertebrates, plants and microorganisms Selye's special stress theory is still applicable. Stressors, in a general sense, then are according to Barrett et al. (ref.6) all perturbations applied to a system (a) which is foreign to that system, or (b) which is natural to that system but applied at an excessive (or suboptimal) level. This definition, however, does not include the disadvantage or possible advantage associated with distress or eustress in the special stress theory. It is, however, more convenient to handle, because any level of perturbation of given length and intensity can be examined or modelled this way. In toxicology including the effect of high energy radiation, the term dose is used, to describe the dose-effect relationship of a given toxic substance (or radiation impact) which is the product of concentration c and time of exposure t:

$$D = c \cdot t \tag{4a}$$

with a proportionality between dose D and effect W:

$$W \sim D = c \cdot t \tag{4b}$$

It was soon found that this relationship does not hold, in general, as e.g. Stratmann (ref.7) has pointed out in his study of the effect of SO_2 on spruce trees, suggesting a more general relationship:

$$W \sim c^\alpha \cdot t^\beta \tag{5}$$

where $\alpha > 0$, $\beta > 0$ are real positive numbers, deviating from 1.

If ecotoxicological experiments are to be achieved within the halflife of a Ph.D. student, often higher doses than in natural ecosystems have to be applied, with the time of exposure and the time of bioaccumulation beyond a threshold lying between seconds and several months. Sheehan (refs.8-10) has summarized the pollutant impact on the individual, by describing the bioaccumulation to effect both the behavioral and biochemical response after a given threshold has been reached (Fig.4a). Stress on the individual level leads via the altered performance to an impact on the population of the species affected. Changes on the community and ecosystem level are then consequences of the stressed individual and populations (Fig.4b). However, it should be noted that even naturally occurring perturbations of terrestrial ecosystems can be very sudden and of high intensity: a forest fire, a devasting hurricane, an extreme frost or insect calamity destroying within minutes to days a patch of ecosystem in the order of ha to several thousand ha. Man's perturbations of terrestrial ecosystems differ in intensity, areal extent and duration over several orders of magnitude.

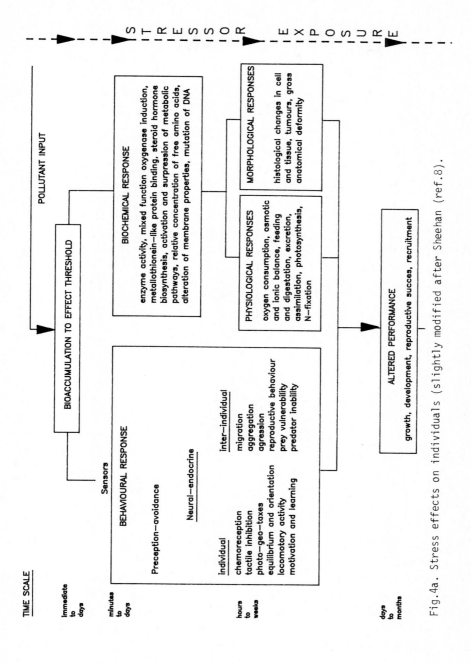

Fig.4a. Stress effects on individuals (slightly modified after Sheehan (ref.8).

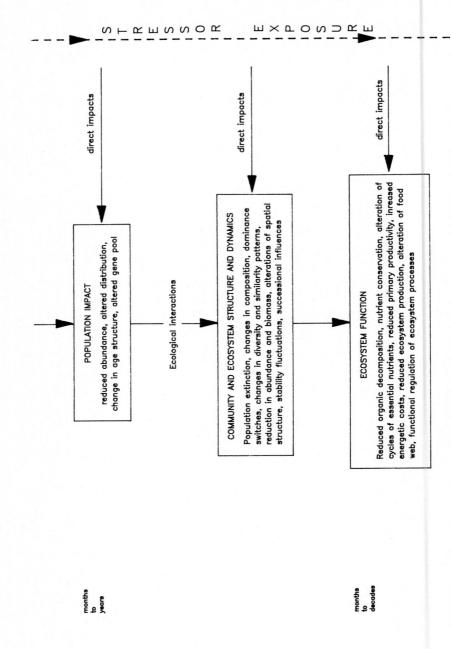

Fig.4b. Stress effects on populations, communities and ecosystems (slightly modified after Sheehan (ref.9).

4.3 The stressed terrestrial ecosystem: changes in structure and functions

In our present understanding of land ecosystems, three types of stress events (natural or man-made) may be distinguished:

1a) <u>Natural catastrophies</u> with lethal stress to most living components of the ecosystem (e.g. forest fire, hurricane, extreme frost, vulcanic eruptions). These events are often of patchy nature allowing the exchange with surrounding ecosystems. Granted this, one can often observe a good recovery through a series of successional stages to a new climax state, the recovery time depending both on the type of system and environmental conditions.

1b) <u>Man-made catastrophies</u> with lethal stress to most living components of the ecosystem (e.g. permanent deforestation, other land use changes, ...). The areal extent is often larger and the environmental constraints persist often beyond the time of catastrophy. The primary impact on the living components of the ecosystem can affect the soil and water space to a serious extent, leading e.g. to nutrient leaching and soil erosion.

2a) <u>Natural environmental conditions</u> which, by unfavourable superposition, lead to a population explosion of a particular species (insect calamities) and community extinction (forest community) <u>with lethal stress to a part of the components of the living community</u>. Recovery time will also here depend on type of ecosystem, areal extent, and persistance of environmental stress.

2b) <u>Acute impact by man</u>, either selective extinction of one or several species (e.g. insecticides, herbicides, hunting) or accidental killing of parts of a living neighbourhood by chemicals of high concentrations. Recovery time will be long if the soil matrix is affected to a serious extent, especially when chemicals are foreign to the natural metabolism of microorganisms.

3a) <u>Chronic stress</u> through fluctuations of naturally occurring environments (e.g. water or nutrient stress for plants over a series of years).

3b) <u>Chronic stress</u> imposed by man on ecosystems through small, but effective changes in the air, soil and water environment acting over a long time. Air pollution is certainly one example, where plants have been exposed not only to acid rain and dry deposition of sulphur and nitrogen compounds, but also to a surrounding atmosphere containing gaseous pollutants, which are taken up along with the CO_2/H_2O exchange process of photosynthesis.

It, however, should be pointed out, that there are man-made environmental changes, from which part of the plant communities far from industrial centers may profit, such as e.g. through the fertilization effect of excess atmospheric CO_2.

In summary we may state that the lethal, acute and chronic stress either natural or man-made will affect the plant, animal and microbial communities in two ways:

1) A structural change affecting the abundance of different species and their biomass. Decrease in the number of species in an ecosystem has been reported as a useful measure of the severity of stress. The relative susceptibility of specific individual species forms a useful basis for the indicator species concept (tolerant versus sensitive species). Several indices have been suggested to describe the species diversity, a concept which contains both the numbers of species present and the distribution of all individuals among these species (refs.8-11). A change of the diversity index is often observed following a pollution impact, however, the results for different ecosystems are contradictory. Simplification of the spatial structure organisation has been observed as well in several ecosystems, e.g. in mangrove communities susceptible to herbicides (ref.12).
2) Functional change affecting the energy flow and element cycling within the compartments: primary producers (green plants), primary and secondary consumers (herbivors, carnivors, ...) and decomposers of dead organic material (soil, flora and fauna) in interaction with the soil/water/air space.

The dynamic stability of ecosystems and their components is most aptly described within the concept of open systems exchanging material, energy and information with their corresponding environments. Response of these open systems to perturbations either within the system itself or through external action defines the properties of dynamic stability. Although both energy and matter interact, a good fundamental understanding (to a first order approximation) is obtained, if one considers at first only the material interactions (dynamic description within the structures of formal chemical kinetics) or the energetic interactions (compare e.g. Odum (ref.13) and Lugo (ref.14). The approach shown here will be formal kinetics describing coupled subsystems by the input/output scheme of flow equilibria.

5 GENERAL CONCEPTS OF EQUILIBRIA IN MATERIAL SYSTEMS, THEIR RESPONSE TO PERTURBATIONS, AND THEIR RELATION TO THE DYNAMIC STABILITY OF ECOSYSTEMS

5.1 Descriptive properties of the dynamic stability of ecosystems

Within the general concept of stability, a dynamical system is defined as locally stable, if the system upon a small perturbation returns to its original dynamic state (negative feedback). The reference state needs not to be characterized by a constancy of all material variables, a steady or stationary state in sensu stricto, but also can be a periodically varying system, as we have in ecosystems with seasonal or diurnal behaviour. We call a system locally unstable when a small perturbation is amplified, thus removing the system even further from the original equilibrium state (positive feedback). Systems are called globally stable, when the system will return to its original state, upon any perturbation, independent of its extent. Systems may be locally stable but globally unstable, if the perturbation exceeds a certain threshold. Dynamic systems may be characterized by several stationary states, allowing a transition between these states, following a perturbation of sufficient strength. Dynamical states are often compared with a rolling ball on a one-dimensional potential surface. A perturbation of given size can move the ball from the original steady state to a new steady state (Fig.5).

Within the ecological sciences several terms have been suggested to describe the corresponding stability of ecosystems (compare Sheehan (ref.9).

The inertia of an ecosystem is defined as the capacity to resist a stressor action, realized e.g. through the chemical buffering capacity, the self-cleansing capacity, but also through the resistance of organisms to environmental fluctuations and functional redundancy within the system.

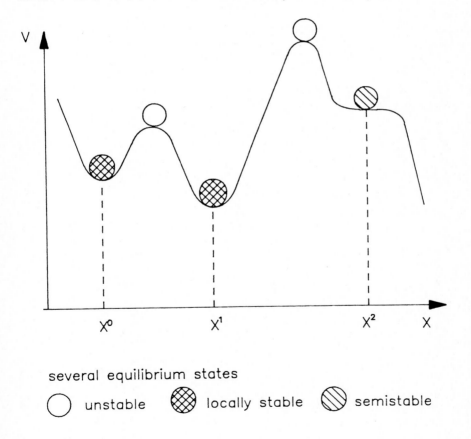

Fig. 5. Potential energy diagram defining several types of stability.

The <u>elasticity of an ecosystem</u> is defined as a measure of its ability to recover within a certain restoration time to its original state, within acceptable limits.

The <u>amplitude</u> is defined as a maximum amount of perturbation from which an ecosystem still can recover.

5.1.1 Hysteresis and malleability

<u>Malleability</u> is defined to be a measure of the ease with which a system is altered to a new stable state. Hysteresis, a term borrowed from the theory of elasticity of solids, is a measure of the degree to which an ecosystem's pattern of recovery is not the reversal of the pattern following the initial stress application.

<u>Persistance</u> is defined as the continuation of the functioning of an ecosystem, including primary production, organic decomposition and nutrient recycling.

5.2 Flux equilibria for a two-compartment system

It has been realized for a long time that the chemical transformation of the species A into the species B is described (1) by a thermodynamical equilibrium state (say A_∞ and B_∞) and (2) by a chemical kinetic rate expression which describes the rate in which an original non-equilibrium situation (say A_0 and B_0 at time $t = 0$) equilibrates. Depending on the "molecularity" of the process, the rates for the forward and reverse reaction can be formulated (ref.1) for the corresponding non-linear process:

$$A \underset{k'_{ba}}{\overset{k'_{ab}}{\rightleftarrows}} B$$

with $R_{ab} = k'_{ab} [A]^\gamma [B]^\delta$ and $R_{ba} = k'_{ba} [A]^\alpha [B]^\beta$ \hfill (4)

where $A = c_A$ and $B = c_B$ denote the concentrations, k'_{ab} and k'_{ba} the specific rates and $\alpha, \beta, \gamma, \delta$ the experimentally determined rate exponents.

In environmental sciences the exchange between two locally separated subsystems (compartments) can be described in a similar fashion, where the rates, R_{ab} and R_{ba} are now identified with fluxes F_{ab} and F_{ba} between the compartments A and B and vice versa and where the concentrations are converted into the mass contents of a particular compartment A or B ($c_A \cdot V_A \rightarrow A$; $c_B \cdot V_B \rightarrow B$). In this way the mass change with time of any given compartment is described by:

$$dA/dt = F_{ba} - F_{ab} = k_{ba} A^\alpha B^\beta - k_{ab} A^\gamma B^\delta \qquad (5a)$$

$$dB/dt = F_{ab} - F_{ba} = -dA/dt \qquad (5b)$$

with F_{ab} and F_{ba} having the same structure as R_{ab} and R_{ba}.

Any internal perturbation is mass conservative, and will lead with a particular response time to the original state, provided the system is globally stable. Any external perturbation, however, can lead to a depletion or increase in mass, which in the steady state is determined by the choice of the parameters $\alpha, \beta, \gamma, \delta$:

$$A_\infty/A_0 = (B_\infty/B_0)^{(\beta-\delta)/(\gamma-\alpha)} \qquad (6)$$

In case of simple, linear donor controlled fluxes ($\alpha = 0$, $\beta = 1$, $\gamma = 1$ and $\delta = 0$), the relative increase in mass in both compartments is equal.

REFERENCES

1. H. Selye, The Story of the Adaption Syndrome, Acta Inc., Montreal, Canada, 1952.
2. P. Raven and P.B. Johnson, Biology Textbook, Times Mirror/Mosby Colledge Publ., St. Louis, 1986.
3. B.L. Bayne, Proc. 9th Eur. Mar. Biol. Symp., 1975, 213-238.
4. A.M. Ivanovici and W.J. Wiebe, in G.W. Barrett and R. Rosenberg (Editors), Stress Effects on Natural Ecosystem, Wiley, Chichester, 1981.
5. D.E. Atkinson, Cellular Energy Metabolism and its Regulation, Academic Press, New York, 1977.
6. G.W. Barrett, G.M. van Dyne and E.P. Odum, BioScience, 28 (1986) 192-194.
7. H. Stratmann, in C. Troyanovski (Editor), Air and Plants, VCH Publishers, Weinheim, FRG, 1985, 182-210.
8. P.J. Sheehan, in Sheenhan et al. (Editors), Effects of Pollutants at the Ecosystem Level, (Effects on Individuals and Populations), Wiley, Chichester, 1984.
9. P.J. Sheehan, in Sheehan et al. (Editors), Effects of Pollutants at the Ecosystem Level, (Effects on Community and Ecosystem Structure and Dynamics), Wiley, Chichester, 1984.
10. P.J. Sheenhan, in Sheehan et al. (Editors), Effects of Pollutants at the Ecosystem Level, (Functional Changes in the Ecosystem), Wiley, Chichester, 1984.
11. J.M. Hellawell, Proc. R. Soc. Lond. Biol. Sci., 197 (1977) 31-56.
12. H.T. Odum, The Effects of Herbicides in South Vietnam. Part B: Working papers, Models of Herbicides, Mangroves and War in Vietnam, National Academy of Sciences, National Research Council, Washington, D.C., 1976.
13. H.T. Odum, in H.E. Young (Editor), Symp. on Primary Productivity and Mineral Cycling in Natural Ecosystems, Univ. of Maine Press, Orono, USA, 1967, 81-131.
14. A.E. Lugo and J.F. McCormick, in G.W. Barrett and R. Rosenberg (Editors), Stress Effects on Natural Ecosystems, Wiley, Chichester, 1981.
15. G.H. Kohlmaier, ISEM Journal, 3 Nos.1-2 (1981) 31-56.

DYNAMIC EQUILIBRIA IN MATERIAL SYSTEMS AND THEIR RESPONSE TO PERTURBATIONS

G.H. KOHLMAIER, A. JANECEK, M. LÜDEKE, J. KINDERMANN, K. SIEBKE and F. BADECK
Institut für Physikalische und Theoretische Chemie
Johann Wolfgang Goethe Universität
Niederurseler Hang, 6000 Frankfurt 50, Federal Republic of Germany

1 STABILITY OF THE GLOBAL CARBON CYCLE WITH PARTICULAR REFERENCE TO TERRESTRIAL ECOSYSTEMS

1.1 Brief description of the carbon cycle

The most important reservoirs of the global carbon cycle are displayed in Fig.1.1. In the atmosphere carbon is present mainly as CO_2 (1987: 347 ppm $\simeq$ 735 Gt C). Due to the Mauna Loa record its concentration change is well known for the past 30 years; its preindustrial value is assumed to lie in the range of 275-285 ppm (latest ice core measurements suggest 278 ppm, 1 ppm $\simeq$ 2.12 Gt).

In the biosphere carbon is the main constituent of all organic compounds (on the average 45% of the dry organic matter). Estimates of the present carbon contents of the living terrestrial biomass lie between 500 and 600 Gt C with the plant biomass constituting more than 99%. Following the death of the organisms (mostly plant biomass) the organic compounds enter into the litter layer. A fraction of this material is respired by the decomposing fauna and microorganisms and enters the atmosphere; another fraction is transformed into long-lived organic compounds, such constituting the humus layer. The total of the soil carbon contents is estimated to be 700-2000 Gt C.

In the oceans carbon is found both as constituent of organisms (mainly phyto- and zooplankton) and dead dissolved and particulate organic matter and as inorganic compounds, mostly in the form of CO_3^{--}, HCO_3^- and $CO_2 \cdot H_2O$ dissolved in the water. The carbon contents of all oceans is estimated to be 38000 Gt C.

1.2 Modelling of the exchange rates

The exchange rate between the atmosphere and the oceans is controlled on the one hand by the chemical equilibria between CO_2, HCO_3^- and CO_3^{--}, such that a relative change of the CO_2 concentration in the atmosphere causes a relative change of only $\sim 1/10$ in the oceans (Revelle buffer factor). On the other hand, the transport processes between the ocean surface layer and the deep sea layer are inhibited kinetically by the existence of a thermocline. Due to this fact the oceans are subdivided into two compartments, the mixing layer M with a depth of

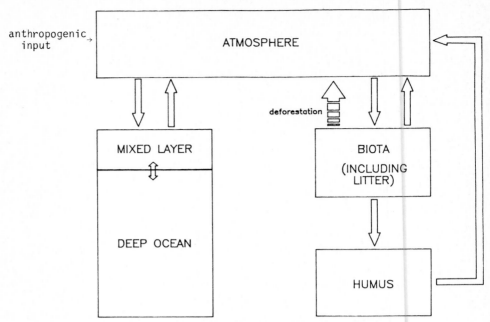

Fig. 1. The most important reservoirs of the global carbon cycle.

∼ 75 m and the deep sea layer D. Owing to its extension the latter could absorb ∼ 85% of excess atmospheric CO_2, while in reality, due to kinetic inhibition, only about 40% are sequestered.

The fluxes between the atmosphere and the biosphere are the net primary production (carbon assimilated by photosynthetic activity minus the autotrophic respiration of the assimilating organisms) on the input side and the heterotrophic and soil respiration on the output side. The heterotrophic respiration encompasses the sum of all processes leading to the transformation of biomass carbon to CO_2 entering the atmosphere without considerable time lags. Here we need to consider the consumption of living biomass by herbivors in the grazing food chain, including their consumption by carnivors on the next trophic level and more importantly so the consumption of litter by the detritus food chain, which makes up more than 90% of the total CO_2 release from ecosystems. As considerable amounts of carbon can be stored in the soils due to a dynamic time delay caused by the long carbon turnover times in the humus layer (50-500 years) the biosphere is subdivided into a living biomass B and soil carbon compartment H (humus). The atmosphere as a link between the two-box reservoirs ocean and biota is modelled by only one compartment A.

For each of the five compartments a balance of the respective inputs and outputs is performed, obtaining a system of coupled ordinary differential equations with the following structure:

$$dA/dt = -F_{am}+F_{ma}-F_{ab}+F_{ba}+F_{ha} \quad \text{(atmosphere)}$$
$$dB/dt = -F_{ba}+F_{ab}-F_{bh} \quad \text{(living biomass)}$$
$$dH/dt = -F_{ha}+F_{bh} \quad \text{(humus layer)}$$
$$dM/dt = -F_{ma}+F_{am}+F_{dm}-F_{md} \quad \text{(mixed layer)}$$
$$dD/dt = -F_{dm}+F_{md} \quad \text{(deep sea)}$$

The F's denote fluxes in (+) or out (-) of the compartment, the indices indicate between which two compartments and in which direction carbon is exchanged (e.g. F_{ba} represents the heterotrophic respiration, transfering carbon from the living biomass compartment B to the atmosphere A).

1.3 The biosphere in the unperturbed system

The submodels for the unperturbed living biomass B and the humus H are formulated in the following way:

$$dB/dt = NPP° \cdot (B/B°)^\delta - (k_{ba}+k_{bh}) \cdot B$$
$$dH/dt = k_{bh} \cdot B - H/\tau_{ha}$$

where $NPP°$ and $B°$ are the initial steady state primary production and biomass, the k's denote specific rate constants, $\tau_{ha} = 1/k_{ha}$ stands for the mean carbon residence time in the humus compartment and δ $(0 \leq \delta \leq 1)$ represents the intensity of coupling between the NPP and the standing biomass B.

It can be seen from the structure of these equations, that there are two equilibrium points:

$$dH/dt = 0 \rightarrow H° = B° \cdot k_{bh} \cdot \tau_{ha}$$
$$dB/dt = 0 \rightarrow k_{ba}+k_{bh} = NPP°/B° = k_b$$
$$\text{with } k_{ba} = (1-\rho) \cdot k_b; \quad k_{bh} = \rho \cdot k_b$$

where ρ is the fraction that enters the humus layer, and $1-\rho$ enters the atmosphere (ρ of the order 5-15%).

Apart from the trivial case of $B° = H° = 0$, a steady state describing the climax state of the biosphere is reached when the influxes are balanced by the effluxes. This steady state represents a flow equilibrium as can be observed in natural unperturbed climax ecosystems, in which the annual integrated net primary production is compensated by the heterotrophic respiration and other losses of living biomass for the compartment B (Fig.1.2) and in which the integrated annual input of dead biomass into the humus compartment is balanced by soil respiration losses. A stability analysis yields that the climax state is a global attractor, independent of the initial values for each of the compartments; provided they are $\neq 0$, the same final state of the biosphere is reached. This behaviour can be demonstrated in a phase portrait, where the development of the system is described by a trajectory in the B-H plane (Fig.1.3).

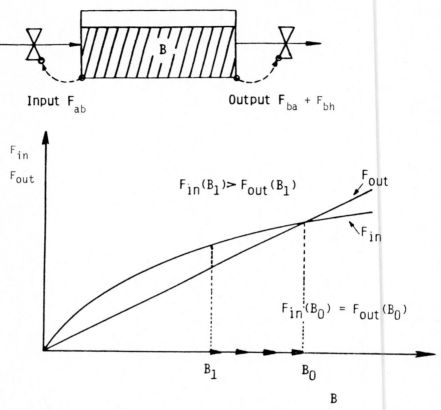

Fig. 1.2. The global attractor: living biomass in its climax state B_0 - any perturbed state B_1 will return to B_0.

1.4 The perturbed system

With respect to the biosphere, two anthropogenic perturbations of the global carbon cycle shall be considered:

1) The burning of fossil fuels, which causes a considerable primary input into the atmosphere. Two different scenarios for the future CO_2 input due to fossil fuel burning are assumed:

   a) an exponential curve fitted to observed values until present time:
   $$F_{f1}(t) = a \cdot e^{b \cdot t},$$
   where a and b are fitting parameters;

b) a logistic growth curve considering the limited fossil fuel resources:

$$F_{f2}(t) = \frac{C_\infty \cdot k_f \cdot \left(\frac{C_\infty}{C^\circ} - 1\right) \cdot e^{-k_f(t-t_0)}}{\left[1 + \left(\frac{C_\infty}{C^\circ} - 1\right) \cdot e^{-k_f(t-t_0)}\right]^2}$$

where C° is the cumulated fossil fuel input at the beginning of indus-

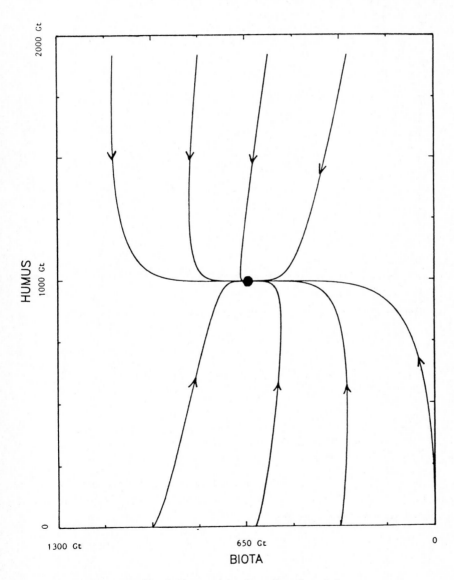

Fig. 1.3. Phase portrait of the combined system B and H.

trialization at time $t_0 = 1860$, here assumed to be 3 Gt C, and where C_∞ are the total fossil fuel resources, here assumed to be 2300 Gt C, while k_f designates the rate of increase of fossil fuel use, here assumed to be 3%.

The rising atmospheric CO_2 concentration interacts with the other two reservoirs of the global carbon cycle: a fraction of the excess CO_2 is absorbed by the oceans; the remaining fraction is likely to influence plant production According to the experimental evidence that the net primary production (NPP) of vegetation is increased with increasing ambient CO_2 partial pressure, the following extended expression for NPP is chosen:

$$NPP = NPP° \cdot (A/A°)^\beta \cdot (B/B°)^\delta \text{ with } \beta > 0 \text{ and } 0 < \delta \leq 1$$

in which $A°$ is a reference (preindustrial) atmospheric CO_2 content and β the CO_2 fertilization factor, which is defined as a relative increase in production for a given relative increase in CO_2.

2) Deforestations in the tropical forests result in an increase of atmospheric CO_2, while at the same time a decrease of the standing biomass and humus is observed. Using the world forest data given by Lanly (ref.1), a model for the different area transfers from tropical forests to logged-over forests, permanent agriculture, pasture and shifting cultivation was developed, as shown in Fig.1.4. From the different densities of carbon in various ecosystems, an aggregated function $G(t)$ can be constructed, which can be considered as a perturbation of the biota compartment. In a detailed model, which was developed by our group, we considered at least 20 age classes of tropical forests, the distribution of which is given in Fig.1.5 for one possible future scenario.

In the simplified "play model of the carbon cycle", presented here, we consider only the five non-autonomous differential equations, which represent some of the essential feedback mechanisms of our carbon cycle studies:

$$dA/dt = -F_{am} + F_{ma} - F_{ab} + F_{ba} + F_{ha} + F_{fi}(t); \quad i = 1,2$$
$$dB/dt = -F_{ba} + F_{ab} - F_{bh} - G(t)$$
$$dH/dt = -F_{ha} + F_{bh}$$
$$dM/dt = -F_{ma} + F_{am} + F_{dm} - F_{md}$$
$$dD/dt = -F_{dm} + F_{md}$$

1.5 Response behaviour to perturbations

1.5.1 Response to singular events

Within the scope of the presented model, two different singular events can be considered: a singular input of CO_2 into the atmosphere (e.g. due to a massive

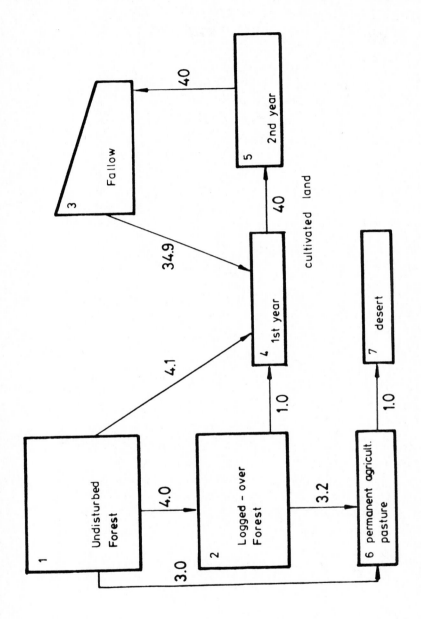

Fig.1.4. Area transfers in tropical forests (10^6 ha), according to Lanly (ref.1).

vulcanic eruption) or a sudden deforestation as may be realized by widespread wild fires or extensive clearing. In Figs.1.6 and 1.7 the system response behaviour to these perturbations is displayed. The response to a singular input of CO_2 into the atmosphere (Fig.1.6) of 0.3 Gt brings about a permanent increase of CO_2 impressively displayed in the deep sea as well as in the ocean surface water, whereas the living biomass and humus, due to the initial peak concentration of CO_2, show a high fertilization effect and the corresponding litter flux to the humus, which, however, is reduced in time, because the atmospheric CO_2 level falls with time. Fig.1.7 displays a carbon transfer within the five compartment system, showing the effect of a massive fire which destroys a fraction of the vegetation ($\sim$ 20 Gt C). At first, atmosphere, surface ocean and deep sea will increase their carbon contents, while the humus is losing carbon, due to the reduced net primary production and along with it a reduced litter production. If the system is given enough time (about 250 years), all reservoirs will eventually equilibrate to the initial state (ref.2,3).

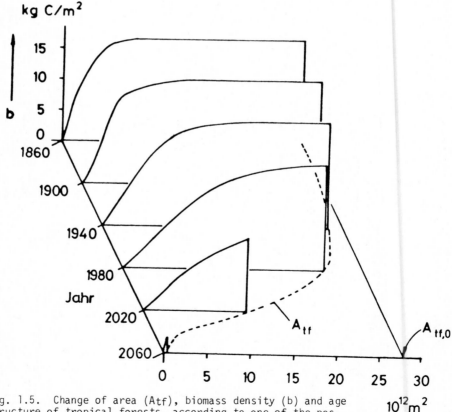

Fig. 1.5. Change of area (Atf), biomass density (b) and age structure of tropical forests, according to one of the possible future scenarios.

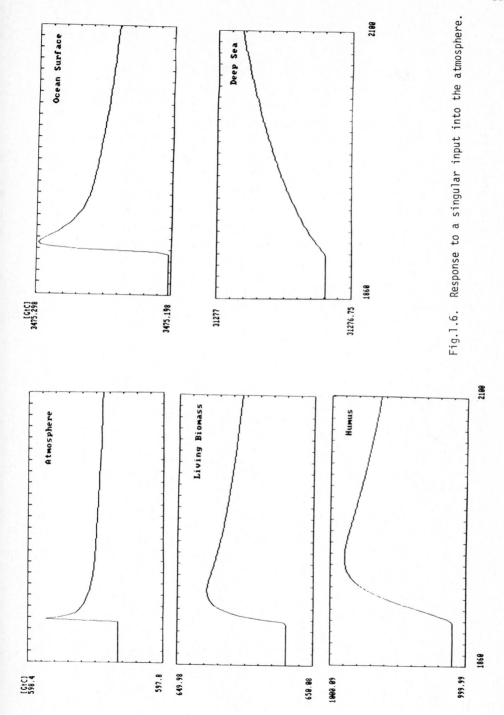

Fig.1.6. Response to a singular input into the atmosphere.

146

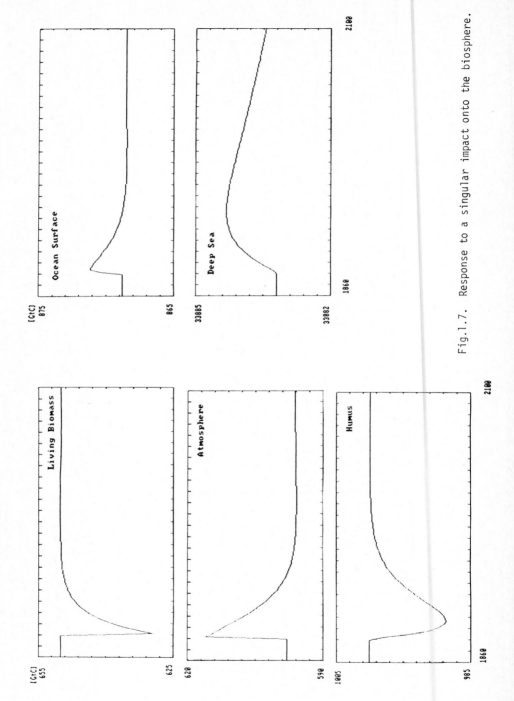

Fig.1.7. Response to a singular impact onto the biosphere.

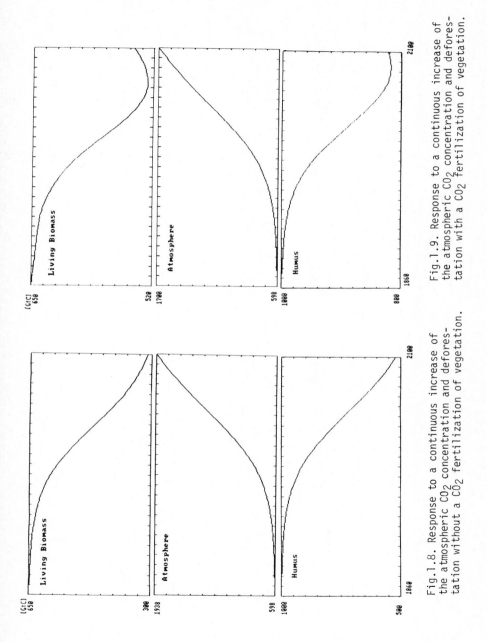

Fig.1.8. Response to a continuous increase of the atmospheric CO_2 concentration and deforestation without a CO_2 fertilization of vegetation.

Fig.1.9. Response to a continuous increase of the atmospheric CO_2 concentration and deforestation with a CO_2 fertilization of vegetation.

1.5.2 Response to continuously increasing fossil CO_2 input into the atmosphere

Figure 1.8 displays the response of the biota-humus system, assuming a deforestation rate, which goes parallel to the fossil fuel consumption and without considering a stimulation of the vegetation by excess CO_2. Introducing, however, as shown in Fig.1.9, a CO_2 fertilization effect, with $\beta = 0.3$, the total living biomass is reduced to a much smaller extent and may eventually even compensate the reduced deforestation rate in the 21th century. Similarly, the humus compartment is reduced by a smaller extent also, because of the assumed close coupling of the two systems.

The simple model displays the main features of the biota-soil interactions with the atmosphere, pointing out the important buffering capacity of terrestrial ecosystems, if deforestation is not maximized and stimulation of the vegetation can still occur within a global atmosphere, which is not totally polluted (ref.4). Extrapolation of man's impact on the vegetation in the future is open to many scenarios, hopefully one which lets survive the biosphere.

2. MODELLING POLLUTANT EXCHANGE BETWEEN PLANT AND ENVIRONMENT: UPTAKE, RESPONSE AND METABOLIZATION OF SULPHUR DIOXIDE BY DIFFERENT LEAF CELL COMPARTMENTS

2.1 Relevance of modelling of air pollutant uptake and metabolization

Nowadays it is the commonly accepted opinion that airborne pollutants are the main causing factor of forest die-back ("Waldsterben"). A number of empirical studies have been conducted to assess the correlations of pollutant concentration in the environment and symptoms of plant damage. In most cases they substantiate harmful effects for relatively high levels of single pollutant concentrations. In real environmental situations, forest die-back is caused by relatively low concentrations of several pollutants. Recent research, therefore, focuses on phenomena like: (1) accumulation of pollutants inside the plants; (2) chronic "invisible" damage; (3) synergistic action of different pollutants; (4) reduction of net photosynthetic production, synthesis of secondary metabolites and their effects; and (5) reduction of tolerance against additional stressors (water stress, extreme frost events, insect infestations, etc.).

In order to explore the mechanisms which lead to the different symptoms of forest die-back it is necessary to take a detailed look into the black box which is located between the causing contamination of the environment and the damage symptoms of the tree. This implies that it is necessary to work on the mechanisms of uptake and metabolization of pollutants by plants and the surrounding soil space. As a first step we look only into the subsystem leaf considering the short range transport and the most important chemical pathways. The computer model, to be presented here, was elaborated to simulate the uptake and metabolization of sulphur dioxide. At the University of Würzburg, Laisk et al. developed a similar model for describing the influence of SO_2 uptake on the pH status within different leaf compartments (refs.5,6).

2.2 Model description

The computer model which is considered here provides a mean to simulate the diffusive transport of sulphur dioxide into the leaves and its subsequent solvation, dissociation, and metabolization.

The main path for the uptake occurs via the stomates. Uptake of gaseous SO_2 across the cuticle plays a minor role because the resistance of the cuticle is about 10^4 times that of the stomatal resistance (ref.7). Besides the stomatal resistance (R_3) the total resistance for the gaseous diffusion of SO_2 into the plant leaves is made up by the resistances of the leaf boundary layer (R_2) and the intercellular air space (R_4) (Fig.2.1). Modelling the gaseous diffusion of SiO_2 is accommlished by adapting the model of Nobel (ref.8) for the uptake of carbon dioxide.

The transfer of SO_2 from the intercellular air space to the liquid phase of the cell walls is governed not only by the solubility of the gas, described by the Henry constant, but also by the degree of dissociation which is dependent on the pH. Given a pH of 5.7 for the liquid phase of the cell walls, stabilized by a buffer, the bulk of the imported sulphur(IV) will occur as hydrogen sulphite HSO_3^- and sulphite SO_3^{--}. S(IV) is transported from the cell walls into the cytosol across the plasmalemma.

As relevant compartments for the transport, metabolization and storage of the sulphur compounds to cytosol, the cloroplast and the vacuole are considered in the model. The concentration within compartments and the fluxes across the membranes are calculated for each of the three S(IV) species (SO_2, HSO_3^-, SO_3^{--}) separately. The permeabilities of the plasmalemma, chloroplast membranes and tonoplast for $SO_2 \cdot H_2O$ exceed the permeabilities for the dissociated species manifold. In addition the respective degrees of dissociation of S(IV) in each compartment are mediated by the compartmental proton concentration (acidity). The compartmental buffer

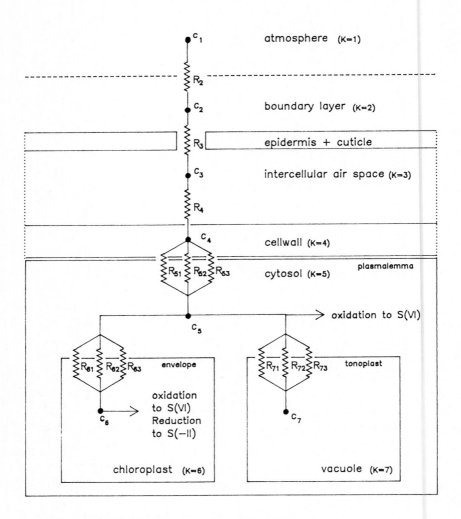

c_K: SO_2 or S(IV)–concentration in compartment K
R_{Ki}: Resistance between compartment K−1 (respectively K−2) and K for the molecule species i

i=1: $SO_2 \cdot H_2O$
i=2: HSO_3^-
i=3: SO_3^{--}

Fig.2.1. Simplified leaf anatomy and S(IV) flux model.

systems are simulated as one single buffer which stabilizes the respective pH-values at 7.4 (in the cytosol), 8.0 (in the chloroplasts), and 5.8 (in the vacuole). Optionally the model can be run assuming an ideal buffer (infinite buffer capacity) or a limited buffer capacity.

Apart from these processes governing the uptake of SO_2, the overall intracellular concentrations of the S(IV) species are influenced by the rates of the metabolizing reactions. Oxidation of S(IV) to S(VI), i.e. sulphate and reduction of S(IV) to S(-II) are reflected in the model. Sulphur of the oxidation state S(-II) is a component of amino acid synthesis or can be alternatively emitted as H_2S or $(CH_3)_2S$ up to a certain degree.

A mathematical outline of the current model is given in Fig.2.2. Under the assumption of a very fast equilibration between the outside SO_2 concentration and the liquid phase of the cell wall (steady state assumption $dc_k/dt = 0$ for $k = 1,2,3,4$), we can solve for c_5 (concentration of S(IV) in the cytosol), c_6 (concentration of S(IV) in the chloroplast) and c_7 (concentration of S(IV) in the vacuole). Diffusion of gaseous sulphur dioxide and transport of the dissolved S(IV) species across the biomembranes are described in terms of Fick's law of diffusion considering specific resistances R_{ki} of gaseous and dissolved SO_2, HSO_3^- and SO_3^{--}. The oxidization of S(IV) to S(VI) - mediated by radical chain mechanism - is modelled as a reaction of first order whereas the reduction of S(IV) to S(-II) - a series of enzyme catalyzed reactions - is modelled with a Michaelis-Menten term. The transport of SO_2 into the cytosol is governed by an effective resistance R_{eff} in which the high plasmalemma resistance of 80 s cm^{-1} becomes nearly negligible since the surface area of the plasmalemma is about 11 times higher than the leaf surface area and since the concentration of dissolved S(IV) species in the cell wall is much higher than the corresponding outside concentration of SO_2. When considering the concentration gradient between gaseous SO_2 of the surroundings, c_1, and the concentration of total dissolved S(IV) species in the cytosol, c_5, we need to multiply c_5 by the factor g, which reflects the accumulation of S(IV) in the liquid phase as a result of solubility and dissociation. The partition coefficients X_i are used to characterize the relative amounts of the different S(IV) species in the respective compartments.

The parameters used in the model runs (cell compartment dimensions, resistances, dissociation constants, buffer concentrations, oxidation and reduction rates) have been adapted from Garsed and Read (ref.11), Pfanz et al. (refs. 12,13), Laisk et al. (refs.5,6) and Seel (ref.14).

2.3 Results

Starting with a hypothetical system, comprising two gaseous compartments of different SO_2 concentrations, we will consider some factors determining the

1. S(IV) concentration in the cytosol:

$$\frac{dc_5}{dt} = \frac{1}{d_5}\left[\frac{c_1 - g\cdot c_5}{R_{eff}} - \frac{A_6}{A}\sum_{i=1}^{3}\frac{X_{5i}\cdot c_5 - X_{6i}\cdot c_6}{R_{6i}} - \frac{A_7}{A}\sum_{i=1}^{3}\frac{X_{5i}\cdot c_5 - X_{7i}\cdot c_7}{R_{7i}}\right] - ko_5\cdot c_5 \quad (1)$$

with: $\quad g = X_{41}\cdot Hv^{-1}\cdot \sum_{i=1}^{3}\frac{X_{5i}}{R_{5i}}\cdot\left[\sum_{i=1}^{3}\frac{X_{4i}}{R_{5i}}\right]^{-1}$

and $\quad R_{eff} = R_2 + R_3 + R_4 + X_{41}\cdot Hv^{-1}\cdot\left[\sum_{i=1}^{3}\frac{X_{4i}}{R_{5i}}\cdot\frac{A_5}{A}\right]^{-1}$

2. S(IV) concentration in the chloroplasts:

$$\frac{dc_6}{dt} = \frac{A_6}{d_6\cdot A}\sum_{i=1}^{3}\frac{X_{5i}\cdot c_5 - X_{6i}\cdot c_6}{R_{6i}} - ko_6\cdot c_6 - \frac{Red_6\cdot c_6}{Kmred_6 + c_6 + [SO_4^=]_6} \quad (2)$$

3. S(IV) concentration in the vacuole:

$$\frac{dc_7}{dt} = \frac{A_7}{d_7\cdot A}\sum_{i=1}^{3}\frac{X_{5i}\cdot c_5 - X_{7i}\cdot c_7}{R_{7i}} \quad (3)$$

4. Ratio of $SO_2\cdot H_2O$ to S(IV) as determined by pH:

$$X_{K1} = \left[1 + \frac{K_1}{[H^+]} + \frac{K_1\cdot K_2}{[H^+]^2}\right]^{-1} \quad (4)$$

5. Ratio of HSO_3^- to S(IV):

$$X_{K2} = \left[\frac{[H^+]}{K_1} + 1 + \frac{K_2}{[H^+]}\right]^{-1} \quad (5)$$

6. Ratio of SO_3^{--} to S(IV):

$$X_{K3} = \left[\frac{[H^+]^2}{K_1\cdot K_2} + \frac{[H^+]}{K_2} + 1\right]^{-1} \quad (6)$$

7. H^+ concentration in compartment K taking into account the influence of H^+ released during formation of HSO_3^-, SO_3^{--} and SO_4^{--} and the buffer of the compartment K:

$$[H^+]_K = E_K\cdot\frac{[HA]_K + X_{K2}\cdot c_K + 2\cdot X_{K3}\cdot c_K + 2\cdot[SO_4^{2-}]_K}{BF_K - ([HA]_K + X_{K2}\cdot c_K + 2\cdot X_{K3}\cdot c_K + 2\cdot[SO_4]_K)} \quad (7)$$

Fig. 2.2. Model equations of S(IV) transport ans S(IV) metabolization (including reduction to S(-II) and oxidation to S(VI) within different compartments of the leaf.

./.

Legend to symbols used in Fig. 2.2.

Indices:
i: molecule species,
i=1: $SO_2 \cdot H_2O$, i=2: HSO_3^-, i=3: SO_3^{--}
K: compartments
K=1: atmosphere, K=5: cytosol, K=6: chloroplast, K=7: vacuole

Variables:
A_K/A: ratio of envelope area of compartment K to leaf area,
BF_K: buffer concentration in compartment K (mM),
c_K: S(IV) concentration in compartment K (mM),
d_K: volume of compartment K per leaf area (cm),
E_K: equilibrium constant of buffer in compartment K (mM),
$[HA]_K$: concentration of protonated buffer anions in compartment K (mM),
Hv: partition coefficient of SO_2 between water and atmosphere
K_1: constant of first step of dissociation $SO_2 \cdot H_2O \rightleftharpoons HSO_3^- + H^+$ (mM),
K_2: constant of second step of dissociation $HSO_3^- \rightleftharpoons SO_3^{--} + H^+$ (mM),
ko_K: oxidation coefficient in compartiment K (1/s),
Kmred: Km for S(IV) reduction (mM),
R_{Ki}: resistance for diffusion of the molecule species i between the compartments K-1 and K (s/cm),
Red_c: max. velocity of S(IV) reduction (mM/s),
$[SO_4^{2-}]_K$: sulfate concentration in the compartment K (mM),
t: time (s),
X_{Ki}: mol fraction of molecule species i in compartment K.

equilibrium concentration inside a system (= compartment 2) which is subjected to a pollutant, SO_2, abundant in the surrounding atmosphere (= compartment 1). Assuming, at first, that both compartments can only accommodate gaseous species, the flux of the pollutant into the system will be determined by the concentration gradient and the resistance for the pollutant's diffusion, respectively, according to Fick's second law. Assuming that the concentration of the pollutant in the environment is constant in time, the concentration in compartment 2 will rise until thermodynamic equilibrium is established. Under these circumstances the concentration inside compartment 2 will be equal to the environmental pollutant concentration.

Assuming in a second step that compartment 2 is constituted by a liquid phase, some additional factors arise which govern the level of the internal equilibrium concentration. The pollutant's solubility and dissociation must be taken into consideration. Because of the high solubility of SO_2 in water when thermodynamic equilibrium is established, $[SO_2 \cdot H_2O]$ in solution will be 40 times $[SO_2]$ in the gaseous phase. Additionally, the total S(IV) concentration will be higher than [$SO_2 \cdot H_2O$] because according to Henry's law only the dissolved gas is equilibrated with the pollutant in the gas phase but not the hydrogen sulphite and sulphite species. Any pollutant molecule transported into compartment 2 will be subjected to dissociation. The equilibrium of the different S(IV) species will be determined by the actual pH. On the other hand any dissociation step will release a proton. As a consequence the pH decreases and the dissociation equilibrium is changed in favour of the undissociated species. Overall S(IV) is again accumulated in compartment 2: 40 fold because of the solubility and another 300 fold because of dissociation.

In a third step the results of computer simulations of SO_2 uptake have to be evaluated. In these experiments, diffusion of the pollutant has been simulated applying the model equations described above. Metabolization of S(IV) is not yet taken into account. Unlike the system constituted of one gaseous and one liquid phase described above, we now deal with several internal compartments in which the acidity is controlled by buffer solutions. Fig.2.3 presents the results of two experiments:

a) assuming an ideal buffer;
b) assuming limited buffer concentration (6.2×10^{-2} M). Under these conditions a thermodynamic equilibrium is reached too. With a pollutant concentration of 0.5×10^{-6} mM in the ambient air, the equilibrium concentration of S(VI) in the cytosol reaches 6 mM with unlimited buffer concentration and 4 mM with limited buffer concentration. That means S(IV) is enriched 10 million times. A high level of a potentially poisonous compound is built up inside the plant cells.

As it can be concluded from these results, the action of cellular buffers bears conflicting results: on the one hand the buffer protects the plant from running out of the physiological range of pH-values in which the enzymes are stable and can operate optimally. On the other hand the stabilization of cellular pH hinders the ratio of undissociated to dissociated S(IV) species to be increased with decreasing pH. As a consequence the equilibrium concentration for intracellular S(VI) is much higher than it would be in an unbuffered liquid phase.

The increase in equilibrium concentration by buffering can be demonstrated by comparing the data for the systems with infinite and limited buffer concentrations, respectively (see Fig.2.3). The ideal buffer stabilizes the pH-value in the cytosol at 7.4. With a buffer concentration of 6.2×10^{-2} M the pH-value in the cytosol is lowered to 7.2 leading to a reduction in the equilibrium concentration of S(IV) by 33%.

In cell compartments with different buffer systems, the equilibrium concentration of S(IV) is altered correspondingly. In the more alkaline compartments

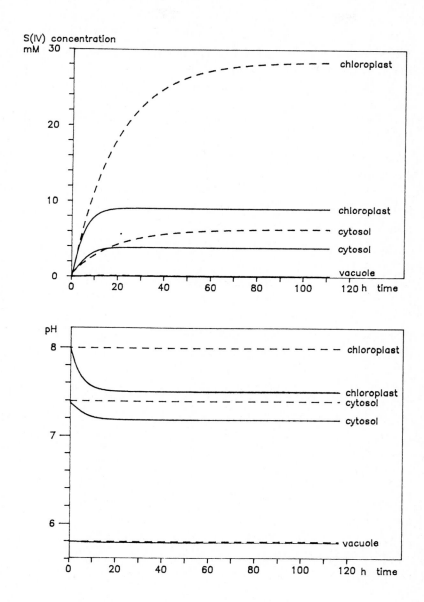

Fig. 2.3. Diffusion of S(IV) into a leaf cell from an atmosphere containing 0.5×10^{-5} mM sulphuridoxide.
--- pH is assumed to be constant;
—— pH is assumed to change due to a limited buffer capacity.

(e.g. chloroplasts, pH 8.0), the mole fraction of $SO_2.H_2O$ is lower and, therefore, the total S(IV) concentration higher than in more acid compartments. Consequently, the equilibrium concentration of S(IV) in chloroplasts is higher than in the cytosol (28 mM vs 6 mM) and much lower in the acidic vacuoles (pH 5.7). Proton stress and potential toxic effects brought about by S(IV) alter with the alkalinity of the respective compartment.

In still another computer experiment we proceeded even one step closer to the real situation inside plants. In addition to diffusion, solution, dissociation and effects onto acidity, now elimination of S(IV) is simulated. Oxidation of S(IV) and transport of sulphate ions together with two protons from chloroplasts and cytosol into the vacuole are modelled.

As can be seen from the results presented in Fig.2.4, the steady state concentrations of S(IV) in the cellular compartments differ significantly from the concentrations given for a system without oxidation. High oxidation rates result in low equilibrium concentrations for S(IV). It was 4 mM for the cytosol in the previous experiment and decreases to 2.8 mM with an oxidation coefficient of 2×10^{-5} s^{-1}, to 0.9 mM with an oxidation coefficient of 2×10^{-4} s^{-1}, and to 0.1 mM with an oxidation coefficient of 2×10^{-3} s^{-1}. Compared with the previous experiment the enrichment of S(IV) in the cytosol is reduced. Without oxidation it was enriched 10 million times, while a value of 5×10^5 is obtained for the highest oxidation coefficient.

In contrast to the previous experiments, in the current experiment the thermodynamic equilibrium for the S(IV) concentration is not established any longer. The S(IV) concentration is now governed by a steady state equilibrium. Apart from the factors discussed up to now, in addition the relation of influx (diffusion) and efflux (elimination reactions) now affects the level of the equilibrium concentration for S(IV). Alterations of the resistances for diffusion and the reaction constants of the elimination reactions will cause alterations in the steady state equilibria.

The model presented here additionally allows to simulate reduction of S(IV) to S(-II) and diurnal changes in stomatal conductance, acidity of chloroplasts and rates of elimination reactions. Up to now it does not comprise possible reactions of S(IV) outside the cells, particularly in the cell wall, and long distance transport of the several sulphur species out of the leaves. But even on the current level of model development the results presented above prove that the effective concentration of pollutants in plant cells and cell compartments are dependent on the internal state of the plant.

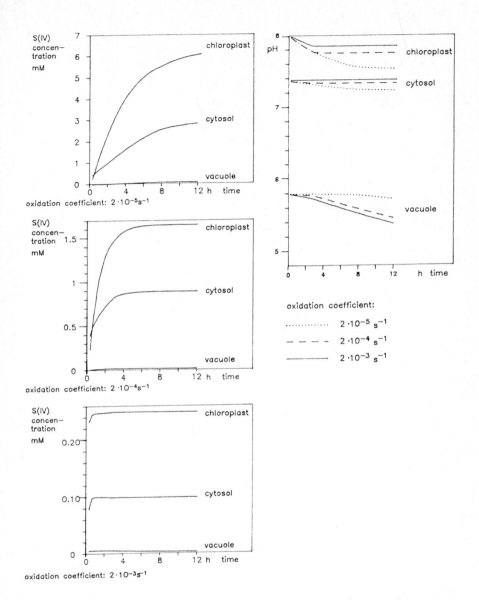

Fig. 2.4. The S(IV) concentration and pH in the compartments assuming an oxidation of S(IV) to S(VI) in the cytosol and the chloroplasts; furthermore, a transport of protons and sulphate from the chloroplasts and the cytosol into the vacuole.

REFERENCES

1. S.P. Lanly, Tropical Forest Resources, FAO, Rome, 1982.
2. G.H. Kohlmaier, Radiat. Environ. Biophys., 19 (1981a) 67-78.
3. G.H. Kohlmaier, ISEM Journal, 3 No.1 (1981b) 31-56.
4. G.H. Kohlmaier, H. Bröhl, E.O. Siré, M. Plöchl and R. Revelle, Tellus 39B (1987) 155-170.
5. A. Laisk, H. Pfanz, M.J. Schramm and U. Heber, Planta, 173 (1988a) 230-240.
6. A. Laisk, H. Pfanz and U. Heber, Planta, 173 (1988b) 241-252.
7. K.J. Lendzian, in M.J. Koziol and F.R. Whatley (Editors), Gaseous Air Pollutants and Plant Metabolism, Butterworths, London, 1984, pp.77-81.
8. P.S. Nobel, Introduction to Biophysical Plant Physiology, W.H. Freeman & Co., San Francisco, 1974.
9. J.W. Anderson, in P.K. Stumpf, E.E. Conn (Editors), Biochemistry of Plants a Comprehensive Treatise, B.J. Miflin (Editor, Vol.5), Amino Acids and Derivatives, 1980, pp.203-223.
10. J.A. Schiff, in A. Lauchli and R.L. Bieleski (Editors), Encyclopedia of Plant Physiology, A. Pirson and M.H. Zimmermann (Editors, Vol.15A), Inorganic Plant Nutrition, Springer, Berlin, 1983, pp.401-421.
11. S.G. Garsed and D.J. Read, New Phytol., 99 (1977) 583-592.
12. H. Pfanz, E. Martinoia, O.L. Lange and U. Heber, Plant Physiol., 85 (1987a) 922-927.
13. H. Pfanz, E. Martinoia, O.L. Lange and U. Heber, Plant Physiol., 85 (1987b) 928-933.
14. F. Seel, Grundlagen der analytischen Chemie unter besonderer Berücksichtigung der Chemie in Wässrigen Systemen, Verlag Chemie, Weinheim, 1965.

ECOLOGICAL ASSESSMENT OF THE DEGRADATION AND RECOVERY OF RIVERS FROM POLLUTION

R.W. EDWARDS
University of Wales Institute of Science & Technology
Cardiff & Welsh Water Authority

1. INTRODUCTION

Methods were devised almost a century ago to describe the assemblages of organisms associated with particular stages in the downstream recovery of rivers from pollution by sewage (refs.1,2). To what extent have we progressed from this early ecological assessment of spatial degradation and recovery downstream of sewage discharges? We now need to describe in ecological terms the impact of the ever increasing range of pollutants on river systems, in both spatial and temporal terms, recognising that polluted rivers rarely maintain a steady state, pollutant loads, and the factors which influence their impact, varying in time. How effectively can we describe dynamic ecological responses to pollution stress? What steps have been made to bridge the gap between description and understanding? How far has our empirical description and comprehension of causal relationships produced a capability to predict responses of river ecosystems and so adjust pollution control strategies to protect these ecosystems? - or at least those parts of their structure and function regarded as important?

To seek answers to such questions the chapter will be divided into three sections. The first explores the "state of the art" with respect to methods of ecological assessment and, increasingly bound up with these, the range of experimental approaches which assist ecological interpretation. The second highlights some of the contemporary changes taking place in patterns of pollution, particularly those which require us to re-examine our traditional approaches to pollution investigation. The third describes some case studies which reveal the strengths and weaknesses of our current approaches.

2. METHODS OF ECOLOGICAL ASSESSMENT

It has often been expressed in varying forms that analytical ecology, which leaves nature to do its own experiments (sometimes stimulated by actions from man), whilst strong on realism, is weak on predictiveness - relying on correlative interpretation of complex variability for its conclusions. On the other hand, man's experimental models of ecological systems are capriciously apt to induce conclusions of little ecological relevance. Bridging the gap between these two approaches, in part by reducing their individual weaknesses and in part by adopting a variety of methods, has been a feature of recent years.

2.1 Describing ecological structure and function

Structure and function can be considered at different hierarchical levels - individual, population and community - although the "community" in most structural descriptions of pollution effects is highly constrained taxonomically (e.g. fish) or by standard sampling procedures (e.g. net sampler). Most attention has been focused on community descriptions, in part because it is generally considered that subtle effects will be detected which might remain unnoted by more detailed studies of individual species, subsequently found relatively insensitive to the imposed pollutant stress. Nevertheless, the ecological literature abounds with autecological studies of aquatic organisms and appropriate methods are available to study many population parameters.

In fact, it might be argued that stress is most easily detected at lower hierarchical levels - particularly that of the individual organism - for each ascending level of organization has an adaptive resilience which reduces the impact of stress. However, this adaptiveness is probably more evident at community level with respect to function rather than taxonomic structure.

At community level it could be argued that the maintenance of function is far more important than taxonomic structure. At population level, however, particularly where species are exploitable resources (e.g. fish) or have particular conservation value, structure is relevant.

There have been several recent reviews of methods to describe community structure and function, the most notable being those mentioned in (refs.3,4,5). In structural terms two approaches have traditionally be adopted:
a) to ignore the qualitative information within data sets and use only numerical information about numbers of species (or frequently taxa of varying status) and numbers of organisms within a sample; these are commonly referred to as <u>diversity indices</u>;
b) to develop <u>biotic indices</u> which are either in the form of numerical scores or named classes and are based upon apparent sensitivities of species or larger taxa to pollution, almost invariably by sewage; such sensitivities are generally derived from field data of previous associations of organisms with pollutant discharges rather than responses obtained from experimental studies of pollutant stresses.

Some indices (e.g. Chandler) have combined these quantitative and qualitative approaches.

In recent years multivariate statistical procedures of classification and ordination have introduced a third and powerful tool for the detection of biological patterns and associated environmental factors.

2.1.1 Diversity indices

Several diversity indices are simply of the form:

$$\frac{\text{function } S}{\text{function } N}$$

where S and N are the number of taxa and individuals, respectively, within a sample. Though many indices have been produced (ref.5) it is clear that they contain little community information. Furthermore, values of such indices are not independent of sample size (ref.6) and even where, throughout a survey, a standard sampling area has been adopted, errors may arise where species-sampling area relationships differ between sites or sampling times. A further problem arises from the spatial variability of overall (and species) density of organisms: confidence limits on the density of benthic organisms (the only ones providing the spatial acuity required for most pollution studies) are unacceptably wide unless very large numbers of samples are taken (refs.7,8). Such sampling and processing resources are rarely available, even in research studies.

Some indices describe the distribution of individuals between taxa, Preston's log-normal distribution and those based on Information Theory (particularly the Shannon-Wiener index) being most widely used (see ref.5). These indices, irrespective of claims of their theoretical value, contain more information about the community, namely species richness and the distribution of numerical abundance of individual species, than those described earlier. However, these two signals sometimes need separation and indices of evenness and redundancy have been developed solely to express this numerical distribution.

The Shannon-Wiener index has been grossly and uncritically used, some authors claiming specific index values are indicative of levels of pollution (ref.9). Taxonomic discrimination, differing widely between investigations, is so critically important in determining index values (ref.6). Furthermore, the logarithmic base (2,e and 10) used in calculating index values varies with investigator and is frequently not explicit. These and other factors (refs.6,10) make numerical comparisons between studies of dubious value.

Edwards et al. (ref.8) compared the values derived for a range of diversity indices from a pollution survey of a river in South Wales and concluded that:
1) if the survey data were only used to generate diversity indices, taxonomic identification to family was adequate, the correlation between index values at family and species level being very high; and
2) the values of different diversity indices at specific sites were highly correlated and that, therefore, most were redundant.

This conclusion is obvious from the similarity in form of the indices. However, indices such as the Shannon-Wiener index, based on numerical distributions of component species, provide information not expressed by simpler functions and are, therefore, not highly correlated with them. In contrast, data provided by Wilhm (ref.11) suggest that in some situations all indices generate values which are highly correlated (ref.5).

The theoretical basis for the application of diversity indices, namely the presumed relationship between community diversity and ecological stability, has been explored by several authors (see ref.5) and the ecological processes underlying such a relationship, where one exists, remain contentious. It has been suggested that trophic diversity (in contrast to taxonomic diversity) within a community might more adequately reflect stability (ref.12). However, testing this hypothesis in aquatic communities is not possible until we can define more widely the trophic status of component species in squatic systems and then express their contribution to energy flow or similar processes.

2.1.2 Biotic indices

Biotic indices have generally been derived either for associations of "attached" micro-organisms (e.g. Saprobic Index of Kolkwitz and Marsson (ref.1)) or of benthic invertebrates (Trent Biotic Index of Woodiwiss (ref.13)), and both schools have their strong adherents, based on tradition and disciplinary expertise as much as the intrinsic merits of the associations. Where pollutant discharges are chronic point sources, both groups of organisms provide spatial discrimination, but with changing patterns of pollution, dominance shifting towards more intermittent sources, long recovery times provide easier detection and in this respect benthic invertebrates may have advantages.

These groups of organisms, or species within them, are used as indicators of pollutional states but such indicators are not equally sensitive to all forms of pollution and so discrimination between different forms of pollution is theoretically possible. However, most biotic indices have been developed in relation to sewage discharges and have little diagnostic value elsewhere. With more sophisticated community analysis based on multivariate statistical techniques to define the groups and to suggest their links with aspects of water quality (refs.14,15), together with an extension and improvement of experimental ecotoxicological studies (refs.16,17) to confirm these links, then the biotic index concept offers some prospects for future development. Some of these indicators will be geographically limited although many micro-organisms are cosmopolitan. A further problem, not yet adequately explored but discussed briefly later in this chapter, is the genetic adaptation of species to pollutant exposure and the consequent variation in resistance of indicator species.

Washington (ref.5) in a detailed and critical analysis of biotic indices lists ten groups, each consisting of up to five closely associated indices.

A few of these indices, including the Chandler Biotic Score (ref.18), combine elements of the biotic and diversity indices. With this system the total score of the sample depends in part on the number of species recorded and their numerical abundance (divided into five classes from present to very abundant). As a consequence, any stress which reduces diversity, affects the score. Headwater streams, although generally unpolluted, are frequently physically stressed by highly variable flows and temperatures and have low species diversity. To accommodate this problem a variant of the Chandler Biotic Score has been developed which essentially expresses the average sensitivity of species in a sample to pollution. This Average Chandler Biotic Score is considered best by the "benthic invertebrate" adherents, being, in particular, sensitive to mild and moderate pollution levels (ref.19) and unaffected by seasonal rhythms (ref.10).

2.1.3 Multivariate techniques and their impact on community description and prediction in relation to water quality

In the last decade or so, several multivariate statistical techniques have been applied to complex data sets of stream fauna. Classification methods have grouped those sites of similar faunal composition and those species which commonly occur together in space or time (refs.20,21). A critical comparison of these classification methods when applied to a faunal survey of the polluted River Ely (South Wales) was made by Learner et al. (ref.22). Most of these early studies were restricted to single river catchments or limited geographical areas but Wright et al. (ref.23) extended his study to 268 sites - all apparently unpolluted - on British rivers. Furthermore, using ordination methods, the relationships between faunal distributions and environmental variables were explored.

Moss et al. (ref.14) have further developed the procedure which uses environmental data to predict the probabilities of macro-invertebrate taxa occurring at unpolluted river sites in British rivers. Using only five physical and chemical variables (distance from source, mean substrate particle size, total oxidized nitrogen, alkalinity and chloride), the fauna at sites could be predicted reasonably accurately, except for rare species which were not commonly found in samples. Thus, it is possible to compare the observed fauna of polluted sites, with the fauna which would be present in the absence of pollution (predicted from easily established environmental properties) and to obtain some measurement of difference between these faunas using distance measurements (e.g. Jaccard's index, Euclidean distance) as numerical expressions of pollutional stress. Of perhaps greater potential importance is the opportunity provided by the technique for qualitative faunal comparisons which are potentially diagnostic in assessing the type of pollution and even its temporal history. Whilst this pollution "finger printing" seems attractive, its full value will not be realised, except in a prospective way, until we know more about the ecotoxicological responses of organisms to pollutants. Furthermore, the predictive environmental variables identified by Moss et al. (ref.14) for unpolluted sites, include two (Cl, oxidized N) which are heavily influenced by sewage (Cl, oxidized N) and agricultural run-off (oxidized N)!

Weatherley and Ormerod (ref.15) using similar classification methods to Wright et al. (ref.23), but at 18 sites on streams of the Tywi catchment in mid-Wales subject to ecological damage from acidification, established three site groups which corresponded to streams draining conifer afforested catchments, acidic moorland streams and circum-neutral moorland streams. These groups were consistent whether spring or summer samples were used or whether riffles, margins or both habitats were sampled. Relationships between site-groups, each with distinctive faunal components and key environmental varaibles, were examined using multiple discriminate analysis. Using three variables (mean aluminium and hardness concentrations, catchment area) it was possible to classify all sites into their correct site group using combined riffle and margin collections in Spring. 94% sites were correctly placed from margin-only collections in Summer.

Whilst these analyses show the crucial importance of water quality to faunal distribution, both hardness (through its buffering effect on pH) and aluminium being associated with acidity, extreme pollution episodes are known to be critically toxic and the relationship between such extreme events and mean water quality still needs to be established.

2.1.4 Functional indices

Matthews et al. (ref.4) reviewed methods used to measure community and population functions and at community level they distinguish three aspects of functional organization: energy flow, material cycling and regulation. The most common *in situ* measurements have related to energy flow, either the measurement of oxygen fluxes (in open or enclosed systems) or the assimilation of labelled organic nutrients, the latter generally requiring enclosure. Odum (ref.24) proposed the measurement of community photosynthesis and respiration, the rates and ratios of these processes being indicative of the health of the ecosystem. In open systems, account needs to be taken of oxygen exchange between the river and the atmosphere in calculating metabolic rates (ref.25) and day-to-day variations in rates caused by light and temperature fluctuations create logistical problems in seasonal integration although oxygen, temperature and light recorders have eased this difficulty (ref.26). Benthic processes dominate overall oxygen fluxes in all but the deepest rivers despite standard equations which ignored both photosynthesis and benthic processes (ref.27) being used for decades. From a study by Boyle and Scott (ref.27) of oxygen fluxes in a shallow river polluted by paper mill effluent which induced the growth of Sphaerotilus, it can be concluded that over a four-month period about 90% of the total respiration ($\equiv 8.8$ g O_2 m^{-2}d^{-1}) was benthic and that even in this polluted river the P:R ratio was more than 30%.

The development of benthic enclosures was principally for respirometric purposes (see Bowman & Delfino, ref.29) but their use could be extended to studies concerning the utilization of organic (ref.30) and inorganic (ref.31) substances whose planktonic uptake by micro-organisms has been widely investigated.

2.2 Experimental simulations of pollution

The earliest experimental studies of pollution effects were short-term toxicity tests, generally with fish or Daphnia species. A sophisticated methodology was developed to analyse the pattern of mortality in such tests (see ref.17), reliable dosing systems have been designed for flow-through systems (ref.32) and the effects of environmental variables, such as pH, hardness and temperature on toxic response were soon recognised and incorporated into testing procedures (ref.33). Despite these achievements, the ecological relevance of results obtained from short-term, single-species, single-pollutant toxicity tests was questioned, because:
a) poisons were rarely present singly and predicting joint action was difficult
   - particularly when concentrations were known to fluctuate widely;
b) temporal extension to chronic exposures, embracing whole and multiple life-cycles was rarely undertaken;
c) the viability of organisms depended not only on direct pollutant action but also on ecosystem response to the pollutant.

2.2.1 Single-species tests

a) Joint action of pollutant mixtures. Whilst recognising the potential potentiation and antagonism within mixtures of poisons (see ref.34), Southgate (ref.35) proposed that, for certain combinations of poisons, toxicity was additive, that is:

when $$\frac{C_a}{48hLC_{50}a} + \frac{C_b}{48hLC_{50}b} \ldots = 1$$

then 50% of fish will die in 48h, a and b being component poisons and C being their concentration. This empirical hypothesis, tested by Lloyd (ref.36), Herbert (ref.37) and others using two and three component mixtures, was generally supported. Later, making this assumption of additive action, Lloyd and Jordan (ref.38) could explain about 70% of the toxicity of 80% of industrial sewage effluents by measuring concentrations of about 9 toxic substances, with the residual toxicity presumably due to other poisons or to potentiation of the detected poisons. Studies were extended to determine the pattern of mortality of fish kept in cages for short periods in industrial rivers and generally such mortalities conformed to expectation. All these early tests were carried out with rainbow or brown trout for which good toxicity data were available.

Edwards and Brown (ref.39) attempted to relate the reported distribution of fish in a large industrialized river catchment (R. Trent) with water quality assuming additive toxicity. The 50 percentile value for the sum of fractions of sites containing fish was 0.1 $48hLC_{50}$ and of fishless sites 0.3 $48hLC_{50}$. Alabster et al. (ref.40) developed this approach within the Trent catchment and were able to define, with considerable precision, the water quality boundary, in statistical terms, between sites with and without fish, the 50 percentile being 0.28 $48hLC_{50}$. They were also able to suggest which poisons might most economically be removed for fish recovery to occur.

The conclusions reached in the Trent catchment were tested in a small river, the Willow Brook, receiving wastes from a large steel works (ref.40). In general the water quality boundary was similar with only marginal fisheries existing where the 50 percentile $48hLC_{50}$ was > 0.28.

That such simple relationships were found was surprising in view of the theoretical reservations about extrapolation from short-term tests and most toxicity studies over the past thirty years or so have sought to detect the lowest concentrations which affect a test species or test "community" directly through experimental procedures. Several reviews describe the multiplicity of options (see Sprague (ref.34) and Buikema et al. (ref.17)).

b) <u>Chronic exposure and life-cycle toxicity tests</u>. One of the developments was to extend the pollutant exposure period to whole life-cycles and so identify particularly vulnerable life-stages. Despite their relatively long life-cycles, most of these studies have been carried out with fish, principally because of their direct commercial or sporting value, adequate knowledge of their ecology and culture needs (with the advantage of developing genetically similar test organisms), and an early erroneous belief that they represent the "top of the food chain" and so act as effective protective surrogates for the aquatic community.

Arising from these long-term studies has been the development of the concept of application factors which can be applied to short-term lethal concentrations to derive maximum allowable toxicant concentrations (MATC). Where this relationship is not known for a toxicant or test species, values for application factors varying from 0.1 to 0.002 have been adopted (ref.41). In a recent study of the toxicity of more than twenty organic and inorganic chemicals to the fish <u>Cirrhina mrigala</u>, using the sensitive larval stage to derive $96hLC_{50}$ values, application factors within a very limited range of 0.036 - 0.079 were calculated (ref.42). The use of application factors has been critically reviewed by Kenaga (ref.43).

Whole life-cycle tests, frequently incorporating measurements of growth-rate and reproductive success, might be regarded as a generally satisfactory basis for establishing MATC values, especially when test conditions physically simulate natural conditions (see Fremling and Mauck (ref.44) for burrowing mayflies). However, normal behavioural responses (ref.34) and predator-prey interactions are generally suppressed in these studies and supplementary investigations of greater experimental complexity are sometimes required. Where poisons accumulate and may be transferred through food-webs, bioaccumulation factors within tissues may also be incorporated in test procedures (ref.17).

Although whole life-cycle toxicity tests with fish species are commonplace, multi-generation tests are comparatively rare (ref.45) and, as a consequence, our appreciation of the extent of genetic adaptation to pollutants is inadequate, particularly is this so for benthic invertebrates where, with the exception of insects with their effective dispersal stage, factors operate to increase the opportunity for selection to take place. Klerks and Weis (ref.46) review evidence for genetic adaptation of aquatic organisms to heavy metals and conclude that unequivocal evidence of adaptation is not widespread in metazoans. This is partly because many investigators, although comparing the resistance of populations from clean and polluted habitats, have not distinguished between physiological acclimation and genetic adaptation through comparative tests on laboratory-reared progeny. Nevertheless, genetic adaptation has been recorded for some representatives of all major metazoan groups except the insects and fish.

A large data base has accumulated on the tolerance of fish species to a wide range of toxicants under a variety of environmental conditions, data that are reliable in predicting whether fish will survive (provided that organisms on which they feed are not adversely affected) but the data base for other organisms which form the basis of field-based biotic indices is far less satisfactory. Until this anomaly is rectified the diagnostic value of field assessments of polluted rivers will remain limited. Sufficient information is available, however, to demonstrate that fish are not reliably the most sensitive organisms to pollutants (ref.47) so making water quality standards based on their responses inadequate in maintaining "ecological integrity" as defined by Cairns (ref.48) and that invertebrate species can vary enormously in their sensitivity to some poisons. Williams et al. (ref.49) reported that the variation in sensitivity of eight macro-invertebrate species to cadmium was $> 6 \times 10^3$ whilst that to ammonia was only four.

Field bioassays using fish species have been carried out for many years to provide protection for water intakes from rivers or to monitor effluent quality (ref.50). Increasingly sub-lethal responses, capable of being translated into electrical signals within the test system, are being used (ref.51).

Organisms, other than fish, are now being used as "bioassays" in field situations. Brown (ref.52) describes the planting of the snail Hydrobia jenkinsi in cages in a river system polluted by copper, chromium and other heavy metals to detect intermittent discharges. From the pattern of deaths in the field and toxicity data derived from laboratory studies of copper and chromium, Brown concluded that field mortality could not be explained solely in terms of the concentrations of these metals. Brown suggested that by using complementary species responding to different poisons one might identify key pollutants and, optimistically, that one might even develop strains of the parthenogenetic Hydrobia jenkinsi with specific responses to different pollutants. Surely, learning the profile of responses of components of natural communities would be more realistic and useful in diagnosing pollutant incidents.

In marine situations, field bioassays have developed further and the "scope for growth" of the bivalve, Mytilus edulis, calculated from a series of physiological measurements, is now used to monitor water quality. Underlying this test procedure is the principle that poisons, even when present in non-lethal concentrations, cause metabolic responses which have energetic costs (ref.53).

2.2.2 Ecosystem responses and multi-species tests

If one is seeking to maintain "ecological integrity", the single species test is likely to prove inadequate - although this is still the established dogma (ref.48). There is no universally sensitive indicator species and where the toxicity of a variety of pollutants has been tested with several species, the ranking of species sensitivities generally varies from pollutant to pollutant and although "relative tolerance indices" have been calculated for test species (ref.54), these will vary with the range of poisons tested (ref.49).

Secondly, biotic interactions may either decrease (ref.55) or increase (ref.56) the effects of pollutants established from single-species tests. Bioaccumulation in food chains is one of the more obvious interactions which has had major effects on predator populations, particularly where prey has been more accessible through behavioural changes (ref.57).

Microcosms, defined by Cairns (ref.16) as "reduced scale models of natural ecosystems or portions of natural systems housed in artificial containers kept in a controlled or semi-controlled environment" have not been widely used in freshwater studies. The simplification and miniaturization of these systems cause instability with so-called replicates behaving differently, scale problems influence nutrient cycling, vertical migrations, etc., and the omission of large organisms from microcosms creates a degree of artificiality which undermines interpretation of their ecological relevance.

External channel systems have increasingly been used for pollution studies in recent years (ref.4). The effects of several pollutants including sewage (ref.58), metals (ref.59), pesticides (ref.60) and suspended sediment (ref.61) have been studied. Both pulses and continuous discharges of pollutants have been used. Frequently, replication has not been possible and variations in ecosystem response between sections of channels have been marked. Sometimes this patchiness has been attributed to shading or to water quality changes down channels (ref.60). Such channel systems are rarely sufficiently large to accommodate fish, and insects, because of their rapid colonization, frequently dominate the benthic fauna, particularly where natural water, containing drift, does not provide the channel flow.

It is but a short step to experiments in natural stream channels where ecological relevance is maximal but replication rarely possible, except where small areas are treated with such pollutants as fine sediment (ref.62). Some experimental additions have been of long duration (refs.55,63,64) whilst others have been short pulses (ref.65). Experiments in natural streams where pollutant addition is over a long period are usually designed to discriminate the effects of single pollutants commonly present in mixtures: such experiments also facilitate the maintenance of constant pollutant concentrations, a condition rarely achieved elsewhere with both effluent varying in flow and strength and with stream flows changing. Experiments in simulating episodic pollution events are particularly useful because good descriptions of such episodes and of their effects are rare as will become clear later in this chapter. Ormerod et al. (ref.65), mimicking a pulse of stream acidity, exposed adjacent lengths of river to low pH and to low pH with aluminium so that contrasting effects could be assessed through changes in drift and benthic populations of invertebrates during the pulse, and through toxicity tests of both fish and selected invertebrates within the stream lengths. With captive fish it was possible to undertake physiological observations throughout the experiment. In later studies (ref.66) a bank-side fish monitor recorded patterns of swimming activity and opercular movements of captive fish; measurements of photosynthetic activity and nutrient uptake by periphyton were included.

3. PATTERNS OF DEGRADATION AND RECOVERY

3.1 Degradation

Changes brought about by pollution can either be viewed from the speed with which they take place or the type of community which results from them. From studies of accidental spillages and experiments in streams involving the release of toxic materials, biological responses are generally rapid, both through mortality and physiological impairment or through emigration (drift).

For example, the effects of an oil spillage into the River Sirhowy, South Wales (ref.67) were detected within several hours of the release through the mortality of fish (Salmo trutta, Noemacheilus barbatulus, Cottus gobio) and benthic invertebrates (mainly Baetis rhodani). In a field experiment to simulate an acid rain episode (ref.65) involving the lowering of a stream pH to 4.2 and, further downstream, increasing the concentration of aluminium to 0.5 mgℓ^{-1}, mortalities of invertebrates and fish had occurred within a few hours and the drift response of some invertebrates, particularly B. rhodani, was almost immediate.

In channels exposed to several heavy metals (Co, Cu, Zn), Eichenberger et al. (ref.59) recorded a sharp reduction in photosynthetic oxygen production a few hours after the introduction of pollutants and within a few days the diatomaceous periphyton had died and become detached.

The ecological significance of emigration on exposure to adverse water quality conditions remains equivocal. Most observed emigration is downstream, as drift, and organisms continue to be exposed to adverse conditions associated with the downstream dispersion of the pollutant, improvement only occurring with pollutant dilution, decay and dispersal. With some forms of pollution, such as deoxygenation, ecological damage may be very local, suitable conditions being restored downstream by surface aeration - a rapid process in fast shallow rivers (ref.25). In such situations, drift may confer individual advantage but population advantage of such catastrophic drift associated with even transient pollution incidents is dubious. With fish the options are wider and movement upstream, into tributaries, or downstream at speeds exceeding pollutant passage are all possible. Detection and avoidance of low concentrations of many poisons have been recorded by several authors (see refs.34,68,69).

Qualitatively, the kinds of ecological response of different types of river resulting from common forms of pollution (particularly sewage) have been described extensively. For example, Woodiwiss (ref.70) related biotic score values to water quality determinands appropriate for organic pollution. Weatherley and Ormerod (ref.15) described the associations of invertebrates in acid hill streams and produced equations relating these associations to water quality determinands, pH or total hardness and aluminium being the most significant. Scullion and Edwards (ref.71) noted the correspondence between the comparative sensitivity of selected invertebrates to low pH, as demonstrated in laboratory experiments, and their distribution in acid streams: they also confirmed the particular sensitivity of some insect species, especially Ephemerella ignita and Rhithrogena semicolorata, to siltation (ref.72). Williams et al. (ref.49) described the correspondence in sensitivity of short-term toxicological responses of Gammarus pulex and Baetis rhodani to zinc with their distribution in catchments polluted by this metal. There were, however, anomalies, e.g. Rhithrogena semicolorata, and they suggest that an improvement in their test procedures to include chronic exposures could well have improved agreement between laboratory results and field distributions.

In summary, many investigators have found, from field studies, similar qualitative responses to particular pollutants and these responses have broadly been supported by laboratory experiments of comparative sensitivity. Employing field data alone, some authors are now using the close correspondence between invertebrate associations and water quality in a predictive manner: nevertheless, the toxicological base for such predictions is still weak. For certain fish species toxicological data are more reliable.

3.2 Recovery

Where discharges of pollutants into rivers are continuous and achieve a constant dilution after mixing, the only processes which bring about a downstream amelioration of effects are: dilution from tributaries, decomposition, settlement or adsorption, and volatilization.

The concept of recovery is solely a spatial one, each zone being "steady-state" unless the response or organisms changes, in the short-term by physiological acclimation or in the long-term by genetic adaptation. This "steady state" is classically described for organic pollution both in terms of the kinetics or organic oxidation (ref.27) and the ecological response of communities to organic pollution (refs.1,2,73). It is now apparent that the simple kinetic description by Streeter and Phelps frequently underestimates the severity of oxygen depletion because of benthic processes (ref.25). Nevertheless, the downstream ecological succession has been verified in numerous investigations.

But to what extent can we regard rivers, their quality and their communities as steady state at any location?

Precipitation patterns influence the discharge and passage downstream of pollutants in a number of ways:
- percolation and removal of materials from soils and tip material;
- discharge of storm sewage direct to streams;
- dilution of sewage within sewers;
- dilution of effluents by increased river flow;
- transport of particulates through deposition (low flows) and scour (high flows).

Phase transfer occurs for soluble materials as well as particles. Volatilization is only significant for relatively few pollutants (e.g. NH_3 at high pH) but adsorption is of wider importance. With metals, adsorption is adequately described by the Freundlich adsorption equation, the Freundlich constant being related to grain size and organic content (ref.74). Particularly for those metals which are organically bonded, desorptive processes are slow and subsequent redistribution is essentially dependent on downstream sediment transport at high flows.

In addition to changing precipitation and river flows, further instability in water quality is caused by changing effluent quality, generally related to seasonal temperature fluctuations: in sewage treatment the concentration of ammonia in effluents is particularly affected by seasonal factors (ref.75).

As a consequence, most rivers have highly variable concentrations of pollutants in time as well as in space. However, most emphasis has been placed on spatial recovery, little attention having been given to temporal responses to water quality changes. The zonation of recovery is best described for sewage of course (refs.1,2) but adequate descriptions are available for other forms of pollution (coal industry (ref.71), heavy metals (refs.76,77).

Temporal variations have either remained undetected because of the infrequency of chemical and, more particularly, biological sampling, or suppressed, because of the practice of combining data to produce annual site mean values and regard short-term variations as "noise" rather than "signal" (ref.71). The conditions to facilitate temporal recovery, in biological terms, are not solely determined by water quality, for the retention of pollutants in sediments etc. prevents normal recolonization. As a consequence, there will be an inertia in the recovery process not solely determined by biological immigration.

The importance of analysing temporal variation in water quality is increasing for as loads of toxic materials discharged in effluents are reduced (and most pollution control programmes concentrate on major chronic sources), the intermittent pollution sources increase in importance. Decisions to restock rivers with fish, based on the reduction in pollutant base-loads or on average quality are fraught with disappointment and when mass mortalities occur they are attributed as "pollution incidents" yet these incidents are often no more frequent and severe than before chronic sources were reduced. They merely reflect the inadequate attention paid to the effects of removing chronic discharges of pollutants on peak rather than mean concentrations.

Not all studies have concentrated on spatial patterns of pollution. Bird (ref.78), for example, analysed metal pollution in a South Wales river receiving both effluents containing nickel, chromium and iron and drainage from tips of metallurgical waste containing zinc, copper, cadmium and lead. Concentrations of all metals were highly variable and could be related to: (1) the river's available dilution (particularly important for metals in effluents); (2) antecedent river flows; (3) river water pH; and (4) prevailing run-off processes; the last three factors were relevant to the release and transport of trace metals from tips. In the case of chromium and iron, transport was dependent on deposition and scour associated with varying river flows.

The emphasis on spatial recovery in field studies has been reflected in the orientation of experimental studies, particularly in single-species toxicity tests, towards predictions of chronic effects; this applies even with the conversion of short-term tests to MATCs through the use of application factors.

3.3 Special factors applying to episodes

Although the escalation in the detection of pollution pulses or episodes is, in part, associated with the removal of background chronic pollution from many rivers, some sources of intermittent pollution have increased in significance. For example, the storage of large volumes of organic farm wastes (e.g. cattle slurries and silage liquor) has increasingly caused problems as the scale of animal production has grown (ref.79). Similarly, the incidence of pulses of acid stream water has increased with increasing atmospheric acidity and as more catchments become denuded of neutralizing bases (ref.80).

The extreme of temporal instability in water quality is the single incident or episode of pollution and there are special conditions and factors which apply (and to a lesser extent where any temporal variation in water quality occurs):

a) ecological conditions before the incident and the concentration-time exposure to pollution, can rarely be defined;
b) attenuation of the pollutant pulse occurs downstream, within the bed and laterally across the stream so exposing organisms to different conditions compared with steady-state releases;
c) recovery of individuals, populations and communities can occur after the incident whereas factors involved in this recovery process do not apply when considering spatial recovery.

a) <u>Definition of preconditions and pollution load</u>. Rarely have detailed ecological surveys been carried out immediately before incidents, unless of course under experimental conditions (ref.65) and one normally relies on a spatial upstream "control" site for comparison. More importantly, precise descriptions of time-concentration exposures or even loads (area under time-concentration curves) are rarely possible.

The author, from data supplied by the Welsh Water Authority of a formaldehyde pulse passing downstream from an industrial spillage in a Welsh river, calculated that about 200 kg had been spilt. This compared with a declared loss of only 30-60 kg. The discrepancy was later resolved and due to differences in analytical techniques in two laboratories. It would have been possible from the changing shape of the concentration-time curve as it passed downstream to calculate longitudinal dispersion parameters and move the pulse upstream (or back in time) to the site of the spillage. This technique has been used by Whitehead et al. (ref.81).

b) _Attenuation of pollutant pulses_. From tracer studies in several Welsh rivers it has been possible to calculate transit times and dispersion for soluble contaminants through a wide range of flow conditions (Welsh Water Authority, personal communication): this was deemed necessary to forecast the consequences of spillages of toxic materials. For the R. Ely near Cardiff under dry weather flow conditions, comparisons were made of the downstream concentration profiles of tracer discharged (1) at a constant rate continuously and receiving only dilution from tirbutaries; and (2) discharged as a one-hour pulse at the same rate, receiving dilution from tributaries and achieving longitudinal dispersal. About 13 km below the discharge point dispersion reduced the peak concentration by 67% compared with that from a continuous discharge. In comparison decay (assuming a first order reaction with normal BOD constant) would have reduced the peak by only 27%.

Such dispersion occurs across the stream, and in backwaters the reduction of peak heights can be very pronounced, particularly near the site of discharge (but below the mixing zone). Similarly, within sediments attenuation of pollutant pulses could have significant effects on the benthic fauna, protecting those living at depth or migrating there during adverse conditions. Ormerod et al. (ref.65) measured the vertical distribution of aluminium in interstitial water within sediment during a 24h aluminium pulse and found decreasing concentrations with depth (to 45 cm) but compensatingly longer residence times. The significance of this attenuation in relation to recovery following pollution episodes needs far more detailed consideration.

c) _Biological recovery processes_. Because of the focus of toxicological studies on needs to predict the effects of "steady-state" chronic exposure, even with short-term tests, post-exposure responses - either death or recovery from sub-lethal damage - have received little attention. The balance between the importance of lethal concentration and the lethal dose needs urgent reappraisal in aquatic toxicological studies, particularly as the significance of chronic pollutant sources is reduced in many rivers. Even the recognition that concentrations of pollutants within rivers are generally variable has not induced many toxicological studies involving exposure to variable concentrations to test the adequacy of the current practice of concentration averaging in predicting ecological effects.

Abel (ref.82) exposed the crustacean, _Gammarus pulex_, to a range of concentrations of the pesticide lindane for periods of 1 to 10^3 min before transferring to clean water. Animals died for up to 3×10^4 min after the initial dose. For a given concentration of pesticide the duration of exposure needed to kill animals was far less than would be suggested using conventional test procedures. For example, at 0.1 mgℓ^{-1} only 3h exposure was required to kill 50% _Gammarus_ within the 3-week observation period whereas the conventional median lethal time (LT_{50}) for the same lindane concentration was 20h. In similar studies with copper (ref.83), even wider discrepancies between exposure time and LT_{50} were recorded.

Pascoe and Shazili (ref.84) and Holdway and Dixon (ref.85) in studies using fish have also drawn attention to the inadequacy of current test procedures to evaluate episodic effects. Pascoe and Shazili (ref.84) exposed young stages of rainbow trout (Salmo gairdneri) to cadmium: at 1.0 mg Ca ℓ^{-1} exposure for only 32 min was necessary for post-exposure mortalities whereas the LT_{50} for this concentration was 1900 min or x60 the lethal exposure period. Holdway and Dixon (ref.85) showed that short pulses of methoxychlor, used to control the pupal emergence of Simulium from rivers, had adverse effects on reproductive parameters (egg production, time to spawning) as well as hatchability and incidence of deformities in offspring even when exposure was limited to 2h at 0.25 $mg\ell^{-1}$ with juvenile flagfish (Jordanella floridae) only 8d old. This exposure compares with a 96h LC_{50} of about 38 $mg\ell^{-1}$.

Where test responses are of sub-lethal effects it is similarly important in relation to episodic events to establish whether recovery occurs after the exposure period. Some studies have demonstrated such recovery. For example, Brafield and Matthiessen (ref.86) exposed the fish Gasterosteus aculeatus to zinc solutions until respiratory stress symptoms developed; they observed redovery 5-10h after transmittance to clean water. Similarly, Asellus aquaticus immobilized during exposure to phenol, regained mobility when restored to clean water, the rate of recovery being dependent upon exposure time, exposure concentration and temperature (Green et al., in press).

Alabaster et al (ref.40) reviewed early studies on the effects of fluctuations in concentration of pollutants on toxicity, testing the hypothesis of the adequacy of concentration averaging. For several poisons such as ammonia, zinc (and mixtures of these), DDT, and an organophosphorus pesticide, it was concluded that mean concentrations over the total exposure period adequately described mortality even when pulses, in some instances, alternated with clean water. In one instance fish were exposed to two poisons, phenol and cyanide, at concentrations equivalent to their 48h LC_{50} for alternate 4h periods; again toxicity could be adequately described by averaging conditions. Examples were, however, given of poisons, e.g. the herbicide Reglone, cadmium, where averaging would grossly underestimate toxicity, irreversible damage being caused by very brief exposures to given concentrations.

A further study, conducted by Siddons et al. (ref.87) of the effects of brief exposures of the brook trout (Salvelinus fontinalis) to aluminium at low pH - a situation which frequently mimics reality - confirms that averaging substantially underestimates toxicity. Fish exposed to aluminium for only two out of six days over a total exposure period of 24d died as rapidly as those exposed continuously.

Invertebrate studies have been few. Aldridge et al. (ref.88) described the effects of intermittent exposure of freshwater mussels to inorganic suspended solids, a situation occurring naturally in waterways with boat traffic. The frequency of exposures to suspended solids was critically important: when exposed to solids every 3h mussels reduced their metabolic rate to compensate for feeding impairment but when exposed to solids every 0.5h catabolic pathways were affected and non-proteinaceous stores (as indicated by O:N ratios) were used almost entirely.

Immigration and population growth processes are crucially important in determining the pattern of river recovery when pollution ceases either after brief or long period. There are, however, other factors related to the receptiveness of the system to immigrants which are also relevant. Eichenberger et al. (ref.59) found that in channels previously exposed to metals their recolonization with diatoms was suppressed by the extensive stands of the metal-tolerant filamentous alga Hormidium. Sediments sometimes remain polluted for long periods after discharges have ceased and prevent the successful reproduction of gravel-spawning fishes such as salmonids (ref.89).

Williams and Hynes (ref.90) compared the importance of various immigration routes (drift, upstream migration, vertical movement from the hyporheos, aerial sources) to areas of the stream bed artificially denuded of fauna. They used traps which only allowed colonization via one of these routes. At the end of a 28d period (June) of colonization, control areas had densities equivalent to the sum of those established on different areas open to only one of the immigration routes. Within that period 39, 17, 18 and 26% of the total fauna were derived from drift, upstream migration, hyporheos and aerial sources, respectively, making drift by far the most important route at this time of the year. Mayflies and mites invaded primarily from drift (which represented almost 70% of the total immigration); bivalves, ostracods and amphipods moved principally upstream; leeches invaded from the hyporheos and aerial sources were particularly important for the Chironomidae. However, the areas denuded of fauna and occupied by traps were only 60x30 cm, so extrapolation of these results and those of other workers using similar techniques (ref.91) to immigration processes affecting substantial lengths of rivers should be resisted. Hyporheol components may not survive, particulary chronic, pollution, and upstream migration rates for all animals except fish will be slow. As a consequence, drift from tributaries and upstream areas, together with aerial sources for insects (through egg deposition) are probably the most significant invasive routes. Microorganisms are probably extensively distributed in catchments, either as spores or vegetative cells, and can be carried into streams through run-off and percolation processes.

Recovery depends not solely on immigration but also through the establishment and growth of populations constrained by the normal processes of competitive interaction. One might expect "r" strategists to dominate initial colonization phases and several workers have observed the explosive growth of microorganisms, usually algae, both in artificial channels and natural streams, in the initial period before grazers have become established (ref.60). These grazers and other functional feeding groups frequently take only a few weeks and rarely more than a few months to re-establish normal population densities except where residual pollution problems occur within the stream bed (ref.92). Frequently these initial invaders are chironomid species and, as suggested earlier, the importance of invasion, either through drift or aerial flight stages, could remain important, particularly for mobile species.

4. SELECTED CASE STUDIES OF THE RECOVERY OF RIVERS FROM POLLUTION

There are many descriptions of the recovery of rivers from chronic pollution. In large rivers there is usually a multiplicity of sources and pollution abatement programmes often take years or even decades to complete. As a consequence, ecological recovery can rarely be attributed to specific improvements in water quality. However, for the Thames Estuary, a few major polluting loads of sewage were primarily responsible for the main problem - oxygen depletion of the estuary, and improvement, as evidenced by fish invasions, was dramatic (ref.93). But such instances are few.

In smaller catchments comparatively short programmes of pollution control are likely to result in more clearly defined responses to specific actions. These small systems are also particularly vulnerable to accidental spillages of pollutants, dilution being low. This does not imply that large rivers, like the Rhine, are immune from such damage (ref.94). Because of their advantages for comprehensive study and the comparative clarity of ecological response, case studies selected for description here are, by international standards, small rivers. All are rivers in Wales and demonstrate that, within a small geographical area, examples of a wide variety of problems can be found and, hopefully, lessons learned.

4.1 River Ebbw Fawr (South Wales) - recovery from sterility

The R. Ebbw Fawr is a river only about 40 km long, flowing down one of the industrial coalfield valleys of South Wales with a major steel works near its headwaters and, until recently, widespread coalmining throughout the catchment producing washery effluents and tip run-off containing high concentrations of suspended solids. Although most of the population of 160,000 within the catchment is located near the estuary, urban development (much of it being 19th century) extends to the headwaters. The main river, which has two major industrialized tributaries and several small unpolluted ones, was essentially abiotic for much of its length for about a century until the period 1970-7 when three improvements took place:

1) The major steel works, the effluents from which had been responsible for the acutely toxic conditions in the river (pH 3-12; DO 0-50% saturation; NH_3-N 2-5 $mg\ell^{-1}$; Pb 1.3 $mg\ell^{-1}$; Zn 1.3 $mg\ell^{-1}$; Cu 0.2 $mg\ell^{-1}$) received a new and comprehensive effluent treatment system. Anaerobic conditions in the river had been caused by the oxidation of ferrous sulphate discharged to the river and wide pH fluctuations occurred through inadequately controlled neutralization of waste acid, using lime.

2) Coal washeries were improved to reduce suspended solid discharges to the river.

3) Storm sewage discharges were reduced by the construction of a new trunk sewer of larger capacity.

The major damage had been caused by the toxic action of steel-works effluents, the other sources of pollution requiring attention to facilitate more complete recovery. During the commissioning period of the waste treatment plant for the steel works (1970-3) marked variations in effluent quality occurred which limited biological recovery. Fish were present only in the tributaries and in situ toxicity tests, with trout and other species, in the main river confirmed that mortality was rapid, particularly in the upper reaches which received no dilution from major tributaries (Fig.1). By 1973, when a marked increase in survival times occurred, the water quality had improved such that the median value of calculated toxicity did not exceed 0.1 48h LC_{50} for brown trout (see ref.33), the dissolved oxygen concentration always exceeded 50% air saturation (except immediately below the effluent) and pH was in an acceptable range (6-9). Fish invasion into the lower reaches of the main river was almost immediate and by 1975 five species were present (Salmo trutta, Gasterosteus aculeatus, Noemacheilus barbatulus, Anguilla anguilla, Platichthys flesus), the latter two invading from the sea. By 1980 further invasion upstream had occurred and two more species (Phoxinus phoxinus, Cottus gobio) were recorded.

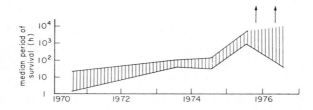

Fig.1. Improvement in survival times of caged trout in the Ebbw Fawr, 1970-76.

With respect to trout (and perhaps other species), embryo development does not occur in the main river because of the quality of sediment. Turnpenny and Williams (ref.89) showed that most ova planted in substrate of the main river died, in contrast to those planted in clean tributaries. In addition to the occlusion of gravel by coal particles, sediments contained high concentrations of toxic metals although the toxicological significance of these metals may be limited by their low solubility at normal pH. It seems that further dramatic improvements - such as the maintenance of self-sustaining trout populations - is unlikely until colliery waste materials, largely derived from tip run-off, are more effectively controlled. Until that has been achieved, trout stocks will be dependent on stocking or on tributaries which are mostly small, produce very few fish and are often culverted near the main river so limiting upstream access to migrating spawners from the main river.

The recovery of bacteria, plants and invertebrates was even more rapid than that of fish. Until 1972 sewage fungus covered much of the stream bed, but only in the middle and lower reaches. Colonization by diatoms and green algae was evident during that year and had extended to upstream reaches by 1974 (Fig.2). Nevertheless, when the survey terminated in 1975 there were still absentees from the flora downstream of the steel works, notably higher plants and mosses, which were present upstream. In situ tests with Cladophora and the mosses Rhynchostegium riparioides and Fontinalis antipyretica confirmed the continued

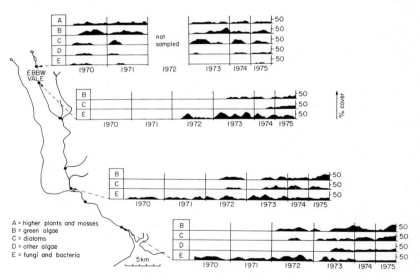

Fig.2. Seasonal changes in percentage cover of groups of plants and bacteria at sites above and below Ebbw Vale during 1970-75. In the period 1970-73 surveys were carried out in spring, summer, autumn and winter; in 1974 and 1975 only spring and autumn surveys were conducted.

but decreasingly damaging effects of water quality during the priod 1970-74.

Macroinvertebrates, particularly naidid and tubificid worms, chironomid midges and the ephemeropteran, Baetis rhodani, were present at downstream sites late in 1970. These groups progressively colonized upstream, being present throughout the river by 1973. Nowhere, however, did their abundance match that in the "clean" headwaters (Fig.3).

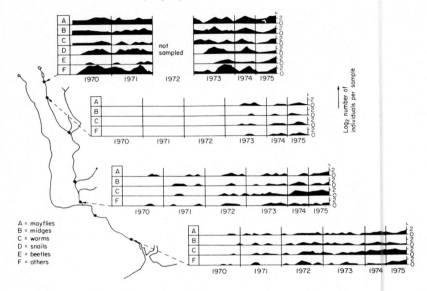

Fig.3. Seasonal changes in the abundance of macroinvertebrates at sites above and below Ebbw Vale during 1970-75.

Both biotic (Trent) and diversity (Shannon-Wiener) indices described the early stages of recovery of the river but this is hardly surprising with such a dramatic change in water quality. These indices suggested that the lower reaches were approaching the quality of the headwaters by 1975 and by 1979 a wide range of faunal groups was present, including several molluscs, crustaceans, leeches and trichopterans.

Several conclusions may be drawn from these investigations:
1) Once soluble toxins are removed, biological recolonization can be extremely rapid, that of plants and invertebrates preceding that of fish.
2) Where toxicity data were available, as with fish, there was agreement between chemical quality and in situ mortality tests with caged fish.
3) The predictions relating to fish distribution based on water quality (ref.40) proved correct.
4) Faunal and floral recovery could be dmonstrated using both biotic and diversity indices.

5) Residual pollution problems prevent the full recovery of the river, particularly the development of self-sustaining trout populations, these problems being exacerbated by barriers to migration.

More detailed descriptions of the recovery of the R. Ebbw are available (refs. 89,95,96).

4.2 River Taff (South Wales) - migratory salmonids respond to water quality

The River Taff is also a small river, only about 70 km in length, and with 4×10^5 people living within its catchment. Its upper reaches were drowned almost a century ago to build reservoirs providing water, principally for Cardiff which is located in its lower reaches. Industrialization, particularly through iron-making, was well advanced at the beginning of the 19th century, damaging all but its headwaters. By the mid-century the coal industry had also developed extensively, polluting several of its tributaries and the middle reaches of the main river. Inadequate sewage treatment and sewer systems, and the establishment of industries associated with coal carbonization, the latter producing toxic effluents, caused a major decline in water quality and damage to river ecology.

Edwards et al. (ref.97) described the biology of the river in 1969, confirming that the benthic fauna was impoverished in the lower reaches and on major industrial tributaries; at that time, although several fish species were present in the rivers of the upper catchment, the main river in its lowest 20 km was devoid of salmonids and sections were seemingly fishless. In 1979, however, small numbers of sea-trout (Salmo trutta) and salmon (Salmo salar) were caught by anglers in the lowest reaches at Cardiff and both species have been observed in these lower reaches every year since, being caught in ever increasing numbers. In 1983 electrofishing also revealed the presence of juvenile salmon, suggesting that some spawning had occurred in the catchment since 1980. Further juveniles of both sea-trout and salmon were caught in 1984. As spawning is not likely in the lower reaches, the presence of juveniles suggests that at least a few fish had ascended a series of weirs in the lower section of the river, built around the end of the 19th century to provide water for developing industries. Such weirs had no fish passes, for migrating fish were almost certainly absent from the river when they were built.

The reason for the return of migrating salmonids to the river around 1979 is apparent from changes in water quality (ref.98). All pollution parameters showed a dramatic improvement at this time, particularly oxygen and ammonia (Fig.4) and water quality conformed to the criteria defined in the EEC freshwater fish directive for salmonids (with the marginal exception of nitrite). These changes were achieved by improvements in sewage treatment and the fortuitous decline in production of a gelatin factory.

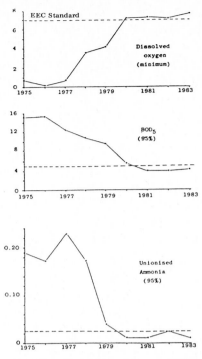

Fig.4. Comparison of annual minimum D.O. and 95 percentiles for BOD and NH_3 (mgℓ^{-1}) in the lower Taff some 1975 to 1983 with criteria for salmonids in the EEC freshwater fish directive.

Surveys over the past 16 years have shown that despite these dramatic changes in both water quality and fishery status, the benthic invertebrates have not changed appreciably, when information is summarized in the form of biotic indices (ref.99), since the studies of Edwards et al. (ref.97) in 1969. Nevertheless, some faunal changes have been recorded by Bent et al. (ref.100). Edwards et al. found only three species of plecopteran in 1969, whereas in 1985, 13 species were recorded. Leuctra fusca, despite a restricted site distribution, was the only abundant stonefly in 1969; its range has now been extended and it represents an indicator species separating major groups in a site classification based on benthic invertebrates (ref.100).

Four such site groups were distinguished. Upper sites on all tributaries were characterised by undepleted, pollution-sensitive taxa, mid-tributary and main reaches by intermediate or depletion fauna and lower main river sites were invariably typified by a pollution-tolerant fauna. These site groups could be classified in terms of water quality, principally ammonia, BOD and suspended solids concentrations, with a high degree of success (71-79% in spring and 55-83% in summer survey).

The following conclusions may be drawn from the case study:
1) The response of migratory salmonids was sensitive and quick to changes in water quality. Spawning and successful rearing to smolts was not delayed after adults were able to ascend the river.
2) The improvement in water quality coincident with the return of migratory salmonids was closely described by criteria of the EEC Fish Directive.
3) Further improvement of the migratory salmonid fishery is dependent on removing physical barriers to migration and on preventing over-fishing of the small and vulnerable stocks.
4) The benthic fauna did not respond temporally to changes in water quality as dramatically as fish, although the main determinands of spatial groupings were ammonia, BOD and suspended solids. The residual occlusion of the river bed with coal solids in the lower reaches could be preventing further improvement
5) Established biotic indices did not detect some fauna changes, e.g. increased diversity and distribution of plecopterans.

As a foot-note to this study, invertebrate samples were taken from sites upstream and downstream of collieries in four river catchments, including the R. Taff, before and after the one year miners' strike of 1984 when coal mines were closed. Although improvements occurred, particularly where upstream sites were of good quality and provided a source for colonization, these were limited and suggested that residual factors (e.g. past occlusion of the river bed, storm sewage) were responsible for the lack of response (ref.101).

4.3 R. Twymyn (N. Wales) - residual metal pollution and delayed recovery

The R. Twymyn is a tributary of the R. Dovey, a renowned sea-trout river in North West Wales. The area was actively mined, principally for lead and zinc, until the early 20th century although on this tributary mining ceased in the late 19th century. The recovery of rivers in such areas of metalliferrous mining has been described by several authors (refs.102,103), who have shown that several decades are required to re-establish a diverse invertebrate fauna and that fish species are often slow to recover, particularly migratory salmonids.

The R. Twymyn is selected because of recent descriptions (ref.104) which include several aspects of chemical and biological recovery. The major source of metal pollution derives from percolation of the headwaters through mine spoil where the annual mean zinc concentration is about 1 $mg\ell^{-1}$; within 17 km this concentration has been reduced tenfold providing a gradient of concentration to study faunal distributions (Fig.5). Lead also leaching from the spoil was similarly diluted.

Invertebrate taxa increased from 15 immediately downstream of the mine-spoil

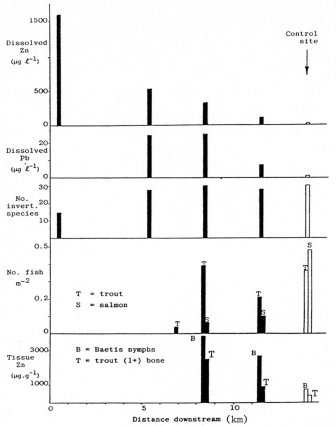

Fig.5. Distribution of zinc, lead, invertebrate taxa, fish density and zinc bioaccumulation in trout bone and Baetis nymphs in the R. Twymyn downstream of an abandoned mine. Values for a control stream are shown for comparison.

to around 30 within 5 km, remaining constant for a further 12 km despite further decreases in metal concentrations. No species of fish was found within 7.5 km of the mine-spoil although downstream at least four species were recorded (S. trutta, S. salar, N. barbatulus, A. anguilla) at reduced densities compared with a control site. Zinc concentrations in both invertebrate species and tissues of brown trout (1+) were found to decrease downstream. Of particular significance was the calculated rate of upstream penetration of trout since 1884 when the mine closed. In the early years after cessation of mining this penetration was equivalent to about 2 km.decade^{-1} but now it has appreciably reduced.

It may be concluded from this study that:
1) Toxic leachates can suppress ecological recovery for several decades.

2) Invertebrates may regain reasonable species diversity more quickly than fish populations recover.
3) Bioaccumulation acts as a useful indicator of residual metal pollution within the catchment.

4.4 Acid streams of Llyn Brianne - complex causation and rehabilitation needs

Llyn Brianne reservoir was constructed in 1969 in the headwaters of the R. Towy in mid-Wales to supply water by river regulation. The decline in the migratory fisheries of the catchment during the 1970s was provisionally ascribed to ineffectiveness of facilities for migrating adults and to changing water quality downstream consequent upon storage. However, investigations in the late 1970s established that young salmonid fish died in several influent streams to the reservoir and that these streams were ecologically impoverished (ref.105). The chemical classification of these streams (Table 1) was shown to be associated with biological properties and land-use (moorland or coniferous forest).

TABLE 1
Biological status in relation to hardness, pH and aluminium concentration (taken from ref.105). M = moorland; F = conifer forest.

Class	Mean hardness ($mg\ell^{-1}$ $CaCO_3$)	Mean pH	Mean sol. Al conc. (μ equiv. ℓ^{-1})	Salmonid abundance ($n.m^{-2}$)	No. invert. taxa
1(M)	> 10	> 6.0	$\leq$ 15	0.6-1.8	} 60-78
11A(M)	8-10	5.5-6.0	$\leq$ 15	0.5-1.7	
11B(F)	8-10	5.0-5.5	25-38	0	23-37
111A(M)	6-8	5.0-5.5	15-20	0.1	46
111B(F)	6-8	4.5-5.0	28-50	0	23-37

More recent investigations have described the detailed chemistry and biology of the streams and are assessing the effects of land-management using potentially ameliorative treatments such as liming and modified tree planting (ref.80). Benthic invertebrate communities have been classified and associated with chemical parameters, particularly total hardness and dissolved aluminium (ref.15); this association facilitates predictions of ecological changes that might ensue from water quality improvements. From mortalities of caged trout it was established that streams where LT_{50}'s < 15d were fishless whilst those where fish survived for longer generally had self-sustaining populations. This relationship suggests that average water quality and quality during storm events and snow melt (when chemical conditions are often most severe) are correlated. Aluminium concentrations were particularly important in determining survival times.

From these links between stream chemistry and biology, Ormerod et al.

(personal communication) have attempted to forecast the biological consequences of future changes in atmospheric inputs and land-use options using a hydro-chemical model of acidification of ground water. The realism of the model was tested by reconstruction changes in atmospheric deposition and land-use over the past 140 years in three contrasting streams of the catchment (two soft water (< 2 mg Ca ℓ^{-1}), one of which is moorland (MS) and the other planted with Sitka spruce (FS); one harder water (4 mg Ca ℓ^{-1}) on moorland (MH). 140 years ago all streams would have contained fish. The model predicted that MH would continue to support fish whereas both soft streams (MS and FS) would suffer a progressive reduction in trout production with time, FS to the point where stocks would become extinct and MS where stocks would be marginally sustained. These predictions are very close to reality (see Table 1).

The model was then used to predict the effects of differences in future atmospheric inputs of acid (100% and 50% of 1984 inputs) and of alternative land uses (moorland and Sitka spruce). With respect to trout, under forest-cover and both continued and reduced acid inputs, fish would be absent, survival times being < 10 days in all cases. Under moorland only, deposition reduction would prevent further deterioration and salmonid stocks would be maintained (Fig.6).

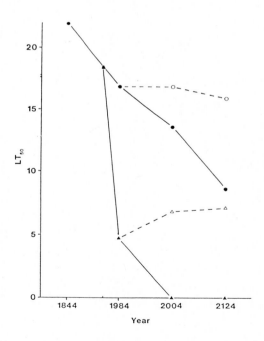

Fig.6. Predicted survival of trout (in days) in a soft water moorland stream on the R. Towy from 1884 to 2124. In one prediction the catchment is afforested in 1958 (▲). Dotted lines show the effect of reducing acid inputs by 50% from 1984.

Predictions of changes in invertebrate assemblages were also made. For forest cover on soft water streams, a rapid shift to a community dominated by stoneflies, even when deposition was reduced by 50%, was predicted.

Reservations about these predictions were recognised by the authors and some of the assumptions will need to be tested and refined further as the responses of catchments to land-treatment become clearer. Nevertheless, they point to the dominating influence of land-use in soft-water areas and suggest that under forest-cover substantial reductions in atmospheric inputs of acid merely retard the rate of deterioration but that with soft moorland streams a 50% reduction would prevent further deterioration and preserve viable, if marginal, fisheries.

The following may be concluded from the study:

1) Empirical models relating fish and invertebrate distributions to water quality parameters have been developed. Land cover has proved crucially important, principally in determining aluminium concentrations which contributed greatly to toxicity.
2) Forecasting the effects of changes in atmospheric inputs has again highlighted land-cover as a critical factor and suggested that expenditure on acid removal alone is of dubious value in forest areas.
3) With the time-scale of response of catchments to changes in pollution loads, modelling, despite its dangers, is the only feasible way to predict change.
4) In view of the importance of biological responses to episodic events, patterns of variability and their relation to critical life stages (ref.69) need further study.

5. CONCLUSIONS

In the introduction to this chapter, several questions were posed relating to the application of our knowledge to environmental management and to the evolution of our comprehension in the processes of degradation and recovery of rivers from pollution. These questions were followed by a review of analytical and experimental methods which have been applied to pollution studies, comments on the changing patterns of pollution which require a re-examination of research priorities and surveillance procedures, and case-studies which expose our strengths and weaknesses in both comprehension of systems behaviour and of the scientific support for environmental management.

5.1 Environmental management

In retrospect it seems that too much effort has been devoted to the development of various diversity and biotic indices and to their application - without recognition of their strengths and weaknesses. Diversity indices, in particular, embody very little information about communities and even that is generally sensitive to sampling effort, seasonal factors and, most importantly, taxonomic penetration.

Biotic indices, incorporating comparative sensitivities of taxa to pollutants have inherently more potential although their selection and use has been geared, almost entirely, to organic pollution with the usual symptoms of low dissolved oxygen, an availability of organic particles and the presence of ammonia. Where only descriptions of the impact of chronic point-sources of pollution are needed these indices can be simplified to provide information on gross changes in response to river quality: simplification by reducing taxonomic penetration and omitting detailed information of numerical abundance are the two most cost-effective steps. As the scores of these various indices are sometimes closely correlated (ref.99), the choice is not closely constrained although their sensitivity to seasonal factors and sampling procedures needs to be recognised. These simple indices are increasingly adopted for national surveillance networks (refs.106,107).

The development of multi-variate statistical techniques of classification and ordination, which facilitate the grouping of both sites and taxa by defined procedures and relate these groupings to environmental variables, represents a potentially powerful step forward in description, in locating the cause of ecological deformation and in modelling effects of environmental change. However, to achieve this potential, taxonomic penetration to species level for a wide spectrum of organisms is required. Brooker (ref.99) has shown that some groups which are taxonomically difficult can be omitted with little effect on site classification. This emphasis on the importance of detailed taxonomy represents a reversal of current trends with the application of simple biotic indices and requires a revision in the deployment of resources for monitoring. Furthermore, identifying the causes of ecological deformation in any instance will be difficult until the ecotoxicological data base is extended to a range of those organisms used for surveillance.

Functional responses of river ecosystems to pollution have not been widely used apart from oxygen consumption measurements of river water (e.g. BOD_5). The poor development of functional measurements stems from the importance of the benthos to the functioning of most rivers and the difficulty in making benthic measurements. Several benthic enclosures have been designed, particularly for measurements of oxygen flux, and their use could be greatly extended to examine a wide range of metabolic activities. Such functional descriptions do not avoid the need for structural descriptions because some species are important as resources or as problems.

Toxicological studies will continue to cover a wide range of procedures from single-species short-term tests to simulated ecosystems. If more diagnostic value is to be achieved by biological surveillance of rivers, life-cycle and multi-generation tests with relevant organisms must be developed further, the latter to provide a more reliable basis for assuming the genetic stability of species in their response to pollutants. Out-door channel systems, despite difficulties of replication and the aberrant composition of communities which can develop within them, will continue to contribute where fine discrimination is needed of the response of systems to chronic discharges or to pollution episodes, but in the latter case interpretation of recovery processes is inevitably suspect.

Bioassay tests conducted both in the field (ref.52) and in the laboratory (ref.108), have proved useful in detecting effects which might otherwise have remained undetected with simple mixtures of known pollutants. These bioassays have increasingly used a diversity of biochemical, physiological and behavioural responses to provide a more realistic assessment of damage.

The pattern of use of these analytical and experimental tools needs to respond to the increasing importance of temporal variability in pollutant concentrations, caused particularly by the removal of chronic point-sources. The remaining inputs, diffuse sources and accidental episodes introduce considerable temporal variability, the first responding to hydrological factors and the second generating single pulses. Both our surveillance and toxicity testing procedures have been oriented towards the steady state which, even with chronic point-sources, rarely exists. Recovery processes, as a consequence, have only been conceived in a spatial sense where factors such as immigration can be ignored. In the case of episodes the reservoir of organisms in the hyporheos may be important but, whatever the immigration route, seasonal factors will dominate early stages of the pattern of recovery. In studying the downstream effects of accidents the additional attenuation process of longitudinal dispersion needs to be considered for in many river systems its effects in reducing peak concentrations may be greater than either tributary dilution or degradation.

Case studies have revealed that where pollutants are removed from a system and do not remain associated with the benthos, recovery can be very rapid. In recovery from the most toxic conditions (see R. Ebbw), microorganisms and invertebrates become re-established faster than fish but where, as on the R. Taff, an adequate range of food organisms is present, it would seem that fish respond more dramatically than benthic invertebrates to improvements in water quality. However, this example from the R. Taff may be a consequence of the residual pollution, i.e. coal particles being predominantly benthic in their location and affecting the organisms which live there rather than in the water column.

Even when industries stop generating pollutants many forms of pollution do not disappear quickly. This applies particularly to mining where waste materials are tipped. Where tip material is inert, problems arise from transport of particles. Even with coal wastes chemical processes can oxidise sulphides and

generate sulphuric acid and, with metalliferous mining, acid problems can be
further exacerbated by the leaching of toxic metals. Patterns of rainfall affect
rates of chemical transformation and the hydrological pathways through waste
material so creating temporal variability in water quality.

Pollution from acid rain has some similar elements, with acid leaching cations from soils - in some cases toxic ones such as aluminium. But where cation exchange capacity from Ca and Mg is limited the recovery, once acid loads are reduced, will be driven by geological weathering unless exchange capacity is replaced by liming. There are seemingly unique aspects to acid rain pollution, particularly the profoundly modifying effects of land-use and the residual problems caused even by overland flow during snow-melt or major storms, for although low in aluminium, pHs can be extremely low (< pH 4) at these times. Despite these complexities of action, progress has been made in modelling effects and forecasting the consequences of change in pollution load and land management.

5.2 Scientific understanding of degradation and recovery processes

Our comprehension of the processes underlying degradation and recovery have been hampered by the separation of research into pollution from the mainstream of ecological theory and, to some degree, from more fundamental aspects of toxic action.

Ecological separation has been far from complete, for stability theory has contributed to the persistence of diversity indices in pollution monitoring and recent multivariate analyses of polluted rivers have contributed to the commentary on the river continuum concept (ref.109). Recent studies of the effects of pulses of pollutants, such as aluminium and low pH, on the behaviour of organisms, either through drift or vertical migration in the benthos, have also drawn upon more basic investigations of drift behaviour (ref.65). Nevertheless, direct chemical-biological interactions have dominated interpretation in pollution studies, trophic and other biological interactions only being considered where introduced pollutants are transferred through food-webs or obviously affect system behaviour. The conclusions of Wright et al. (ref.23) would, however, suggest that the macro-distribution of invertebrates can be explained in these chemical terms.

The major fields of experimental studies of pollution have not been so obviously separate from mainstream toxicological studies except that orientation has been towards effects of equilibrium concentrations rather than doses, so recovery as a physiological process has been neglected and behavioural responses have only been of interest in establishing lowest effect thresholds. Abel (ref.83) and others have recently stressed the need to take more account of these factors. In some senses, aquatic pollution studies have reached a remarkable degree of sophistication (ref.34), particularly in consideration of the biochemical action of poisons such as cadmium, the joint action of pollutants and the establishment of MATCs. In recent years theoretical ideas relating to the effects of poisons on energy for growth (ref.53) have impinged and increasing consideration is now being given to genetic selection as a process of adaptation to aquatic pollutants.

6. ACKNOWLEDGEMENTS

I wish to thank staff of Welsh Water Authority and the Water Research Centre who suggested case material and to Dr. S.J. Ormerod and Dr. D. Pascoe who made comments on an early draft of this paper. Figures 1, 2 and 3 were taken from "Ecology of European Rivers" by kind permission of Blackwell Scientific Publications.

REFERENCES

1. R. Kolkwitz and M. Marsson, Ber. dtsch. bot. Ges., 26A (1908) 505-519.
2. R. Kolkwitz and M. Marsson, Int. Rev. Hydrobiol., 11 (1909) 126-152.
3. E.E. Herricks and J. Cairns, Wat. Res., 16 (1982) 141-153.
4. R.A. Matthews, A.L. Buikema, J. Cairns and J.H. Rodgers, Wat. Res., 16 (1982) 129-139.
5. H.G. Washington, Wat. Res., 18 (1984) 653-694.
6. B.D. Hughes, Wat. Res., 12 (1978) 359-364.
7. P.R. Needham and R.L. Usinger, Hilgardia, 24 (1956) 383-409.
8. R.W. Edwards, B.D. Hughes and M.W. Read, in The Ecology of Resource Degradation and Renewal, Blackwell Scientific Publ., Oxford, 1975, pp.139-156.
9. J.L. Wilhm and T.C. Dorris, Bioscience, 18 (1968) 477-481.
10. P.M. Murphy, Environ. Pollut., 17 (1978) 227-236.
11. J.L. Wilhm, J. Wat. Pollut. Control Fed., 39 (1967) 1673-1683.
12. R. Macarthur, Ecology, 36 (1955) 533-536.
13. F. Woodiwiss, Trent Biotic Index of Pollution, 2nd Quinquennial Abstract of Statistics relating to Trent Watershed, Trent River Authority.
14. D. Moss, M.T. Furse, J.F. Wright and P.D. Armitage, Freshwat. Biol., 17 (1987) 41-52.
15. N.S. Weatherley and S.J. Ormerod, Environ. Pollut., 46 (1987) 223-240.
16. J. Cairns. Wat. Res.. 15 (1981) 941-952.
17. A.L. Buikema, B.R. Niederlehner and J. Cairns, Wat. Res., 16 (1982) 239-262.
18. J.R. Chandler, Wat. Pollut. Control, 69 (1970) 415-421.
19. D. Balloch, C.E. Dames and F.H. Jones, Wat. Pollut. Control, 75 (1976) 92-114.
20. M.P. Brooker and D.L. Morris, Freshwat. Biol., 10 (1980) 437-458.
21. P. Clare and R.W. Edwards, Freshwat. Biol., 13 (1983) 205-225.
22. M.A. Learner, J.W. Densem and T.C. Iles, Freshwat. Biol., 13 (1983) 13-36.
23. J.F. Wright, D. Moss, P.D. Armitage and M. Furse, Freshwat. Biol., 14 (1984) 221-256.
24. H.T. Odum, Limnol. Oceanogr., 1 (1956) 102-117.
25. R.W. Edwards and M. Owens, in G.T. Goodman, R.W. Edwards and J.M. Lambert (Editors), Ecology and the Industrial Society, Blackwell Scientific Publ., Oxford, 1965, pp.149-172.
26. M.P. Brooker, D.L. Morris and R.J. Hemsworth, J. Appl. Ecol., 14 (1977) 409-417.
27. H.W. Streeter and E.B. Phelps, Bull. US Publ. Hlth. Serv., No.146 (1925).
28. J.D. Boyle and J.A. Scott, Wat. Res., 18 (1984) 1089-1099.

29 G.T. Bowman and J.J. Delfino, Wat. Res., 14 (1979) 491-499.
30 J.E. Hobbie and R.T. Wright, Limnol. Oceanogr., 10 (1965) 471-474.
31 R.H. Monheimer, Can. J. Microbiol., 20 (1974) 825-831.
32 D.W.J. Green and K.A. Williams, Lab. Pract., 32 (1983) 74-76.
33 V.M. Brown, Wat. Res., 3 (1968) 723-733.
34 J.B. Sprague, Wat. Res., 4 (1970) 3-32.
35 B.A. Southgate, Q.Jl. Pharm. Pharmac., 5 (1932) 639-648.
36 R. Lloyd, Ann. Appl. Biol., 46 (1961) 535-538.
37 D.W.M. Herbert, Ann. Appl. Biol., 50 (1962) 755-777.
38 R. Lloyd and D.H.M. Jordan, J. Proc. Inst. Sew. Purif. (1963) pp.167-173.
39 R.W. Edwards and V.M. Brown, Wat. Pollut. Control, 66 (1967) 3-18.
40 J.S. Alabaster, J.H.N. Garland, I.C. Hart and J.F. de L.G. Solbë, Symp. Zool. Soc. London, 29 (1972) 87-114.
41 National Technical Advisory Committee, Water Quality Criteria, Report to the Secretary of the Interior Federal Water Pollution Control Administration, Washington, 1968, pp.1-234.
42 S.R. Verma, I.P. Tonk, A.K. Gupta and M. Saxena, Wat. Res., 18 (1984) 111-115.
43 E.F. Kenaga, in K.L. Dickson, A.W. Maki and J. Cairns (Editors), Analyzing the Hazard Evaluation Processes, American Fisheries Society, Washington D.C., 1979, pp.101-111.
44 C.R. Fremling and W.L. Mauck, in A.L. Buikema and J. Cairns (Editors), Aquatic Invertebrate Bioassays, American Society for Testing and Materials, Philadelphia, 1910, pp.81-97.
45 D.A. Benoit, E.H. Leonard, C.M. Christensen and J.T. Fiandt, Trans. Amer. Fish. Soc., 105 (1976) 550-560.
46 P.L. Klerks and J.S. Weis, Environ. Pollut., 45 (1987) 173-205.
47 E.E. Kenaga, Environ. Sci. Technol., 12 (1978) 1322-1329.
48 J. Cairns, in R.K. Ballentine and L.J. Guarraia (Editors), The Integrity of Water, U.S. Environmental Protection Agency, Washington D.C., 1977, pp. 171-187.
49 K. Williams, D. Green and D. Pascoe, in D. Pascoe and R.W. Edwards (Editors), Freshwater Biological Monitoring, Pergamon Press, Oxford, 1984, pp.81-91.
50 C. Henderson and Q.H. Pickering, J. Am. Wat. Wks. Ass., 55 (1963) 717-720.
51 J. Cairns and W.H. van der Schalie, Wat. Res., 14 (1980) 1179-1196.
52 L. Brown, Wat. Res., 14 (1980) 941-947.
53 P. Calow, Invertebrate Ecology: a Functional Approach, Croom Helm, London, 1981.
54 W. Sloof, Aquat. Toxicol., 4 (1983) 73-82.
55 P.J. Eisele and R. Hartung, Trans. Amer. Fish. Soc., 5 (1976) 628-633.
56 J.R. Robert, D.W. Rodgers, J.R. Baily and M.A. Rorke, Polychlorinated biphenols: biological criteria for an assessment of their effects on environmental quality, Ntl. Research Council, Ottawa, Canada, 1978.
57 R.W. Edwards, H. Egan, M.A. Learner and P. Maris, J. Appl. Ecol., 1 (1964) 97-117.
58 A.J. Watton and H.A. Hawkes, Wat. Res., 18 (1984) 1235-1247.
59 E. Eichenberger, F. Schlatter, H. Weilenmann and K. Wuhrmann, Ver. Internat. Verein. Limnol., 21 (1981) 1131-1134.
60 M. Yasuno, Y. Sugaya and T. Iwakuma, Environ. Pollut. (Series A), 38 (1985) 31-43.
61 D.M. Rosenberg and A.P. Wiens, Effects of sediment addition on macrobenthic invertebrates in a Northern Canadian river, Wat. Res., 12 (1978) 753-63.
62 J.M. Culp, F.J. Wrona and R.W. Davies, Can. J. Zool., 64 (1986) 1345-1351.
63 C.E. Warren, J.H. Wales, G.E. Davis and P. Sonderoff, J. Wildl. Mgmt., 28 (1964) 617-660.
64 R.H. Hall, G.E. Likens, S.B. Fiance and G.E. Hendrey, Ecology, 61 (1980) 976-989.
65 S.J. Ormerod, P. Boole, C.P. McCahon, N.S. Weatherley, D. Pascoe and R.W. Edwards, Freshwat. Biol., 17 (1987) 341-356.
66 S.J. Ormerod, N.S. Weatherley, P. French, S. Blake and W.M. Jones, Ann. Soc. R. Zool. Belg., 117 suppl.1 (1987) 435-447.

67 Welsh Water Authority, Macroinvertebrate Studies of the River Sirhowy Following an Acute Oil Pollution Incident in March 1984, 1985, SE/16/85.
68 D.S. Cherry and J. Cairns, Wat. Res., 16 (1982) 263-301.
69 J.M. Gunn, Environmental Biology of Fishes, 17 (1986) 241-252.
70 F.S. Woodiwiss, Chem. Ind., March 1964.
71 J. Scullion and R.W. Edwards, Freshwat. Biol., 10 (1980) 141-162.
72 P.M. Nuttall and G.H. Bielby, Environ. Pollut., 5 (1973) 77-86.
73 H.B.N. Hynes, The Biology of Polluted Waters, University of Liverpool Press, 1960.
74 F. Tada and S. Suzuki, Wat. Res., 16 (1982) 1489-1494.
75 A.M. Bruce and H.A. Hawkes, in C.R. Curds and H.A. Hawkes (Editors), Biological Filters, Acad. Press, London & New York, 1983, pp.1-111.
76 B.A. Whitton and P.J. Say, in B.A. Whitton (Editor), River Ecology, Blackwell Scientific Publ., 1975, pp.286-311.
77 J.F. de L.G. Solbé, in J.S. Alabaster (Editor), Biological Monitoring in Inland Fisheries, Applied Science Publishers Ltd., London, 1977, pp.97-105.
78 S.C. Bird, Environ. Pollut., 45 (1987) 87-124.
79 W.R. Howells and R. Merriman, in J.F. de L.G. Solbe (Editor), Effects of Land Use on Freshwaters, Ellis Horwood Ltd., Chichester, 1986.
80 Welsh Water Authority, Llyn Brianne Acid Water Project, J.H. Stone (Editor), 1987.
81 P.G. Whitehead, R.J. Williams and G.M. Hornberger, J. Hydrol., 84 (1986) 273-286.
82 P.D. Abel, Freshwat. Biol., 10 (1980) 251-259.
83 P.D. Abel, Prog. Wat. Tech., 13 (1980) 347-352.
84 D. Pascoe and N.A.M. Shazili, Ecotox. Env. Safety, 12 (1986) 189-198.
85 D.A. Holdway and D.G. Dixon, Can. J. Fish. Aquat. Sci., 43 (1986) 1410-1415.
86 A.E. Brafield and P. Mathiessen, J. Fish. Biol., 9 (1976) 359-370.
87 L.K. Siddons, W.K. Siem and L.R. Curtis, Can. J. Fish. Aquatic Sci., 43 (1986) 2036-2040.
88 D.W. Aldridge, B.S. Payne and A.C. Miller, Environ. Pollut., 45 (1987) 17-28.
89 A.W.H. Turnpenny and R.W. Williams, J. Fish. Biol., 17 (1980) 681-693.
90 D.D. Williams and H.B.N. Hynes, Oikos, 27 (1976) 265-272.
91 A.L. Sheldon, Oikos, 28 (1977) 256-261.
92 K. Gose, Jap. J. Ecol., 18 (1968) 147-157.
93 M.J. Andrews, in P.J. Sheehan D.R. Miller, G.C. Butler and Ph. Bourdeau (Editors), Effects of Pollutants at the Ecosystem Level, Wiley J. and Sons, 1984.
94 D. MacKenzie, New Scientist, 112 (Nov. 27th) (1986).
95 A.W.H. Turnpenny and R. Williams, Environ. Pollut. (Series A), 26 (1981) 39-58.
96 R.W. Edwards, P.F. Williams and R. Williams, in Ecology of European Rivers, Blackwell Scientific Publ., Oxford, 1984, 83-111.
97 R.W. Edwards, K. Benson-Evans, M.A. Learner, P.F. Williams and R. Williams, J. Inst. Water Pollut. Control, 71 (1972) 144-166.
98 G.W. Mawle, A. Winstone and M.P. Brooker, Nature in Wales N.S., 4 (1985) 36-45.
99 M.P. Brooker, in D. Pascoe and R.W. Edwards (Editors), Freshwater Biological Monitoring, Pergamon Press, 1984, pp.25-33.
100 E.J. Bent, J.M. Dack and K.R. Wade, The Environmental Quality of the Taff Catchment 1985: Biology, Welsh Water Authority (SE/6/86).
101 N. Reynolds, Macroinvertebrate Studies to Assess the Effects of the 1984 National Mineworkers Strike on the Biological Quality of five South Wales Rivers, Welsh Water Authority SE/15/86.
102 R.D. Laurie and J.R.E. Jones, J. Anim. Ecol., 7 (1938) 272-289.
103 J.R.E. Jones, J. Anim. Ecol., 18 (1949) 67-88.
104 K.T. O'Grady, Minerals and the Environment, 3 (1981) 126-137.
105 J.H. Stoner, A.S. Gee and K.R. Wade, Environmental Pollution (Series A), 36, 1984, 125-157.

106 N. de Pauw and G. Vanhooren, Hydrobiologia, 100 (1983) 153-168.
107 C.A. Extence, A.J. Bates, W.J. Forbes and P.J. Barkam, Environmental Pollution, 45 (1987) 221-236.
108 C.L.M. Poels, M.A. van der Gaag and J.F.T. van der Kerkhoff, Wat. Res., 14 (1980) 1029-1035.
109 S.J. Ormerod and R.W. Edwards, Freshwater Biol., 17 (1987) 533-546.

BIOMANIPULATION OF AQUATIC FOOD CHAINS TO IMPROVE WATER QUALITY IN EUTROPHIC LAKES

R. DE BERNARDI

C.N.R. - Istituto Italiano di Idrobiologia - 28048 Pallanza, Italy

It is widely accepted that lake ecosystems may be divided into two components, the biotic and the abiotic, which interact through reciprocal feed-back mechanisms. Consequently changes in one component will induce corresponding changes in the other. A second basic ecological concept is that biological communities are structured along trophic hierarchies - food chains - which direct energy·flow within the ecosystem (ref. 1) and thus influence virtually every aspect of ecosystem function.

However, since each species has its own biological and ecological characteristics, food chains of different structure or species composition will accomplish their function in different ways and with different efficiencies.

These simple beliefs have fostered much of the great interest in the structure of biological communities and in the response of community structures to environmental change.

Despite these common elements in ecological thought, the awareness that each species exist only within its limits of tolerance (ref. 2) and that all systems are now subject to ever increasing loads of a vast array of foreign substances has encouraged a reductionistic and unidirectional view in which an organism is simply the pawn of its environment. A given community structure is therefore considered only in terms of the resistance of the biota to prevailing abiotic conditions, ignoring completely the importance of biological interactions. Alternatively, one encounters the no less reductionistic and naive study that considers a single organism or species in the laboratory and then extrapolates those results to all communities and all biotic conditions forgetting that, in nature, all individuals occur only as members of a population and the populations only as components of a community.

Although Lindemann's (ref. 1) trophodynamic view has contributed greatly to our quantitative understanding of ecosystem function, one must still agree with

Hutchinson (ref. 3) that the role of different species of organisms in structuring or organizing an ecosystem must be more carefully considered: the character of a biological community is determined by the characteristics of its component populations and a community's homeostasis is the direct result of the summed interactions of these populations.

The aim of this paper is to consider community structure as a result not just of the abiotic environment, but especially of biotic interactions, like predation and competition, which connect different members of trophic hierarchies. Furthermore, I will provide evidence that community structure plays an important role in controlling the abiotic environment.

Lake ecosystems can be subdivided into a series of well defined but highly interacting subsystems. Among these, the pelagic environment is doubtless the most peculiar. Most environments display complex geometries scored by sharp physical gradients and discontinuities; the open water of lakes contains a relatively simple vertical gradient in light, temperature and chemistry with the result that it is one of the most uniform of all environments.

A further peculiarity of the pelagic environment is that its primary producers (unicellular or colonial algae) are very small relative to the size of the planktonic herbivores. These herbivores, in turn, are much smaller than many of their own predators. Thus because pelagic predators are often much larger than their prey, the prey are completely open to predator attack. In addition, despite differences in the mechanisms of food gathering (ref. 4) all planktonic herbivores compete for food particles in the same size range (about 1 to 15 μm) (ref. 5). Thus, in the pelagic environment, the biotic interactions of competition and predation may rank among the most important factors determining and regulating community structure.

That the eutrophication is, without any doubt, the most important reason for lacustrine environment deterioration is widely accepted. Similar consensus exists on the fact that causes of eutrophication must be recognized in an increased availability of nutrients (mainly P) for algae growth. However, this awareness, in some circumstances, leads us to ignore that lakes are ecosystems and like any ecosystem are complex in structure.

Moreover, interactions among these structures are an important factor for lake ecosystem functioning, so that it is unproductive to consider them as single elements. For the same reasons it is a reductionistic point of view to consider eutrophication as a binomial phosphorus-algae as in many circumstances it was. In such a vision of the process not only the structural functional complexity of the lake is ignored but also methods of lake recovery that do not necessarily fit only this binomial are not taken into consideration.

Hrbáček, Desortovà and Popovsky (ref. 6) showed that the phosphorus-chlorophyll relationship of Dillon and Rigler (ref. 7) held for central European lakes too, but they also found that departures from this relationship could be explained by reference to the effects of fish predation. Analogous conclusions were drawn (ref. 8) from studies in Lake Harriet near Minneapolis (ref. 9). Finally, Lynch (ref. 10) in (ref. 8) found that fish density in enclosures with similar phosphorus levels was positively related to algal density and negatively related to water transparency.

This body of evidence leads to two conclusions: first, biological factors play an important role in determining community structure, and thereby also influencing physical and chemical characteristics of the environment; second, the food habits and population dynamics of organisms near the top of the trophic pyramid (ref. 11) are important factors in this process. The first conclusion is scarcely novel. In 1948 Omer-Cooper (ref. 12) stated that biological factors are the most important of ecological factors normally acting on lakes and ponds and that the appearance of a new species in an animal community easily produces a more marked effect than any normal physical or chemical variation. However, such an opinion, until recently little known and generally ignored, is now gaining some well deserved recognition. It is perhaps less widely accepted that the structure of the food chain is more sensitive to variations near or at its vertex than those at its base (ref. 13). By extention, this suggests that each trophic level is influenced more by the level immediately above it than by that immediately below.

While predation and competition determine the species composition of the community, the effect of an increase in food or nutrients is often translated at a first step only into a numerical increase in the species already present (ref. 14).

This is shown in the process of the lake enrichment and the biological changes this process entails.

In an oligotrophic lake, algal production is not usually limited by zooplankton grazing but by the availability of nutrients, primarily phosphorus and/or nitrogen. The zooplankton in such systems is limited in its turn by predators. If such a lake receives an increase amount of nutrients, an increase is noticed in both production and biomass of phytoplankton, particularly diatoms and green algae.

A eutrophic lake produces much more plant material than can be used by the herbivores. This surplus accumulates in the lake until decomposed by bacteria. This leads to a progressive decline in dissolved oxygen in the deeper strata, and in extreme cases, to anoxia and accumulation of nutrients released from the sediments to these deep layers. In the surface water, an increase in pH is observed following the consumption of CO_2 by photosynthesis. These variations in the chemical environment provoke modifications in the biological community. Stenotypic fish species such as salmonids and coregonids are gradually replaced by more tolerant cyprinids; in the phytoplankton, green algae are supplanted by blue-green which gain more advantage in competition for CO_2 and nutrients at higher pH values. The zooplankton populations also undergo a profound structural change passing from a dominance of calanoid copepods to small cladocera and rotifers. More generally speaking at this level of the food chain a gradual shifting to smaller species is observed.

This represents a very important point for the energy budget in the lake environment.

As smaller-sized species are less efficient in utilizing the available food particles the eutrophication determines a larger gap between primary production and algae biomass utilization by herbivorous zooplankton.

There are two main reasons for the shift toward smaller herbivores in the planktonic community of eutrophic lakes: first a variation in the fish community with a dominance of planktophagous species, selectively preying upon larger zooplankters; and, second, the reduced possibility presented by these last to avoid clogging of the filtering apparatus by filamentous and colonial algae that progressively dominate phytoplankton communities. It must be noticed in this respect that formation of filamentous or large colonies by planktonic algae has been interpreted as an evolutive strategy to avoid grazing by filter-feeders.

The result of the eutrophication process is thus a decrease of the efficiency of energy transfer along the pelagic food chain with more important losses to the detritus chain (Fig. 1).

Eutrophication can be, in this respect, considered a sort of self-intoxication of the lake system which destroys the pre-existing equilibrium and establishes a new one within a different biological community.

The term "self-intoxication" seems justified because, although the initial cause of the process is external (an increase of nutrient availability) the process is furthered by the activity of the biological community of the eutrophying lake. In other words, an increase in nutrients initiates internal processes which stabilize the new trophic state.

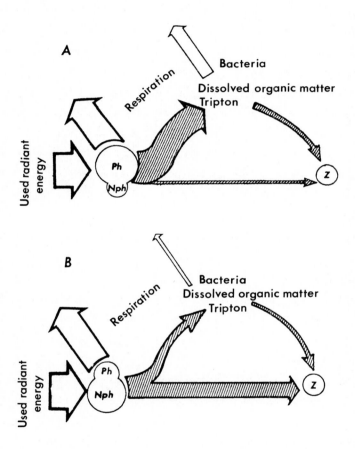

Fig. 1. The schematic diagram of main paths of energy flow between the producers and consumers level in the pelagial of eutrophic (A) and oligotrophic (B) lakes. As trophic level of lake increase a contemporaneous increase of energy flowing directly to bacteria is observed.
Ph=phytoplankton; Nph=nannophytoplankton; z=zooplankton (ref. 15).

For these reasons attempts to re-establish the original trophic status only by reducing the nutrient level often do not produce the desired results in a reasonably short time.

A typical eutrophic community tends to maintain its identity by counteracting attempts at purification. In such cases, manipulation of the food chain could prove a useful and efficient mechanism for reducing the most obvious drawbacks in eutrophy.

In effect, such biomanipulation consists of a shift in controlling factors from that shown as "type A" of "type B" (Table I). A type A community is that usually found in lacustrine environments. The phytoplankton is limited by the availability of algal nutrients, but the second level, the zooplankton, is not limited by food availability but by predation from still higher trophic levels represented by the fish. The fish are in turn limited by food availability. One appropriate intervention in a "type A" eutrophic lake would induce control of the algal community by grazers rather than nutrients. This could be achieved by reducing the population of planktivorous fish (ref. 8) for example by removing all fish from the lake, or by selective poisoning of planktivores (for example antimicine is specially toxic to perch), or by introduction of large fish which prey upon the planktivores and by other mechanisms as well (ref. 17), (ref. 18).

TABLE 1. Principal limiting factors of major components of the limnetic community under two different systems of management. See text for explanation (ref. 16).

Link	Main limiting factor	
	A	B
Phytoplankton	Nutrients (P and N)	"Grazing"
Zooplankton	Predation	Food
Planktophagous fish	Food	Predation

In entomology this technique is not new: it is called "biological pest control" and consists in the use of predatory organisms, parasites, pathogenus or producers of pherormoned to check the development of undesirable insect populations. In limnology it has been called biomanipulation (ref. 19) and may be considered something new; the principles which form their basis are common to both, firstly, a scientific use of biotic interaction among the different organisms which constitute natural trophic chains. It is a matter largely of modifying the degree of interaction between two or more populations, increasing or decreasing it, with the aim of achieving control of the density of a certain species or group of organisms, heightening or lessening in this way the role they play in a given ecosystem. In other words, modification are initiated which, by altering the delicate balance between species, produce between them the "construction" of quantitatively different relationships such that community structures may be achieved whose biological activity acts in accordance with certain demands made by man on the environment.

Thus, while in entomology such a technique has been used from the start to control undesirable populations (the first example of biological control goes back to 1762 with the introduction into the Mauritius Islands of birds to control the density of the locust populations (ref. 20), in limnology this technique is proposed to control the eutrophication process. Although new in limnology, so that further scientific confirmation is required, these techniques appear very promising and have in any case the advantage of suggesting an integrated biological approach for the qualitative restoration of polluted lake ecosystems.

With these techniques, in fact, the problem of the improvement of water quality is faced by considering the lake as a whole, taking into account all its component parts and mechanisms that permit its functioning.

However, it must be emphasized that such interventions carry a certain risk for the existing community and must be seriously controlled. A deeper knowledge of the structure and function of the food chain is essential if the approach is to be made sufficiently precise. An indispensable step in obtaining this knowledge is an accurate taxonomy based not just on the morphology of dead specimens, or on the biology and ecology of the species in isolation, but on their role as members of different biological community and food chains. Such basic research can assume an importance far beyond its scientific interest and may prove essential for effective management of aquatic ecosystems.

Following the pioneering studies by Hrbáček (ref. 21) and by Hrbáček, Dvořakova, Kořinek and Prochazková (ref. 22) demonstrating differences in the zooplankton between ponds with and without fish, a large and active field of research has developed around the theme that fish predation influences zooplankt

on between ponds with and without fish, a large and active field of research has developed around the theme that fish predation influences zooplankton composition. Subsequently, Brooks and Dodson (ref. 5) developed their size-efficiency hypothesis which states that planktonic herbivores compete for

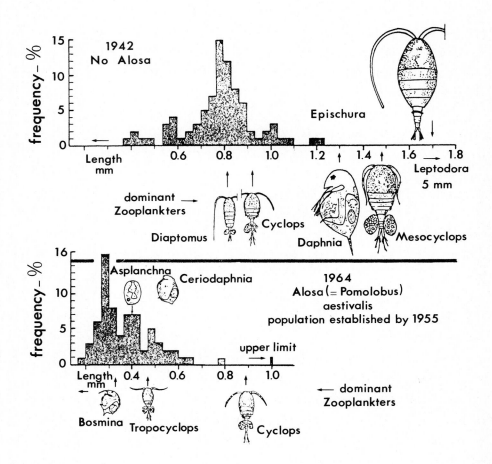

Fig. 2. Composition of the zooplankton population in Crystal Lake, Connecticut, before (1942) and after (1964) the introduction of Alosa aestivalis. Each unit in the histogram indicates 1% of the total sample counted. The largest zooplankton was not represented because of the scarcity of adults. The organism pictured represent the mean length of the smallest mature individuals. Arrows indicate the position of the smallest mature stage of each species on the histogram. The predatory rotifer Asplanchna priodonta is the only non-crustacean included. Other rotifers were not considered. The introduction of planktivorous fish determine to great changes in the zooplankton community structure leading, the dominance of smaller species (ref. 5).

food particles ranging from 1-15 μm in size, and that larger organisms, as well as being able to utilize particles of a greater size, are also more efficient feeders. This may lead to three possible consequences:

A) at a low predation level small planktonic herbivores are eliminated by the competition with those of larger size (dominance of large cladocera and calanoid copepods);

B) At a high predation level small-sized species dominate (dominance of rotifers and small cladocera);

C) at a moderate predation level the co-existence of different sized organisms can occur.

It follows that intervening in planktivorous fish populations by controlling their density, the establishment of a zooplanktonic type A community may be achieved as Brooks and Dodson demonstrate (ref. 5) (Fig. 2), and as may also be concluded from the results of de Bernardi and Giussani (ref. 23) in Lago di Annone (Fig. 3), and of Schindler and Comita (ref. 24); Galbraith (ref. 25) and many others.

These changes at a zooplankton level do not occur without consequences for the biocenotic structure of phytoplankton communities, as well as for the physical and chemical characteristics of the environment, in particular for water transparency and nutrient concentration.

Some examples may give a more complete picture of such theoretical considerations.

Figure 4 shows biomass values and percentage composition of phytoplankton populations in plastic bags with and without fish placed in the eutrophic Swedish lakes Trummen and Bisjön (ref. 26). It appears from this figure that the bags with fish present all the elements that characterize the algal communities of eutrophic environment, that is, high biomass values and the dominance of blue-green algae, together with pH values and low water transparency. On the other hand, in the bags without fish, despite the similar initial level of nutrients, algae are less abundant, pH values are "normal" and transparency is high.

This last environment presents oligotrophic characteristics that contrast with the situation that might have been expected on the basis of nutrient level.

Even clearer evidence may be drawn out from Figure 5 which reports the results of experiments by Lynch (ref. 10) in the United States. Also in this

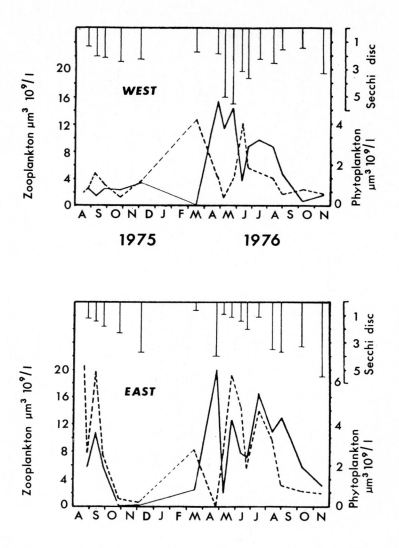

Fig. 3. Seasonal development of the biomass of herbivorous zooplankton (———) and phytoplankton (------) and water transparency (Secchi disc depth) in the east and west basins of Lago di Annone. Higher zooplankton populations lead to lower algal densities and greater transparencies in summer both basins (ref. 23).

case the presence of an increasing number of fish, in environments with a comparable nutrient content, is strictly correlated with a reduction in water transparency and with an increase in algae density. In both the above examples, the relationship between fish and algae is mediated by the zooplankton community

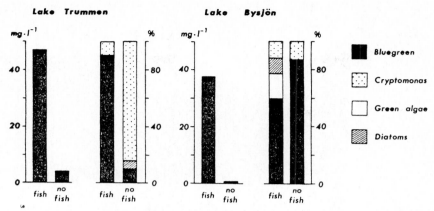

Fig. 4. Biomass (mg fresh weight/litre) and percent composition of phytoplankton in plastic bags with and without fish in two Swedish lakes: Lake Trummen and Lake Bisjön. The presence of fish resulted in an increase in phytoplankton and, in Lake Trummen, dominance by blue-green algae (ref. 26).

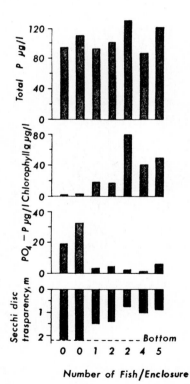

Fig. 5. The effect of increasing fish stock on water transparency, chlorophyll a and phosphate concentrations in plastic bags containing similar concentrations of total P. The evidence for eutrophy is most marked when more fish are present (ref. 10).

which, in the absence or low density of fish, may constitute a dominant factor for phytoplankton control (Fig. 6a, 6b).

Fig. 6a. Effects of planktivorous fish on water transparency measured by Secchi disc in experimental tanks filled with the same water. In the two tanks the Secchi disk is maintained at the same depth. To the left the tank with fish, to the right the tank without fish (ref. 27).

Fig. 6b. Different phytoplankton population densities in the two tanks of the figure 6a - CS2=tanks without fish; HS3=tanks with fish (ref. 27).

More evidence in support of this statement is reported in Figure 7, which refers to the evolution of Daphnia density in Lake Harriet, in relation to total phosphorus and chlorophyll a content. Despite the fact that phosphorus is maintained at an almost constant level in different years, chlorophyll concentration is clearly a function of Daphnia population density (ref. 8).

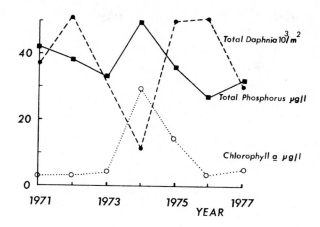

Fig. 7. The effect of Daphnia density on the concentration of total P and chlorophyll a in Lake Harriet 1971 to 1977. Although total P is virtually unchanged, chlorophyll level varies greatly with zooplankton density (ref. 8).

More recently, an analogous phenomenological picture has emerged, demonstrating how the presence of planktivorous fish can produce on the environment a phytoplankton structure very similar to that obtained in environments without zooplankton, and showing also how the latter can even control colonial forms of large size (ref. 28, 29). Figures 8 to 11 and Table 2 report the most salient results obtained with them. In particular, Figures 8 and 9 show a comparison between concentration values of particulate organic carbon (P.O.C.) subdivided in three granulometric fractions (1-25; 1-50; 1-100 μm) in environments with phytoplankton only, with phyto- and zooplankton, and finally with phytoplankton, zooplankton and planktivorous fish. The influence of fish in the sense described before is clear, as it is clear also from Figure 10, in which a relationship is made between the density of the blue-green algae and that of the most important filter feeding zooplankters. It emerges clearly that the latter are capable of controlling the abundance of blue-green algae which, it should be remembered, are those most characteristic of eutrophic environment.

The form assumed by this relationship (Fig. 11) is a clear indication that there are fairly well defined levels of relative density which regulate the interactions between these two groups of organisms.

In reality, the relationships between zooplankton and phytoplankton are much

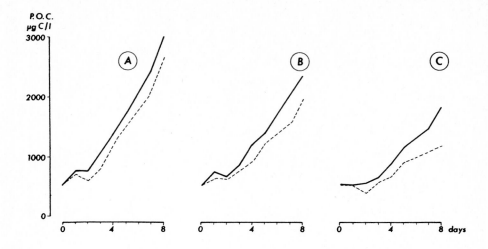

Fig. 8. P.O.C. concentration (µg C/l) in three different granulometric fractions (A = 1-100 µm; B = 1-50 µm; C = 1-25 µm) in enclosures with (- - -) and without (———) Daphnia obtusa (aproximate density 10 ind/l (ref. 29).

TABLE 2. P.O.C. concentration (µg C/l) in three different granulometric fractions in enclosures with and without Daphnia obtusa (ref. 29).

Days	fraction 1-25 µm without Daphnia	with Daphnia	fraction 25-50 µm without Daphnia	with Daphnia	fraction 50-100 µm without Daphnia	with Daphnia
0	537	537	0	0	0	0
1	531	536	241	109	4	83
2	555	403	139	241	80	0
3	655	591	214	199	218	17
4	871	668	331	250	182	264
5	1146	902	242	325	322	248
7	1461	1077	565	544	408	534
8	1815	1189	520	754	682	727

more complex than may be deduced from simple prey-predator relationships, in so far as there is the intervention of problems of size and species specific selection, and this makes it difficult to formulate predictive biological models for the evolution of algal community structure as a consequence of biomanipulati-

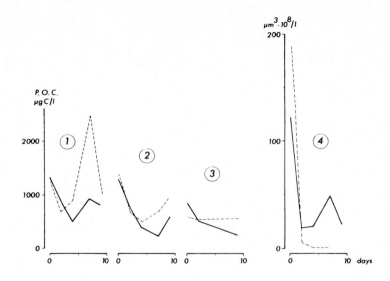

Fig. 9. P.O.C. concentration (µg C/l) in three different granulometric fractions (1 = 1-100 µm; 2 = 1-50 µm; 3 = 1-25 µm) and herbivorous zooplankton biomass ($um^3 \times 10^8$/l) in enclosures with (- - - -) and without (———) planktophagous fish. (ref. 29).

on. For this reason present research is tending towards a clarification of the relevance of this food particle selection, and its consequences for the biocenotic structure of algal communities.

Table 3 refers to a synthesized picture of some pertinent results obtained up to now.

At this point the objection could be made that the results of experiments in biomanipulation, however interesting, are still of a somewhat theoretic, pioneering nature, as they have been obtained in semi-artificial environments. In fact, this is not the case. After their initial phase in semi-artificial environments, these studies are now being transferred to natural environments, and the results so far obtained suggest that the elements acquired are valid also for the real-scale experiments, which highlights the important role that biomanipulation can play in the future recovery of eutrophic lacustrine environments.

To dispel these doubts, it should be sufficient to quote, from among the increasing number of practical real-scale biomanipulation treatments of natural

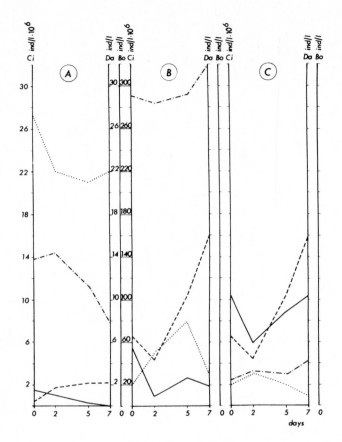

Fig. 10. Temporal variations of blue-green density in enclosures with (---) and without (——) zooplankton, and of the densities of the main filter-feeding zooplankters (Daphnia = ..., Bosmina = .-.-.-). (ref. 29).

environments, those of Andersson's co-workers on some Swedish lakes (ref. 26), or those of Shapiro on American lakes (ref. 8) or, again, to experience acquired by Hrbáček in fish ponds and artificial lakes in Czechoslovakia (ref. 22).

In this connection, the results obtained by Henrikson, Nyman, Oscarson and Stenson (ref. 30) in the Swedish lake Lilla Stockelidsvatten, which has undergone biomanipulation treatment, are very indicative.

Figure 12 illustrates the pluriannual evolution of the average primary production values in this lake after the removal of the fish in 1973. This figure and the next (Fig. 13), illustrating the water transparency values of the

TABLE 3. Patterns of different algal species in enclosures with and without Daphnia obtusa (ref. 29).

Algal species presenting lower density in the bags with Daphnia	Algal species presenting greater density in the bags with Daphnia	Algal species presenting the same density in the bags with and without Daphnia
Anabaena sp. Oscillatoria rubescens Lyngbya limnetica Tot. Cianophyceae Cyclotella comensis Fragilaria crotonensis Tot. Diatomeae Rhodomonas minuta Cryptomonas ovata Tot. Crypthophyceae Scenedesmus sp. Scenedesmus armatus Chlamydomonas sp. Tot. Chlorophyceae	Gemellicystis Oocystis lacustris Micractinium pusillum Mougeotia sp.	Synedra sp. Nitzschia actinastroides Rhodomonas lacustris Tetraëdron minimum Schröderia setigera Pediastrum boryanum

lake before and after the treatment indicate clearly how results obtained in experiments using plastic bags are also repeated on real scale, thus confirming the validity of the plastic bags experiments and, in general, of biomanipulation technique.

I think that the foregoing outline illustrates clearly enough the basic principles and potentiality of these techniques of biological control of the quality of the environment.

However, because the ecological processes involved are complex and delicate, and because the knowledge to be acquired is very far-ranging, these are clearly not techniques that can be applied indiscriminately. Their practical application must be proceeded and constantly accompained by careful ecological research on the structure and function of the specific environments to be treated. In fact, it should be evident that, just as every environment has its own specific structure and function which makes it unique, the biomanipulation treatment which is planned for and applied to it must be equally specific.

This is true even if, among the various advantages which the techniques of food chain biomanipulation for the recovery of polluted aquatic environments

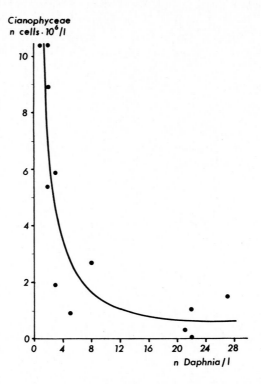

Fig. 11. Experimental relationship between blue-green and <u>Daphnia</u> densities in the different enclosures. (ref. 29).

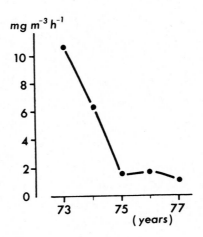

Fig. 12. Lake Lilla Stockelidsvatten: limnetic primary production. Mean mid-day values. The removal of fish by rotenone in November 1973, determine a decrease in primary production. (ref. 30).

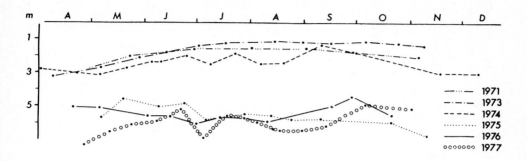

Fig. 13. Lake Lilla Stockelidsvatten: Secchi disc transparency before and after the treatment with rotenone in November 1973. (ref. 30).

offer, there is without any doubt that of being able to reconstruct or, if you like, return to pre-treatment conditions, should the results not be those anticipated, and all this at a relatively modest cost compared to that of other projects.

Still, it must not be thought that biomanipulation is a cure for all the ills afflicting lakes, or that it can always and in any case offer an alternative intervention to those which are now usual. The field of action, the characteristics of environments in which intervention is possible and the already emphasized need for an in-depth ecological knowledge, of necessity limit the field of these techniques. It is however beyond doubt that, once the above conditions are satisfied, the correct application of biomanipulation techniques can form a useful supplement to (and in certain specific cases a substitute) the usual technological interventions for the recovery of the quality of lacustrine waters.

This is particularly important each time the internal loading is sufficient, without further external loading, to maintain the eutrophication level or, again, to speed up the recovery in environments previously subjected to the control of external loading through conventional treatment plants.

REFERENCES

1. R.L. Lindeman, The trophic-dynamic aspect of ecology, Ecology, 23 (1942) 399-418.
2. V.E. Shelford, Animal communities in temperate America, University of Chicago Press, Chicago, 1913.
3. G.E. Hutchinson, The prospect before us, In: Frey D.G. (Editor), Limnology in North America, University of Wisconsin Press, Madison, 1963, 683-690.
4. J. Hrbáček, Competition and predation, relation to species composition in freshwater zooplankton, mainly Cladocera, In: Cairns J.Jr. (Editor), Aquatic Microbial Communities, Garland Publ.Inc., New York, 1977, 305-353.
5. J.L. Brooks, S.I. Dodson, Predation, body size and composition of plankton, Science, 150 (1965) 28-35.
6. J. Hrbáček, B. Desortová, J. Popovsky, Influence of the fishstock on the phosphorus-chlorophyll ratio, Verh.Int.Ver.Limnol., 20 (1978) 1624-1628.
7. P.J. Dillon, F.H. Rigler, The phosphorus-chlorophyll relationship in lakes, Limnol.Oceanogr., 19 (1974) 767-773.
8. J. Shapiro, The need for more biology in lake restoration, Contr. N° 183 of the Limnological Res. Center, University of Minnesota, Minneapolis, Mimeografia, 1978, pp. 20.
9. J. Shapiro, H.O. Pfannkuch, The Minneapolis chain of lakes, A study of urban drainage and its effects, Interim Report n. 9, Limnological Research Center, University of Minnesota, 1973, pp. 250.
10. M. Lynch, Unpublished result, Limnological Research Center, University of Minnesota, In: J. Shapiro, 1978 (1975).
11. S.H. Hurlbert, J. Zedler, D. Fairbanks, Ecosystem alteration by mosquito fish (Gambusia affinis) predation, Science, 175 (1972) 639-641.
12. J. Omer-Cooper, In: Hurlbert, Zedler, Fairbanks, 1972, Proc.Egypt.Acad.Sci. 3 (1948) 1.
13. B.C. Patten, Need for an ecosystem perspective in eutrophication modeling, In: E.J. Middlebrooks, D.H. Falkenborg, T.E. Maloney (Editors), Modeling the eutrophication process, Proc.Workshop, Utah State University, Logan, Utah, 1973, 83-87.
14. T.M. Zaret, A predation model of zooplankton community structure, Verh.Int.Ver.Limnol., 20 (1978) 2496-2500.
15. Z.M. Gliwicz, Studies on the feeding of pelagic zooplankton in lakes of varying trophy, Ekologia Polska, Ser. A, 17, 36 (1969) 663-708.
16. R. de Bernardi, Biotic interactions in freshwater and effects on community structure, Boll.Zool., 48 (1981) 353-371.
17. J.R. Gammon, A.D. Hasler, Predation by introduced muskellunge on perch and bass, I. Year 1-5, Wis.Acad.Sci.Arts.Lett., 54 (1965) 249-272.
18. W.R. Schmitz, R.E. Hetfeld, Predation by introduced muskellunge on perch and bass, II. Years 8-9, Wis.Acad.Sci.Arts.Lett., 54 (1965) 273-282.
19. J. Shapiro, V. Lamarra, M. Lynch, Biomanipulation - an ecosystem approach to lake restoration, Contribution n° 143, Limnol.Res.Center, Univ.Minnesota Mimeografia, 1975, pp. 32.
20. H.F. Van Emden, Pest control and its ecology, Studies in Biology n° 50 Edward Arnold Publ.Ltd.London, 1974, pp. 60.
21. J. Hrbáček, Typologie und Productivität der teichartigen Gewässer, Verh.Int.Ver.Limnol., 13 (1958) 394-399.

22 J. Hrbáček, M. Dvoraková, V. Korínek, L. Prochazková, Demonstration of the effect of the fish stock on the species composition of zooplankton and the intensity of metabolism of whole plankton association, Verh.Int.Ver.Limnol., 14 (1961) 192-195.
23 R. de Bernardi, G. Giussani, The effect of mass fish mortality on zooplankton structure and dynamics in a small italian lake Lago di Annone, Verh.Int.Ver.Limnol., 20 (1978) 1045-1048.
24 D.W. Schindler, G.W. Comita, The dependance of primary production upon physical and chemical factors in a small, senescing lake, including the effects of complete winter oxygen depletion, Arch.Hydrobiol., 69 (1972) 413-451.
25 M.G. Galbraith, Size-selective predation on Daphnia by Rainbow Trout and Yellow Perch, Trans.Amer.Fish.Soc., 96 (1967) 1-10.
26 G. Andersson, H. Berggren, G. Conberg, C. Celin, Effects of planktivorous and benthivorous fish on organism and water chemistry in eutrophic lakes, Hydrobiologia, 59 (1978) 9-15.
27 R. de Bernardi, Verso un approccio ecologico integrato per il controllo delle acque: la biomanipolazione di catene alimentari, Acqua e Aria, 10 (1983) 1075-1084.
28 R. de Bernardi, G. Giussani, E. Lasso Pedretti, Relazioni alimentari fitoplancton-zooplancton in esperimenti di biomanipolazione, In: R. Bertoni e R. de Bernardi (Editors): "Atti V° Congresso Associazione Italiana di Oceanologia e Limnologia, Stresa 19-22 Maggio 1982" (1982a) (in stampa).
29 R. de Bernardi, G. Giussani, E. Lasso Pedretti, Selective feeding of zooplankton with special reference to blue-green algae in enclosure experiments, Paper presented at: "Plankton Ecology Group Annual Meeting, Trondheim, Norway 23-28 August 1982" (1982 b).
30 L. Henrikson, H.G. Nyman, H.G. Oscarson, J.A.E. Stenson, Trophic changes without changes in the external nutrient loading, Hydrobiologia, 68 (1980) 257-263.

LAKE ECOSYSTEM DEGRADATION AND RECOVERY STUDIED BY THE ENCLOSURE METHOD

O. RAVERA
Department of Physical and Natural Sciences, Commission of the European Communities, JRC - 21020 Ispra (VA)

1. INTRODUCTION

The aim of this paper is to discuss the degradation of the aquatic ecosystem due to chemical pollutants, and its possible recovery studied by the enclosure method. Because there are variuos attitudes towards the degradation and recovery of the ecosystems, it seems opportune for me to express my opinion on some aspects of these problems, before discussing the specific subject of this paper.

An ecosystem consists of a community living in a given physical environment. The community is composed of the plant and animal species-populations. The physical environment is composed of a solid phase (soil or rocks in terrestrial ecosystems and sediment in aquatic one) and a fluid phase (air in terrestrial ecosystems and water in aquatic one).

This description, which may be criticized for being excessively superficial, seems to be more practical for the aims of the paper than others which are more complete and problematic "(refs. 1-2)".

The most important characteristics of the ecosystem are the interrelationships between biota and physical environment, the interrelationships among the species populations of the community and the continuos modifications of the biotic and abiotic compartments. Because of these variations, the interrelationships among the species populations and those between biota and physical environment cannot be stable. This is the basis of the ecosystem's evolution.

On the other hand, the homeostatic mechanisms tend to keep the general structure and the functions of the ecosystem stable showing its evolution-rate. The opposing tendencies to change the ecosystem and to keep it stable may be compared to the transmission of the genetic characteristics and to the mutation-rate in a species-population.

Each ecosystem has a natural evolution trend and its rate may be more or less rapid. The more stable ecosystems, if considered on a geological scale, also evolve; for example, the salinity of the sea increases with time and its community today is very different from that of the mesozoic era. There are various natural stresses which may modify the evolution of an ecosystem. For example, in a few days a landslide may transform a river into a lake and an eruption of lava changes a wooded area into a desert in a few hours. Some ecosystems have cyclic variations; for example, temporary pools, during the dry season, become dry areas.

In some centuries or in several thousands of years, depending on the morphometrical and hydrological characteristics of the basin, a lake passes from an oligotrophic stage to a eutrophic stage through a mesotrophic. This natural evolution may be greatly accelerated by human activities, and a medium size lake may be eutrophicated in a few years. Man may vary the evolution trend of the ecosystem, prevent its evolution or produce a regression in its natural succession. For example, man may hinder a valley with a dam and transform it into an artificial lake (impoundment), prevent the succession from a prairie to a forest by using the grass land for pastoral activity or clear a woodland to cultivate crops, that is he may produce a regression from the climax to one of the first stages of the succession. The pollution load may greatly modify the ecosystem, reducing its diversity and productivity, replacing some species with others and altering the community structure and functions.

In conclusion the ecosystem may vary because of natural stresses as well because of the influence of man. The variations of the ecosystem may be easily detected if it is studied for several years, but it is not always easy to separate the influence exerted by man from the effects produced by natural causes. In addition, the concept of ecosystem degradation is often subjective. In fact, the inhabitants of a very polluted area will judge as satisfactory the conditions of a lake (or a forest) which might be considered degraded in a region with a low pollution level. Pollution is only one of the causes of ecosystem modification; other causes are: elimination of animals and plant species, transfer of species from one continent to another, simplification of communities and alteration of the hydrology and geomorphology of an area.

A degraded ecosystem is a system out of place, that is its processes are not balanced; for example, in an eutrophic lake the production rate of organic matter greatly exceeds its mineralization rate; as a consequence, a great amount of organic material is accumulated in the sediments producing oxygen depletion in the deeper waters. Primary production of a water body treated with algicides decreases to a very low level or is abolished with a consequent decrease in abundance of the consumers (herbivors and carnivors).

It is rather difficult to evaluate the degradation level of an ecosystem, except for those very polluted. For example, can a lake eutrophicated (or acidified) by natural causes be considered degraded? The degree of alteration is often established according to the use of the ecosystem resources. This pragmatic concept cannot be accepted from a scientific point of view; for example, the same eutrophic waterbody is considered degraded if it is used for drinking water, but in good condition if the most important use of its water is the irrigation of crops. In addition, the actual degradation level of an ecosystem cannot be judged on the basis of legislation. The legislation applied to protect the environment from chemical pollution varies from one Country to another and is frequently based on general information concerning the toxicity of a series of pollutants tested on a few species kept in laboratory conditions. In addition, the legislation must be applied to all the ecosystems of the same type (for example, lake, river) and, commonly, considerating one compartment of the ecosystem (for example, air, water) and never the global ecosystem and its "personality".

Legislation frequently, concerns only the pollutant concentrations in the effluents and not in the water body receiving them. These criteria, which may be critizised from the ecological viewpoint, are necessary, because legislation must be easily understandable and simple and economic in application and must recommend methods which may produce results in a short time. From the biological control methods commonly used (for example, biotic indices, saprobic classes) a practical and rough idea of the degradation level of a water body (generally lotic ecosystems polluted by non-toxic organic substances) may be obtained. Several approaches have been applied to classify the trophic degree of lake ecosystems such as the classification developed by the OECD Eutrophication Programme "(ref. 3)" and the "Trophic State Indices" proposed by Carlson "(ref. 4)". These approaches are very useful, but corcern only some parameters: for example, water transparency, chlorophyll-a and total phosphorus concentration. The approach by Vollenweider dealing with the probability distribution for trophic categories seems to be the most realistic because it shows the uncertainty in classifying the trophic level of a water body "(ref. 3)".

It is well-known that the toxicity of several pollutants varies in relation to their physico-chemical form and that the form may be modified by the characteristics of the physical environment (for example, pH, alkalinity, concentration and nature of organic and mineral chelating substances and concentration of particulate matter). Some pollutants (e.g. oil, acids) discharged into the aquatic environment in great amounts, may modify its physical and chemical characteristics. The strong acid load decreases the alkalinity and pH values and modified the transfer of several elements from the sediments to the water, producing alterations in animal and plant populations. The biota react to the combined influence of the pollutant and the physical environment, and the latter, in turn, is influenced by the modified biota. Micropollutants (e.g. biocides, heavy metals) are generally present in the aquatic environments in concentrations so low that they have a negligible influence on the general characteristics of the water. On the other hand, these low concentrations may be sufficiently high to alter the structure, biomass and production rate of the community; these biological modifications may have an important influence on the physical environment. For example, the decreased photosynthetic-rate, due to the pollutant, produces a lowering of the pH value, an increase of the nutrients not used by the phytoplankton and an increase of the water trasparency consequence of the decreased algal biomass.

In conclusion, some pollutants (e.g. oil, nutrients) may alter the physical environment and, consequently, the biota; other pollutants (e.g. heavy metals) may influence the biota, which in turn, alters the physical environment. From these considerations the interrelations between pollutant, biota and physical environment are evident and, then, studies on polluted water bodies must always consider these three variables.

Because of the great variability of the ecosystems and the various trends of their evolution, there is great uncertainty in defining the concept of ecosystem recovery. The recovery must be considered in relation to a given ecosystem model, but the choice of this model is very subjective. It might be considered that to restore a lake means to modify its present characteristics to obtain an ecosystem identical to that existing before any human influence. Because the ecosystem continuously evolves, it is very difficult to establish what stage of evolution the lake must reach. In addition, we know very little about the natural pre-impact state of most lakes. Generally, a very clean water body with high transparency is considered the natural state of the lake pre-human impact. This concept is rather utopistic because even lakes uncontamined by man may be, according to their stage of evolution, mesotrophic or eutrophic (e.g. Lake George, Tanzania). Indeed, to transform a naturally eutrophic lake into an oligotrophic one, does not mean to recover it, but to obtain an artificial regression of the lake's natural evolution. The recommendation by the OECD "Eutrophic Programme" to recover the eutrophic lakes at least to the mesotrophic state seems to be more realistic "(ref. 3)". This criterion has been developed overall for economic and technological reasons with the two-fold aim of ameliorating the lake ecosystem altered by man and of increasing the water resources available for various uses. Generally, the recovery of a lake means to reduce man's noxious influences to minimun to improve its general condition. To this aim the causes of lake degradation must be identified and the importance of the anthropogenic impact estimated. Two important concepts related to lake recovery must be considered:

- except for a very few cases, a degraded lake ecosystem spontaneously recovers when the causes of its alteration have been removed. Because the recovery rate is generally rather slow, man's interventions aim to accelerate this natural process.
- the complete recovery of most polluted lakes is impossible, because some species have been eliminated from the community and, if introduced, they need environmental conditions similar to those existing before the lake was polluted. This is not always easily obtained.

An aquatic ecosystem may receive a certain amount of noxious substances (toxics and nutrients) without showing evident effects. This occurs because the load of pollutants is counterbalanced by the mechanisms which eliminate its effects (e.g. through the lake outflow) or neutralize them (for example, by detoxification processes). Above a certain loading of pollutant (depending on the dilution capacity of the ecosystem and the persistance of the pollutant) the first effects are evident. In time, if the pollutant load is permanent, there is a progressive accumulation of the pollutant in the medium (water and sediment), the more sensitive species decrease or are eliminated while the more resistant survive and, generally, increase in abundance. At this stage the ecosystem is completely modified and it attains a new equilibrium, i.e. the resources are used by other species-populations which were scarce or absent in the ecosystem before the pollution. If the toxic load increases above a certain level, the ecosystem collapses and may be considered completely degraded.

From these considerations it seems reasonable that degraded lakes should be recovered to the stage at which no relevant effect produced by pollution is evident, although a certain amount of pollutant might be discharged into the water body.

2. ENCLOSURE TECHNIQUE

The problem of the degradation of an aquatic ecosystem and its recovery is extremely complex. Consequently, from field studies it is practically impossible to evaluate the actual effects produced by the pollutants and to establish a relationship between the concentration of each pollutant in the medium and the effects at population and community level.

Information about the time when the pollution started and the variations of the pollutant loadings during the period preceeding the study is generally scarce. To evaluate pollution damage by comparing a polluted ecosystem with others which are not polluted may be hazardous if the pollution level is not very high. Laboratory experiments are always very simple as compared to the natural ecosystem; in addition, the most important parameters are kept constant, while the most evident characteristic of the natural environment is their variations.

As a consequence, with these experiments the potential noxiousness of pollutants may be accurately evaluated, but the results obtained may be extrapolated to the natural ecosystem with great difficulty.

To acquire better knowledge on pollution effects at community and ecosystem levels, during the last few years, investigations have been carried out with the enclosure method, which represents a compromise between field study and laboratory experiment.

With this method conclusions on the actual effects of a given pollutant may be drawn by replicated experiments carried out in the same ecosystem in different seasons. In addition, these conclusions may be generalized by means of experiments replicated in the same seasons in water bodies with various characteristics.

This technique consists in isolating a part representative of the ecosys-

survival and reproduction of part of the populations due to selective and adaptive mechanisms and the increase of some resistant species-populations favoured by the decrease of predators and competitors less resistant to the pollutant. In a few experiments the pollutant concentration in the enclosed water column is kept constant. Generally, at the beginning of the experiment (or after a few days to permit the adaptation of the enclosed system) a known amount of pollutant is added to the enclosure to obtain a given concentration in the water column. During the experiment the wall of the enclosure, the periphyton growing on it, the seston and the sediments (if these are enclosed with the water column) adsorb a significant percentage of the added pollutant. As a result, pollutant concentration in the water column decreases in time and, consequently, the medium becomes less and less toxic for the biota. The decrease of the pollutant concentration is, obviously, more rapid for the easily degradable substances (e.g. some biocides) and those with a high vapour pressure (e.g. some hydrocarbons). One of the exceptions to the decrease of the pollutant concentration is the continuous flow experiment (Melimex) carried out in eutrophic Lake Baldegg (Switzerland) on the effects of a mixture of metals (Hg, Cd, Cu, Zu and Pb). The treated enclosures received a continuous metal loading by a pump system until to have concentrations equal to the legally tolerant limits "(ref. 30)". Thomas et al. "(ref. 31)" maintained the copper concentration of 10 μg/l constant in an enclosed water column for 28 days by periodic analysis and copper addition.

A clear example of adaptation to the pollutant has been reported by Kuiper and Hanstveit "(ref. 22)". On the 5th day from the beginning of an experiment, a single dose of 4-Chlorophenol (4CP) was added to an enclosure; a second dose of the same coumpound was added on the 32nd day. After the first addition, the degradation rate of the 4CP was extremely slow, but after the second the pollutant decreased in the water column very rapidly indicating the presence of an adapted bacterial flora. This conclusion was confirmed by laboratory experiment. Azam et al. "(ref. 32)" calculated that the heterotrophic activity of the bacterial populations, in enclosures contamined with 1 μg Hg/l and 5 μg Hg/l respectively, decreased to 1% of the activity measured in the control. Because this percentage did not decrease after further additions of mercury, the authors supposed that this was due to the selection of mercury-resistant strains. Accurate experiments confirmed this hypothesis and demonstrated that the increased chelating capacity of seawater was not the cause of the mercury tolerance of the bacteria. As a consequence, the authors concluded that mercury-resistant bacteria may transform the ionic mercury into organic form and then modify the toxicity of this metal to other taxa of the community.

The pollutant modifies the community structure reducing the abundance of some more sensitive species-populations; this reduction may favour their preys and competitors and depress their predators. For example, Kuiper and Hanstveit "(ref. 22)" observed an increase of phytoplankton population density in enclosures contaminated with 1mg/l of 2,4 -Dichlorophenol (DCP). This increase was not due to the stimulating effect of the toxic substance, but to that of the lower grazing pressure by the depressed copepod population, more sensitive to DCP than phytoplankton.

Azam et al. "(ref. 32)" carried out experiments with enclosures to study the effects of low concentrations of mercury (1 and 5 μg/l) on marine bacterial populations. After 19 hours from the contamination a sharp decrease of the bacterial biomass and a reduction of the heterotrophic activity of about 99% were observed. Because the reduction of the biomass was of one order of magnitude, whereas the inhibition of the heterotrophic activity and the viable bacterial counts were of 2 and 3 orders of magnitude, it was probable that the mercury caused a bacteriostatic effect and not a bactericidal one. The recovery of the biomass occurred within 5 to 7 days and that of the heterothophic activity within 4 to 11 days. Consequently, the effects of one dose of mercury on bacterial populations seems to be transient. Gächter and Máreš "(ref. 30)" from enclosure experiments carried out on the effects of metals on phytoplankton observed an initial decrease of the phytoplankton biomass, species number and photosynthetic rate. The community structure was modified and the more resistant species were favoured. At the end of the experiment, higher population density were measured in the treated enclosures, than in the control.

Initial inhibition of phytoplankton biomass and production followed by their recovery was observed by Thomas et al. "(ref. 31)" in enclosed water columns contaminated with copper (10 and 50 μg/l). At the end of the experiment the values of the biomass and production were similar to those calculated for the control. In a second experiment, copper (5 and 10 μg/l) seemed to favour the increase of the phytoplankton population density, but not its production rate. According to the authors during this experiment the microflagellates, resistant to copper, were very abundant. The higher values of the phytoplankton biomass in the treated enclosure, when compared with those of the control, are probably the effect of the zooplankton inhibition and then of the less severe grazing.

The recovery rate varies with the taxon in relation to the seriousness of the damage suffered by it and its reproductive rate. As a consequence, a modified community may be expected even after the pollutant disappears from the medium. Similar results may be obtained when noxious substances (e.g. oil, nutrients) exert a greater influence on the characteristics of the physical environment than on the species, because these react to the modified medium in various ways according to their own sensitivity to the environmental factors.

Sensitivity to a pollutant varies with the species, the physiological and development stage and the physical environment characteristics. For example, if the pollutant is added to the enclosure when most of the population is in the most sensitive development (or physiological) stage, the population density will show a dramatic decrease. The same effect will be obtained if the pollutant is added when the conditions of the physical environment are unfavourable to the species.

Therefore, more general conclusions may be drawn from experiments duplicated in different seasons and in various water bodies because the variations of the physical and chemical characteristics as well as those of the community structure are taken in account.

If the physical and chemical factors of the environment influence the community, this in turn, controls these factors. For example, the oxygen concentration and pH value decrease as a consequence of the phytoplankton depression and the recovery of the latter increases the values of oxygen as well as of pH. A reduction of phytoplankton grazing, produced, for example, by a biocide which does not influence algal populations but only their feeders, enhances phytoplankton growth which decreases the water transparency and nutrient concentration in the trophogenic layer. Consequently, at least at first, the recovered community grows in physical environment rather different from that existing before pollution. These are the reason why it is relatively easy to predict the recovery of a community exposed to moderate concentrations of toxic substances, but it is quite impossible to imagine the characteristics acquired by the recovered community and then by its ecosystem. An example of the complex relationships between pollutant, physical factors and biota was by the results obtained by Lang and Lang-Dobler "(ref. 33)" from enclosure experiments on the effects produced in Tubificid and Chironomid populations by a continuous pollution of a mixture of five heavy metals. Heavy metals significantly reduced the phytoplankton production rate; this depression with the consequent decrease of the sedimentation rate of organic matter, improved the oxygenation at the sediment-water interface. As a consequence of the higher oxygen concentration, an increase of the benthonic population densities in the treated enclosure was observed.

It is quite easy to explain the biomass and production recovery, because they are supported by the nutrient concentrations which are of the same order of magnitude in both the treated enclosure and in the control. In a community where the structure is modified by the pollutant, the nutrients should be utilized by species different from those of the control, but the amount of material and energy utilized should be similar, because the recovered community, generally, attains a biomass value not significantly different from that of the control. In conclusion, after a certain time from the pollutant addition, the biomass and production may not be different from those of the control, although the community structure is modified "(refs. 30-31)". As a consequence, the recovery of a community does not consist in the growth of a community identical to the original one, but in a community quantitatively similar, but qualitatively also very different from that existing before the pollution. This means that in similar physical environments different communities may give rise to functioning ecosystems.

5. REFERENCES

1. O. Ravera, Contributi Centro Linceo Interdisciplinare di Sc. Mat. e loro Applicazioni, 51 (1980) 101-159.
2. O. Ravera, in J.H. Cooley and F.B. Golley (Editors) Trend in Ecological Research for the 1980s, Plenum Pyblishing Corporation, 1984, pp. 145-162.
3. R.A. Vollenweider and J.J. Kerekes, in Restoration of Lakes and Inland Waters, EPA 440/5-81-010 (1981), pp. 25-36.
4. R.E. Carlson, Limnol. Oceanogr., 22 (1977) 361-369.
5. K.R. Solomon, T.Y. Yoo, D. Lian, N.K. Kaushik, K.E. Day and G.L. Stephenson, Can. J. Fish. Aquat. Sci., 42 (1985) 70-78.
6. C.D. McAllister, T.R. Parson, K. Stephens and J.D.H. Strickland, Limnol. Oceanogr. 6 (1961) 237-259.
7. D.W. Blinn, T. Tompkins and L. Zaleski, J. Phycol., 13 (1977) 38-61.
8. O. Ravera, in A. Boudou and F. Ribeyre (Editors) Aquatic Ecotoxicology Fun-

damental Concepts and Methodologies, CRC Press, Boca Raton, USA (in press).
9 R. Gächter, Schweiz Z. Hydrol., 41 (Sep. No. 770) (1979) 169-176.
10 B. Zeitzschel, Oceanology Int. 78, Tech. Sess. B (Biol. Mar. Tech.), 1978, pp. 22-52.
11 J.H. Steele and D. Menzel, Rapp. P.V. Réun. Comm. Perm. Int. Explor. Mer., (1978).
12 J.M. Davies and J.C. Gamble, Phil. Trans. R. Soc. London B, 286 (1979) 523-544.
13 S. Draggan, The Microcosm: Biologiçal Model of the Ecosystem, Techn. Report Monitoring and Assessment Research Centre, University of London, 1980.
14 J.W.G. Lund, Verh. Internat. Verein. Limnol., 18 (1972) 71-77.
15 C.R. Goldman, Limnol. Oceanogr., 7 (1962) 99-101.
16 J.C. Lacaze, Téthys, 3 (1971) 705-716.
17 J. Shapiro, Science, 179 (1973) 382-384.
18 J.J. Lack and J.W.G. Lund, Freshwater Biol., 4 (1974) 399-415.
19 J.C. Gamble, J.M. Davies and J.H. Steele, Bull. Marine Soc., 27 (1977) 146-175.
20 O. Ravera and D. Annoni, in Atti 3° Congresso Associazione Italiana Oceanologia e Limnologia, Sorrento 18-20 Dicembre 1978, 1980, pp. 417-421.
21 D.J. McQueen and D.R.S. Lean, Water Res., 17 (1983) 1781-1790.
22 J. Kuiper and A.O. Hanstveit, Ecotoxicology and Environ. Safety, 8 (1984) 15-33.
23 U. Uehlinger, P. Bossard, J. Bloesch, H.R. Bürge and H. Bührer, Verh. Internat. Verein. Limnol., 22 (1984) 163-171.
24 A. Salki, M. Turner, K. Patalas, J. Rudd and D. Findlay, Con. J. Fish. Aquatic Sc., 42 (1985) 1132-1143.
25 O. Ravera, N. Riccardi and P. Sechi, in Atti 3° Congresso Società Italiana Ecologia, Siena, Ottobre 1987, (in press).
26 P.H. Kerrison, A.R. Sprocati, O. Ravera and L. Amantini, Envir. Technol. Letters, 1 (1980) 169-176.
27 S. Zarini, D. Annoni and O. Ravera, Envir. Technol. Letters, 4 (1983) 247-256.
28 D. Annoni and O. Ravera, in A. Moroni, A. Anelli and O. Ravera (Editors), S.I.T.E. Atti 5, Edizioni Zara, Parma, 1985, pp. 567-572.
29 C.E. Shannon and W. Weaver, The Mathematical Theory of Communication, University of Illinois Press, Urbana, 1963.
30 R. Gächter and A. Máreš, Schweiz Z. Hydrol., 41 (Sep. No. 770) (1979) 228-246.
31 W.H. Thomas, O. Holm-Hansen, D.L.R. Seibert, F. Azam, R. Hodson and M. Takahashi, Bull. Marine Sc., 27 (1977) 34-43.
32 F. Azam, R.F. Vaccaro, P.A. Gillespie, E.I. Moussalli and R.E. Hodson, Marine Sc. Communications, 3 (1977) 313-329.
33 C. Lang and B. Lang-Dobler, Schweiz Z. Hydrol., 41 (Sep. No. 770) 271-276.

THE IMPACT OF HEAVY METALS ON TERRESTRIAL ECOSYSTEMS: BIOLOGICAL ADAPTATION THROUGH BEHAVIOURAL AND PHYSIOLOGICAL AVOIDANCE

H. EIJSACKERS

1. INTRODUCTION

Relative to air and water pollution, soil pollution has been largely underestimated for a long period. The last fifteen years much has come to light, however. On many locations large quantities of waste, partly synthetic organic compounds, appeared to have been dumped. At the moment there are some 7500 recorded sites with polluted soil in The Netherlands. In addition, one has come aware that many substances like heavy metals are non-biodegradable and are leachable only to a limited extent, so that adverse effects may occur in the long term. This has been recognized by the authorities, which has resulted in legislation on soil contamination and sanitation and on manure applications. Much of the required knowledge, however, is still lacking. Therefore various agencies have started research programmes, finally converging on a "Netherlands integrated soil research program", in which 54.000.000 Dutch guilders will be invested in the next four years.

2. RESEARCH REQUIRED

Four main aspects are involved in the soil biological research about pollution and decontamination, viz. original or reference values, biotic and abiotic interactions, dose-effect relations, and methods for assessment and for establishing limits (Table 1). These aspects are further elaborated by Eijsackers (ref. 1), it should be made clear here which basal processes are involved.

In modern society the soil is used for many purposes. It carries society including the plants and animals on which people live. The soil provides useful substances including nutrients for plants and animals and it provides a filter for waste substances. Most of these functions are only possible as a result of natural breakdown processes.

For a continued performance of this variety of functions the mechanisms of the related processes have to be known, including the minimum conditions required. If the constituent processes are known, as well as the organisms involved and the various interactions, it will be possible to achieve optimum utilization of the complete process.

TABLE 1. Soil biology research into soil pollution and sanitation.

Areas of research	Abiotic research	Biotic research
Reference studies	-pedological -physical -chemical	-numbers and activities of soil organisms
Dynamic aspects	-availability of substances	-bio-uptake of substances -fluctuations in numbers and activities
Dose-effect relations	-interactions of physical and chemical processes and of various substances	-parameter level (individual, population, ecosystem) -recovery (recolonization, adaptation)
Models for fixing limits	-substance levels based on exchange reactions	-effect levels based on natural fluctuations -shelter function based on geometric distribution

For this purpose not only the process under abnormal conditions shall be examined, but also the extent of influences, both from natural and unnatural situations. In this way it can be established how sensitive soil types are to various forms of disturbance and pollution.

Disturbance and pollution will bring about changes in a community. Some species can adapt more readily than others. The pollutant then acts as selecting factor. Fundamental questions in the research will be:

1. To what extent can populations resist certain selective effects, such as pollutants, and how strong should selective effects be to bring about changes in the population?
2. How persistent are such changes, especially in cases in which the selective effects are only temporarily active?
3. At the expense of what are such changes (adaptations) effectuated, and what other adaptations are (temporarily) lost?
4. Which strategies are pursued to obtain a necessary adaptation with the lowest possible number of other adaptations being lost?

The resistance referred to under 1. has been described by Orians (ref. 2) and others for ecosystems as "inertia", whereas the (speedy) recovery after a disturbance towards the original situation is called "elasticity". In a similar way it can be investigated how strong the numerical fluctuations of populations are under normal conditions as well as natural stress and pollution stress. This will yield results as to the degree of inhibition (decrease in numbers of organisms or their activities), and to the rate at which the original situation recovers. After a disturbance due to felling, which can more or less be compared with a natural disturbance due to storm, the soil fauna failed to regain its original composition, even after fifteen years (ref. 3). Controlled fires to accomplish silvicultural objectives in woodland of North America, however, appeared to be possible every four or five years without permanent changes in the soil fauna being caused (ref. 4).

An important item in this recovery is recolonization from intact areas. The legislation on soil protection in The Netherlands also considers 'soil protection areas' to be important recolonization centres from where recovery of affected areas can occur. Den Boer (ref. 5, 6) performed research with groundbeetles as to the effects of a heterogeneous environment (suited spots within a much less suited area) and a heterogeneous population on the stability of this population in the total area. Via the spreading of risk, heterogeneity would contribute to the unchanged composition of the population, provided sufficient possibilities remain for exchanges.

On the question of at what cost an adaptation is achieved, the Department of Ecology and Ecotoxicology of the Free University of Amsterdam carries out research in isopods which are exposed to the effects of heavy metals. Isopods that were better adapted to heavy metals, have appeared to be less adapted to desiccation (Van Cappelleveen, oral comm.).

It has not yet been examined what strategies are pursued to achieve an adaptation. On the basis of differences in survival strategies it can be studied how different groups react to stress due to, for instance, heavy metals. For earthworms Satchell (ref. 7) made a classification into r and K survival strategies. Roughly speaking, for the survival of the species emphasis is put on the production of a numerous offspring (r-strategy) or on the survival of the adults (K-strategy). Ma (ref. 8) performed experiments into Cu sensitivity in six species of worms. These provide a first indication of the ED-50 value (the Dose at which 50% of the worms investigated show the Effect searched for) for the cocoon production (an r-property) with r worms being slightly higher. Consequently, the reproduction of this type of earthworm might not be adversely affected immediately. With the K worms the ED-50 for the weight loss is slightly higher. This could imply that the condition of adults of these species deteriorates less in case of Cu stress (Table 2).

TABLE 2. ED-50 for Cu with six earthworm spicies (in brackets: 95% confidence interval (Ma 1983).

r/K*	Species	Cocoon prod.	Mortality	Weight changes
K	A. longa	38(10-147)	>300	395(178-878)
K	A. caliginosa	27(17- 45)	>300	405(213-769)
K/r	A. chlorotica	28(21- 38)	127(90-180)	121(76-194)
r/K	L. rubellus	80(44-147)	205(151-278)	240(171-337)
K/r	L. terrestris		130-200	139(110-175)
r	E. foetida		231(153-348)	285(201-403)

*distinctions based on Satchell (ref. 7).

3. DYNAMIC PROCESSES

Soil biological changes take much time, and abiotic conditions in the soil fluctuate little. Nevertheless, the processes in the soil constitute a very dynamic complex, in which the biological availability of substances is a major aspect. This is determined by the chemical equilibrium between the labile and stabile forms of a chemical compound and by the biological absorption of the substance by soil organisms. As a result of the absorption into soil organisms the quantity of a labile compound diminishes, thus shifting the chemical equilibrium. Because of these dynamic equilibria it is not the total quantity of a substance that conditions the strength of the effect of the substance, but the quantity that is biologically available. The present approach of establishing the biologically available amount of a substance is based on extraction with a range of chemical solvents of increasing strength. This does not render, however, a satisfactory understanding of how the above processes regulate the availability. For some time toxicologists have been working on quantitative structure activity relations (QSAR's) by which a range of related substances are compared to discover which physical and chemical properties ("structure") of these substances correlate with the effects found ("activity"). These relations have hardly been investigated for soil biological mechanisms so far; Ma (ref. 9) did some research on the impact of soil pH, CEC and organic matter content onto heavy metal accumulation in earthworms.

A second dynamic aspect is that of numerical and activity fluctuations in the soil fauna. These fluctuations need be established to determine whether a low number or a reduced activity of a certain species is still within the normal fluctuation pattern. Domsch et al. (ref. 10) studied this aspect for micro-organisms, whereas Eijsackers et al. (ref. 11) did so for the soil fauna (Fig. 1).

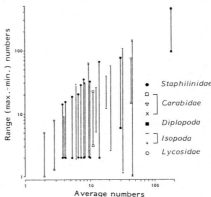

Fig. 1. Seasonal amplitudes (maximum - minimum) of various species of surface fauna related to the average population level (composed on the basis of various references).

In all cases there appears to be a more or less fixed relationship between the average level and the range of lowest and highest numbers. So, for each species with known average level under undisturbed conditions, it can be found what minimum numbers are within the natural fluctuation limits and what are not. If they are not, there has been a disturbance. As a second criterium the ratio between the amplitude of fluctuations and the duration of the fluctuation with regard to the average population level could be examined.

4. DOSE-EFFECT RELATIONS

The Department of Environmental Pollution of the Research Institute for Nature Management has been performing much research on dose-effect relations. The effects of heavy metals have been studied, both in micro-organisms, soil arthropods, and earthworms. In many cases it appeared that the dose-effects relations can be described as logarithmic relations between dose and effect. The dimension of the effect is then established as the dose at which the effect is reduced by 50% with regard to the reference value: ED-50. Although this dose is the most suited from the statistical point of view (minimal variation), it is questionable whether it is the most relevant one from the ecological viewpoint (Figs. 2 and 3). With doses that are too weak to show the effects as measured, it may be expected that several sub-processes are inhibited indeed, which play a role in the effect process. In case of microbial respiration a number of sensitive species will be inhibited at a dose at which resistant species are fully active still, so that the total volume of the respiration is not affected (see also Figure 6). Therefore, it should be better to use the ED-10 value or the No Effect Level.

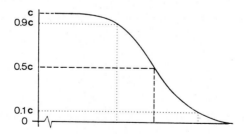

Fig. 2. Relation between dose (Cu) and effect (as percentage of control) (ref. 12).

Figure 3 shows, moreover, clearly the varied responses (with a short and a long lag-phase and a gradual or steep decline) of micro-organisms to copper and zinc in various soil types. This stresses the importance of bio availability of heavy metals in relation to physico-chemical characteristics of the soils.

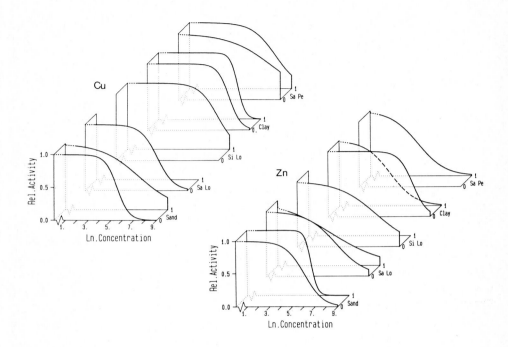

Fig. 3. Dose-effect relations of urease activity in five soil types (sand, sandy loam, silty loam, clay, sandy peat) contaminated with copper (Cu) and zinc (Zn) and measured immediately after contamination (front curve) and one year later (rear curve). The urease level in non-contaminated soil was put at 1.0 (ref. 12).

A basal point to be noted refers to the question at what level of biological organization the parameters have to be selected. This can range from the level of the cell to that of the ecosystem. Political considerations mostly refer to different areas, and here parameters of the highest level seem to be the most relevant (ecosystem/biotic community). For protected species, it is the level of biotic community/population, whereas for explanatory research the population, organism, and cell levels seem to be the most appropriate. For ecosystems the species diversity is often used as a measure to indicate the effect of pollution, especially in water (ref. 13), and different sets of diversity and evenness indices have been developed (Fig. 4). Yet this measure has to be applied with utmost care.

diversity indices	Evenness indices	alternative probability indices	similarity indices
$S \approx \alpha \ln N/\alpha$ $D = (S-1)/\ln N$ $SI = \dfrac{\Sigma n_i(n_i-1)}{N(N-1)}$ $H' = -\Sigma p_i \log_2 p_i$ $MI = \sqrt{\left(\sum\limits_{i=1}^{S} n_i^2\right)}$ $SCI = R/N'$ $TU = 1 - \left(\dfrac{N'}{N'-1}\right)\left(\sum\limits_{i=1}^{k} p_i^2 - \dfrac{1}{N'}\right)$	$\epsilon = \dfrac{S}{S'}$ $J = \dfrac{H'}{\log_2 S}$ $E = \dfrac{eH'}{S}$ $E = \dfrac{eH-1}{S-1}$	$PIE = \left[\dfrac{N}{N-1}\right]\left[1 - \Sigma\left(\dfrac{n_i}{N}\right)^2\right]$ $E(S_n) = \Sigma\left[1 - \dfrac{\binom{N-n_i}{n}}{\binom{N}{n}}\right]$	$I = \dfrac{c}{a+b-c}$ $I = \dfrac{2c}{a+b}$ $I = \Sigma \min(a,b,c,\ldots,n)$ $PS = 100(1 - 0.5\Sigma|p_{ij}-p_{ik}|)$ $BC = \dfrac{\Sigma|n_{1i}-n_{2i}|}{\Sigma(n_{1i}+n_{2i})}$

Fig. 4. Survey of species richness indices based on diversity, evenness, similarity and probabilistic distribution (ref. 13).

When calculating five diversity indices for the entomofauna in grasslands with increasing levels of fertilization, Siepel and van de Bund (ref. 14) found that the degrees for two consequtive years rendered conflicting conclusions with regard to the possible impact of fertilization on species diversity index, whereas a multiple variance analysis showed the constant presence of such impact.

A second approach is thought to be the use of indicator organisms, which are representative of similar organisms, see for instance Ma (ref. 9) with respect to earthworms. The question is, however, in what way these species are "representative". The "functional classification" developed by van Straalen et al. (ref. 15) offers perspectives as it provides an insight into the degree of ecological representativeness of a selected organism on the basis of food choice, reproduction rate, desiccation tolerance.

A third approach via sum parameters, e.g. soil respiration stating the total activity of the organisms, has the disadvantage that the decrease in respiration activity in one species may be compensated for by an increase in another species. Heavy metals do show distinct effects on microbial soil respiration, nitrogen mineralization and nitrification as summarized by Doelman (ref. 16), albeit with a large range (Fig. 5). These data illustrate the different sensitivity of these processes for different heavy metals. The range of concentrations in which either never, sometimes or always inhibition was observed, are for the greater part caused by the soil conditions in which the inhibition was measured. In general lowest effect concentrations were established in sandy soils and highest in clay and organic soils.

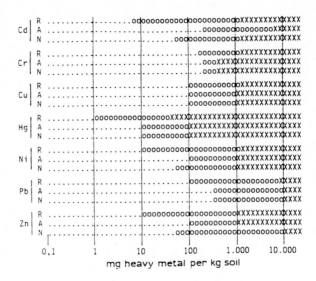

Fig. 5. Summarized literature data on the effects of Cd, Cr, Cu, Hg, Ni, Pb and Zn on soil respiration (R), nitrogen mineralization (A) and nitrification (N) with heavy metal ranges in which respectively never (.), sometimes (o) or always (x) inhibition was recorded (ref. 16).

By studying the consequences of a certain dose-effect relation for an organism considering the related elements of its biotic community, we are able to consider possible effects for higher levels of biological organization. A very illustrative example of this is formed by the research project on the heavy-metal accumulation in forest soil near a copper smelter in southern Sweden (Gusum). A clear decrease in various microbial soil processes went together with a much lower decrease in the total fungal biomass and a steady number of fungal taxa (Fig. 6). This is explained by the shifting composition of the fungal population in which some species even clearly increase. These resistant species, however, do not show a level of activity similar to that of the sensitive species, as indicated by the decrease in the different processes.

Moreover, we have to realize that in this investigation it is assumed that bacteria play a negligible role in these forest soils. Metabolic activities of eukaryotes (fungi) are, however, less diverse than those of prokaryotes (bacteria) and their growth and reproduction rates are lower. It is clear that the explanation of this type of interactions requires the fundamental research into the correlations between biotic and abiotic processes in the soil, as mentioned in the paragraph "Research required".

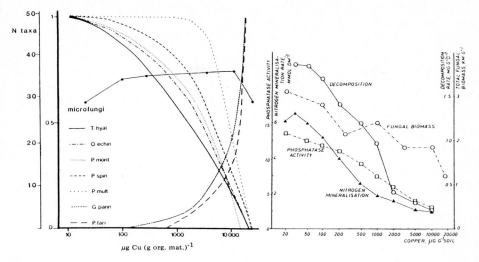

Fig. 6. Number of fungal species and survival of five sensitive and two resistant fungal species (left) compared with (right) development of fungal biomass and general microbial activities (N mineralization, phosphatase activity and decomposition rate) in forest soils with increasing Cu content near a copper smelter at Gusum, southern Sweden (ref. 17).

5. ADAPTATION BY BEHAVIOURAL AND PHYSIOLOGICAL AVOIDANCE

Resistance to heavy metals can be achieved by different mechanisms. For micro-organisms these mechanisms could be, according to Doelman (ref. 16):
- an excretion mechanism to keep the intracellular heavy metal concentration low,
- an oxidation process to change heavy metals to less toxic oxides,
- a trapping mechanism by biosynthesis of intracellular polymers (metallothioneines),
- a binding or precipitation mechanism to the cell wall,
- a biomethylation mechanism.

It is of importance to know how fast, how long, to what extent, and at what costs these mechanisms are started and maintained, as each mechanism requires an input of cellular energy. From different studies it appears that all these factors vary considerably. In experiments with Pb, resistance was observed after two years and the percentage resistant micro-organisms had not changed after three and four years (ref. 18), whereas in a partially sterilized soil the bacteria population became 100% resistant against Cd within two weeks (ref. 19). We have to realize in this respect that the difference in bacterial resistance between unpolluted and slightly polluted soils is larger and more distinct than between slightly and medium or heavily polluted soils. Multiple resistance has been observed in a few cases and suggests a similar site of action or resistance mechanism.

For soil faunal organisms the mechanisms mentioned above will be relevant too. Woodlice (Isopoda), earthworms (Lumbricidae) and springtails (Collembola) have developed a variety of storage and excretion organelles for different heavy metals (Table 3). These organelles protect the animals for the toxic action of heavy metals by storing them but, on the other hand, pose a threat by transfer of accumulated heavy metals in the food chains.

TABLE 3. Storage organelles for heavy metals in different soil fauna groups*.

	storage	excretion
Earthworms		
Ca-gland	Pb Zn	Zn
body wall chloragosomes	Pb Cu Cd Zn Fe	
body wall surface	Cd	
body wall mucoid coat	Zn	Zn
waste nodules	Pb Cd	Pb
intestine		Zn Pb Fe
Woodlice		
hepatopancreas B cells	Fe Zn Pb	
S cells	Zn Cd Pb Cu	Cu
metallothioneines	Cd	
Springtails		
cuticula	Pb	
gut epithelium	Pb	

* compiled from different literature sources.

Another adaptation mechanism for soil faunal organisms is by avoidance behaviour. Both earthworms, isopods and springtails are able to avoid heavy metal contaminated substrates as illustrated by the burrowing pattern of earthworms in a partly with Cu contaminated soil (Fig. 7). Springtails, moreover, are quite capable in distinguishing between polluted substrate and polluted food as observed for springtails by Ohlsson and Eijsackers (ref. 20), whereas isopods distinguish food types (= leaf species) from uncontaminated and contaminated areas (Fig. 8) as observed by Van Capelleveen (ref. 21). By these avoidance mechanisms soil fauna organisms effectively lower their exposition to heavy metals.

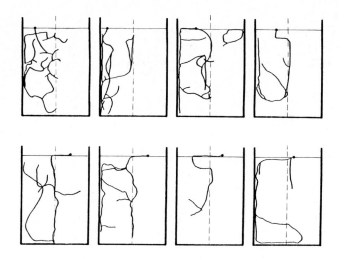

Fig. 7. Burrowing patterns of earthworms (L. rubellus) in thin soil slabs of uncontaminated (left) and partly with 240 mg.kg^{-1} Cu contaminated soil (right). The earthworms were placed on the surface of either the uncontaminated or the contaminated part of the soil.

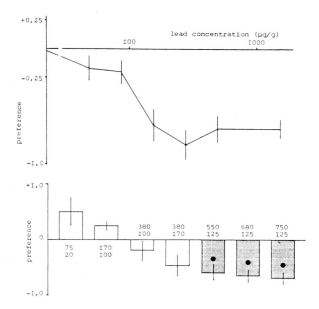

Fig. 8. Preference of P. scaber, when offered a dichotomous choice between A. artificially contaminated and uncontaminated oak litter (mean ± SD) and B. polluted and less polluted common oak leaf (open columns) or a mixture of common oak, red oak and silver birch materials (dotted columns), sampled from the fermentation layer at the roadside (●: significant preference reaction, α = 0.05) (ref. 21).

6. RECOLONIZATION

Recovery is being paid growing attention to in the government policy of The Netherlands. In addition to the many sites of contaminated soil in The Netherlands there is diffuse pollution caused by persistent compounds, such as heavy metals. Because of the persistent character the pollution in the soil will accumulate. For all of these situations knowledge is needed whether the natural functioning of the soil can be restored, possibly after sanitation, for only then one of the fundamental principles of soil protection is being met, which is the potential multi-functionality. Research into recovery shall consider a number of questions:
- the exposure limits above which effects may occur,
- the temporary exposure limits up to what effects are reversible,
- and the extent to what recolonization can occur from outside the area.

Important values are the speed and distance involved in recolonization in places where the soil has been sanitized and put back or where the pollution has disappeared through natural processes. When investigating a fly-ash tip (fly-ash consists of minute silica particles resulting from coal burning in power plants) Eijsackers et al. (ref. 22) found an evident relation among the numbers of earthworms, the age of the fly-ash and the distance from the surrounding reference field from where recolonization could take place (Table 4). Satchell and Stone (ref. 23) found a similar recolonization pattern in ash tips. Small soil animals, such as springtails and mites, are often passively moved over the surface (by wind and water). They can recolonize very quickly as observed on a heap of soil sanitized with a thermical sanitation method (internal RIN-CML-report), though it should be thoroughly verified whether these animals can indeed maintain themselves in the recolonized area.

TABLE 4. Numbers of worms, worm species and cocoons in coal fly-ash of different age and composition and survival of L. Rubellus inoculated on various ash sites (ref. 22).

Ash-age	1-2	3-4	4-5	15-18	control (clay)
Distance from control area (m)	71	48	38	18	
pH	11.1	10.1	8.2	7.9	7.6
N worms.m^{-2}	0	1	50	270	355
N worm species.m^{-2}	0	1	2	4	5
N cocoons.m^{-2}	0	1	675	691	*
Survical percentage of inoculated L. rubellus	0	42.5	32.5	92.5	

*no data available.

CONCLUSIONS

The impact of heavy metals on terrestrial ecosystems depends on the distribution of the metals in the soil. This comprises the spatial distribution which can be rather irregular and enables soil biota to avoid contact with the heavy metal. It comprises also the chemical "distribution" in relation to physico-chemical characteristics which influence the bio-availability of the metals and the resulting uptake, storage and excretion of the metals by soil animals.

Biotic activities like litter breakdown will change the distribution of physico-chemical characteristics of the soil profile and of the heavy metals within the profile. Moreover, storage and excretion processes may change the character of the heavy metal (i.e. biomethylation) and hence its availability. All these processes are mutually dependent as shown in figure 9.

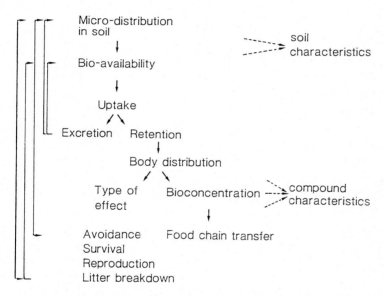

Fig. 9. Processes and characteristics involved in the "flow" of heavy metals through soil organisms in relation to the effects caused.

Because of these different mechanisms (avoidance, storage, excretion) heavy metals will have a very variable impact on soil organisms. Consequently it is not possible to derive general rules for effects of heavy metals or to set fixed limits; effects on soil organisms will depend on type of soil, heavy metal and organism. When using general parameters such as soil respiration or number of species one runs the risk of underestimating the total effect as resistant species may have taken over the activities of negatively affected sensitive species. Moreover, it may obscure the appearance of heavy metal effects. Therefore when studying the effects of heavy metals on an ecosystem level both the functional and structural characteristics have to be included.

ACKNOWLEDGEMENT

The material for this contribution was for a large part collected during a sabbatical leave at the Department of Soil Ecology of the University of Lund (Sweden).

7. REFERENCES

1 H. Eijsackers, Soil biological research into soil pollution, Annual report 1986, Research Institute for Nature Management, Arnhem, 1987, 105-119.
2 G.H. Orians, Diversity, stability and maturity in natural ecosystems, In: W.H. van Dobben and R.H. Lowe McConnel (eds.), Unifying concepts in ecology, Junk, The Hague, 1974, 139-150.
3 V. Huhta, E. Karppinen, M. Nurminen and A. Valpas, Effects of silvicultural practices upon arthropod, annelid and nematode populations on coniferous forest soil, Ann. Zool. Fenn. 4, 1967, 87-104.
4 L.J. Metz and D.L. Dindal, Effects of fire on soil fauna in North America, In: D.L. Dindal (ed.), Soil biology as related to land use practices, Proc. Fourth Int. Coll. Soil Zoology, Syracuse, EPA, Washington, 1980, 450-569.
5 P.J. den Boer, Stabilization of animal numbers and the heterogeneity of the environment: the problem of persistence of sparse populations, Proc. Adv. Study Inst. Dynamics Numbers Popul., 1971, 77-97.
6 P.J. den Boer, Dispersal power and survival: carabids in a cultivated countryside, Miscellaneous Papers 14, Agricultural University, Wageningen, 1977.
7 J.E. Satchell, r-Worms and K-worms: a basis for classifying lumbricid earthworms strategies, In: D.L. Dindal (ed.), Soil biology as related to land use practices, Proc. Fourth Int. Coll. Soil Zoology, Syracuse, EPA, Washington, 1980, 848-864.
8 W. Ma, Meten van nevenwerkingen van bestrijdingsmiddelen en andere stoffen in laboratoriumtoetsingen met regenwormen, Intern RIN-rapport van onderzoek gefinancierd door VoMil, 1983.
9 W. Ma, The influence of soil properties and worm-related factors on the concentration of heavy metals in earthworms, Pedobiologia 24, 1982, 109-119.
10 K.H. Domsch, G. Jagnow and T.H. Anderson, An ecological concept for the assessment of side-effects of agrochemicals on soil micro-organisms, Residue Reviews 86, 1983, 65-160.
11 H. Eijsackers, C.F. van de Bund, P. Doelman and Wei-chun Ma, in press, Aantallen en fluctuaties van het bodemleven, Bodembeschermingsreeks, Staatsuitgeverij, Den Haag.
12 P. Doelman and L. Haanstra, Short- and long-term effects of heavy metals on urease activity in soils, Biology and Fertility of Soils 2, 1986, 213-218.
13 P.J. Sheehan, D.R. Miller, G.C. Butler and P. Bourdeau, Effects of pollutants at the ecosystem level, Scope 22, Wiley, New York, 1984.
14 H. Siepel and C.F. van de Bund, in press, The influence of management on the microarthopod community of grasslands, Pedobiologia.
15 W.M. van Straalen, H.A. Verhoef and E.N.G. Joosse-van Damme, Functionele classificatie van bodemdieren en de ecologische functie van de bodem, Vakblad voor Biologen 65, 1985, 131-135.
16 P. Doelman, Resistance of soil microbial communities to heavy metals, In: V. Jensen, A. Kjoller and L.H. Sorensen (eds.), Microbial communities in soil, Elsevier, London, 1986, 369-384.
17 R. Rühling, E. Baath, A. Nordgren and B. Söderstöm, Fungi in metal-contaminated soil near the Gusum brass mill, Sweden, Ambio 13, 1984, 34-36.

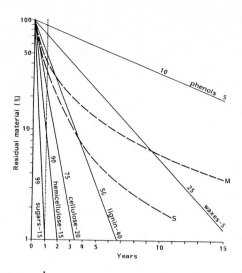

Fig. 1. Breakdown of litter components on the basis of a logarithmic process. The first (underlined) figure with each component denotes the percentage of breakdown in one year after leaf fall, and the second figure the component share by weight in the material. The dotted lines indicate the decomposition curves for all substances for several years in soil horizons with high (S) and low (M) breakdown rates (ref. 2).

Depending on the route of immission the compound may turn up in the upper soil layer of either a natural or an agricultural area or in deeper soil layers. Soil layer and soil type will affect the possible toxic action of the compound as well as the chance to be degraded, because there are significant differences in the composition of soil biota in different soil layers and soil types of which Table 2 and 3 give some examples.

TABLE 2. Vertical distribution of different groups of micro-organisms in the soil ($\times 10^3 \cdot g^{-1}$ soil). (ref. 3)

Depth (cm)	aerobic bacteria	anaerobic bacteria	actinomycetes	fungi	algae
3 - 8	7800	1950	2080	119	25
20 - 25	1800	379	245	50	5
35 - 40	472	98	49	14	0,5
65 - 75	10	1	5	6	0,1
135 - 145	1	0,4	-	3	-

Not only toxic action, but also biodegradation will be affected by the composition of the soil biota.

TABLE 3. Numbers of soil fauna organisms per dm^2 in forest and arable soil (compiled from different literature sources).

	Forest	Arable
Nematoda	100.000	20.000
Rotatoria	5.000	+
Acari	3.000	350
Collembola	2.000	110
Tardigrada	1.000	+
Enchytraeidae	300	110
Diptera larvae	4	1
Lumbricidae	2	2
Millipeda	2	2
Chilopoda	1	+
Isopoda	1	+
Aranea	1	+
Mollusca	1	+

+ present $< 1.dm^{-2}$

4. BIO-AVAILABILITY IN RELATION TO TOXICITY AND BIOCONCENTRATION

4.1 Laboratory QSAR-studies

Bio-availability is a dynamic process of physico-chemical ad/desorption processes and biological uptake-mechanisms. This can be studied by using a range of related chemical compounds with stepwise changed structures and characteristics and testing their biological activities (Quantitative Structure Activity Relationships). Van Gestel and Ma (ref. 4) studied this for two earthworm species and six chlorophenols (3-; 3,4-di; 2,4,5-tri; 2,3,4,5-tetra; 2,3,4,6-tetra; penta) with increasing lipophilicity and adsorption characteristics. Measurements in two different soil types rendered different toxicity-values. By expressing these toxicity-levels in relation to the amounts of chlorophenol available in the soil moisture - with the use of measured adsorption-coefficients - the differences between the two soil types disappeared.

Adsorption and lipophilicity happened to be the key-variables affecting bioaccumulation, toxicity and biodegradation of these compounds (Fig. 2).

Bioaccumulation increases with increasing lipophilicity, and this in turn is related with an increasing toxicity. An increased lipophilicity will affect biodegradability negatively as the better binding of chlorophenols to organic soil substances will protect the phenols for biodegradation.

These laboratory studies, therefore, not only provide data for deduction of QSAR's, but also have practical implications.

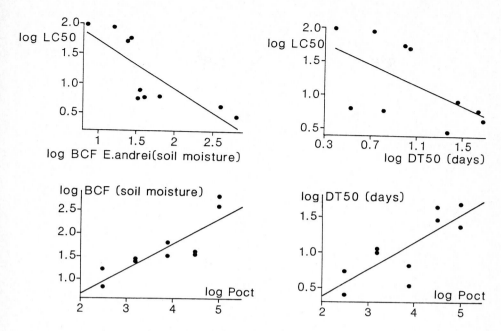

Fig. 2. Bioaccumulation (log BCF soil moisture), toxicity (LC50), and biodegradation (log DT 50) of five chlorophenols in relation to their lipophilicity, and the relation between toxicity and biodegradation for the earthworm Eisenia andrei. (ref. 4)

4.2 Field studies on bioconcentration

In a large waste dump site north of Amsterdam (Volgermeerpolder) Prins and Eijsackers (ref. 5) studied the bioconcentration of different organic compounds from the soil via earthworms to small mammals and birds of prey. The contents of different chlorobenzenes of these items have been summarized in table 4 for one sampling spot in the Volgermeersite, because within the site the amounts of organic compounds can differ considerably.

TABLE 4. Contents of different chlorobenzenes* (mean mg.kg^{-1} dw ± s.d.) of soil, earthworms and mammals (mice) in the Volgermeerpolder, and resulting bioconcentrationfactors (BCF) for the transfer soil-earthworms (s-e) and earthworm - mammals (e-m). (ref. 5)

	1,2,3,4-TeCB*	1,2,4,5-TeCB	QCB	NCB
soil	1.05 ± 0.33	0.44 ± 0.19	0.55 ± 0.2	0.34 ± 0.18
earthworms	0.38	0.08	3.1	6.6
mammals	0.03 ± 0.02	0.004 ± 0.01	0.35 ± 0.33	2.49 ± 1.86
BCF s-e	0.28 - 0.53	0.12 - 0.31	4.21 - 9.01	12.7 - 41.3
BCF e-m	0.03 - 0.15	0.00 - 0.18	0.01 - 0.22	0.10 - 0.66

* 1,2,3,4-Tetrachlorobenzene; 1,2,4,5-Tetrachlorobenzene; Pentachlorobenzene; Hexachlorobenzene.

Bioaccumulation by earthworms from the soil (a layer of about 10 cm organic material with a thin litter layer) is the first major step in the bioconcentration-process. Further transfer to mammals (mice, shrews and voles) is far more limited as can be deduced from the BCF-values for the different benzene-compounds. Moreover, there is a very significant difference in bioaccumulation in relation to the chlorination of the benzenes which will result in an increased lipophilicity ($P_{oct/water}$).

For Hexachlorobenzene the food chain transfer could be further extended by analysis of other food items of herbivore mammals, herbivore/insectivore birds and birds of prey (buzzard). Moreover, a limited sampling program was carried out in the waterways within the site for water and sediment, crustaceans (Daphnia spec.), fish and herons. These sampling data are compiled in Figure 3.

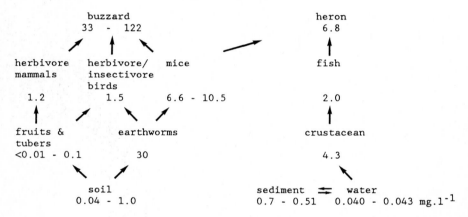

Fig. 3. Hexachlorobenzene-content of different parts of a natural food-system. Water in mg.l^{-1}, soil and sediment in mg.kg^{-1} dry weight, fruits & tubers, earthworms and crustaceans in mg.kg^{-1} fresh weight, all other in mg.kg^{-1} fat weight. (ref. 5)

From the figure it appears that it is not possible to formulate simple general rules about bioconcentration in natural food systems. Depending on food choice, food consumption rate, metabolic rate, and specific degradation mechanisms, the uptake and further transfer will differ. In this situation the largest bioaccumulation was realized by earthworms from the soil. These data illustrate the problems in evaluating food-chain transfer. Generalized transfer-patterns compiled from literature references from different areas, therefore, have to be interpreted with great caution.

5. BIOLOGICAL SOIL SANITATION

5.1 Required conditions for biodegradation

Biodegradation depends on the structure of the compound, the environmental conditions and the presence of biodegrading organisms.

With respect to the structure of the compound it is not yet possible to give general rules. For specific groups like hydrocarbons present in natural oil it has been established that:
- lineair chains degrade better than branched ones,
- short chains degrade better (but chains with less than 10 C-atoms cause increased intoxication, thus hampering biodegradation),
- alkanes degrade better than alkenes,
- aliphatic and aromatic ringstructures degrade slower than chains,
- polycyclic compounds degrade slower.

Moreover, structure will affect physico-chemical characteristics like solubility and affinity to clay mineral and organic matter which influence the availability of a compound and hence its degradability. Besides the structure of the compounds also the quantity is of importance: a high amount will cause intoxication, whereas very low amounts could be "overlooked" by microorganisms.

Further physico-chemical characteristics will stimulate or hamper microbial activity in general and availability of different compounds (and consequently their potential biodegradability) in particular.

Relevant characteristics are:
- temperature: - increased temperatures (10 - 25°) stimulate microbial activity
  - above 25° desiccation will hamper microbial activity and increase adsorption
- moisture content: - low contents hamper biodegradation
- pH: - low pH correlates with high adsorption
- organic matter: - high organic matter content will increase adsorption and increase microbial activity
- clay content: - high clay content will increase adsorption
- texture: - a fine texture influences moisture content and aerobic conditions
- aerobic/anaerobic: - aer/anaerobic conditions will stimulate specific microorganisms and consequently specific biodegradation processes
- electron acceptors: useful acceptors have to be available.

A special characteristic is the distribution of the compound in the soil. Mostly this will be irregular, especially after dumping, and at depths in the soil profile where microbial activities are rather limited (below 75 cm, see Table 2). Moreover, waste compounds may be combined with other substances. In an area in The Netherlands HCH-isomers were dumped packed in chalk resulting in small lumps (ø a few mm). This necessitated a pre-homogenizing procedure in order to make the HCH seizable.

The third group of environmental conditions necessary for biodegradation are the microorganisms. A compound may not be biodegraded because:
- there is no microorganism able to tackle it,
- the microorganism is not present (distribution-problem),
- the necessary enzymes are not induced,
- electrondonors are not available,
- microbial species show competition.

If the quantity of the compound is too low, the microorganisms may not "observe" the compound and not induce relevant enzymes. Only enzymes which are already present and degrade homologous compounds can degrade the compound by co-metabolism or "fortuitous" metabolism. However, this process will be rather slow.

To describe the degradability of a specific compound the DT50 (Degradation Time for 50% of the compound) or DT95 is used. Because a number of variables (solubility, temperature, pH) will influence the degradation rate, the DT50 is not independent of the compound concentration and will not have a constant rate (1st order-reaction). Furthermore, it only considers the first degradation step in a process with possibly far more steps. Very easily degradable compounds (like sugars) will show a 0-order reaction, whereas most pesticides presumably will have a 2nd-order or repeating 2nd-order reaction (Fig. 4).

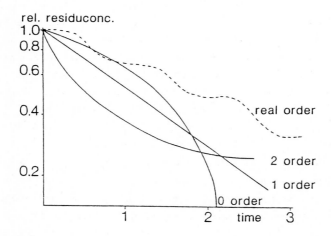

Fig. 4. Type of chemical reactions involved in degradation of pesticides.

Besides, we have to realize that degradation and mineralization are not similar as mentioned above with respect to the first degradation step.

Mineralization means that the compound is <u>completely</u> degraded, whereas in the literature degradation is frequently interpreted as the first degradation step in which the compound looses its original chemical configuration. As quite a number of metabolites have an even enhanced toxicity or persistance relative to the original compound this poses a serious danger for overestimating potential biodegradability.

To illustrate the necessity of an integrated approach of biosanitation of soil polluted with organic compounds two examples of biodegradation (HCH-isomers and Endosulfan) in relation to formation of metabolites and their toxicity for soil biota will be discussed.

5.2 Biodegradation of HCH-isomers

With the production of the pesticide lindane (the γ-HCH isomer) large amounts of the α-HCH isomer together with β-HCH and δ-HCH are formed. These waste-products have been dumped in the surroundings of a pesticide plant in the eastern part of The Netherlands, mixed with chalk to bind the HCH. In order to investigate the possibilities of microbial degradation laboratory- and semi fieldstudies were carried out which were aimed at optimization of the natural degradation potencies. To that extent the heterogeneous distribution of the HCH's within the soil was tackled first by intensively homogenizing, further microbial activities were stimulated by adding nutrients and varying environmental conditions.

From the literature (ref. 8) it was known that degradation occurred at anaerobic conditions. However, this regarded low concentrations, mainly of γ-HCH (lindane), whereas in this situation we were faced with high concentrations (up to 7000 mg HCH per kg soil) and four isomers. Preliminary laboratory experiments revealed that best degradation results were obtained with aerated slurry (degradation rate 14 mg.day^{-1} at 20°C), whereas with anaerobic slurry the rate was 10 mg.day^{-1} (ref. 6). Furthermore inoculation of microorganisms especially "adapted" to degradation of organochlorines did not give better results than optimization of the natural degradation potencies.

Further upscaling of the experiments in 60 liter-containers in a glasshouse showed the advantage of working with slurry for homogenization although it proved to be difficult to maintain aerobic conditions. If so, the α-HCH concentration decreased in 23 weeks in slurry from 400 mg.kg^{-1} to ca. 10 mg.kg^{-1} and in moist soil from 360 mg.kg^{-1} to ca. 60 mg.kg^{-1}. In the first degradation-phase, with the fastest degradation rate (Fig. 5), the degradation process continued also at lower temperatures 12 - 17°C (ref. 7).

In a final experiment under field conditions the low temperatures slowed down the process to a limited extent, in 23 weeks the α-HCH contents decreased from 350 to ca. 100 mg.kg^{-1}. During the whole degradation process a number of microbial features were measured; i.c. microbial respiration, glutamic acid degradation and numbers of specific HCH-degraders.

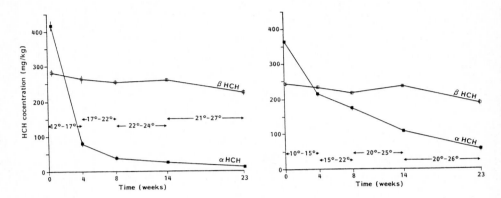

Fig. 5. Degradation of α and β HCH in slurry (left) and moist soil (right) in a glasshouse. For each sampling-interval the range of soil temperatures is given (ref. 6).

These features showed a maximum level at the end of the first degradation phase. This suggests an adaptation of microorganisms to use α-HCH as a substrate, followed by a population increase and a subsequent HCH-decrease. With slurry this adaptation phase did not occur, presumably because of the pre-adaptation during homogenization optimizing the contact between microorganism and HCH. From the laboratory experiments it had already appeared that HCH isomers or metabolites did not intoxicate microorganisms and processes. Preliminary experiments with earthworms about their possible rôle as soil homogenizers and aerators revealed that HCH intoxicated the earthworms and slowed down their burrowing activity, but activity of L. rubellus has a slight positive effect on degradation.

Besides, in our field studies in the Volgermeerpolder α-HCH (soil concentration 0.02 ± 0.03 mg.kg^{-1}) was strongly bioaccumulated by earthworms (BCF 3.08 - 15.4) and bioconcentrated to a limited extent by mammals (BCF 0-0.50), leading to contents of 0.154 and 0.028 ± 0.051 mg.kg^{-1} α-HCH respectively in spot A (similar to table 4). In an adjacent spot (soil concentration 0.45 ± 0.55 mg.kg α-HCH) the earthworm and mammal-contents were far higher, however, with contents of 10.77 ± 4.35 and 7.08 ± 10.31 respectively and BCFs of 6.4 - 150 (!) and 0.03 - 2.7 (ref. 5). This poses a serious problem for soil sanitation as it forces the government to set very low acceptable limits for HCH and consequently necessitates the degradation process to be very effective.

5.3 Biodegradation and ecotoxicity of Endosulfan

Endosulfan is used in mushroom-growing for insect-control. The mushroom-compost with Endosulfan-residues is dumped near growing-houses where Endosulfan can leach into surrounding ditches accumulating in the sediment.

To assess the potencies of biodegradation of Endosulfan, the sediment was thoroughly homogenized and cultured under aerobic and anaerobic conditions. Regularly the sediment and water were tested for toxicity for microorganisms, earthworms, fish and Daphnias (Table 5). Under aerobic as well as anaerobic conditions the degradation passed with the same rate; after 12 weeks 64 and 68% of the original amount resulted respectively, although the amounts of different metabolites differed considerably between these conditions.

The toxicity of the sediment remained about the same during the first six weeks, but at the end of the experimental period the toxicity especially for fish and Daphnias increased. This indicates a gradual release of the different Endosulfan compounds from this rather peaty soil. Because of this soil type the adsorption of Endosulfan will be strong which may also explain the low degradation rate.

TABLE 5. Relative contribution of parent compounds (alpha and beta endosulfan) and different metabolites after 12 weeks degradation and toxicity of these compounds for microorganisms (glutamic acid degradation time), earthworms (% mortality), fish (% mortality), Daphnia (% mortality). (ref. 9)

	aerobic 0	aerobic 12	anaerobic 12		aerobic 0	4	8	12	weeks
alpha	44	8	8	microorganisms	16	-	19	20	
beta	48	20	22	earthworms	13	3	13	9	
lacton	0	0	30	fish	25	25	80	100	
sulfaat	4	32	4	daphnia	0	30	30	60	
ether	4	4	4						

6. CONCLUSIONS

The problem of organic pollutants is not merely a problem of considerable quantities or concentrations of specific pollutants, but also of a specific irregular distribution of these pollutants in the upper-soil (for instance by mixing with other constituents) and in the sub-soil. Therefore there will not always be sufficiently intimate contact between pollutants and soil biota. Although this can be an advantage, as it will limit the extent of intoxication, it will also be a disadvantage by hampering biodegradation processes.

Moreover degradation processes produce metabolites which may be more toxic or persistent than the parent-compound.

Therefore research on the impacts of organic pollutants on the soil system must always be combined with research on degradation processes and vice versa.

Besides, research must include the initial compound as well as the metabolites; references which only mention the disappearance of the initial compound do not really provide data on degradation, or more precisely, mineralization of a compound.

In laboratory studies on related chlorophenols it was shown that toxicity, bioaccumulation with earthworms and degradation of the chlorophenols by microorganisms are closely related and carry back to the availability of the compounds. Adsorptioncoefficient (expressed in the amount of chlorophenol in the soil water) and lipophilicity ($P_{oct/water}$) provided an adequate framework to explain different responses of the different chlorophenols in two soil types. A number of field inventories on bioaccumulation and bioconcentration of organic compounds in soil and terrestrial animals revealed that especially the accumulation of compounds by earthworms from the soil happens to be very efficient. Further accumulation from earthworms to mammals (mice, shrews and voles) is less efficient which may reduce the potential toxic effects of bioconcentration.

Under field conditions too an increased chlorination of phenol- and benzene compounds and a subsequently increased lipophilicity result in an increased Bio Accumulation Factor (BCF).

Optimization of the natural degrading potencies of soil microorganisms proves to provide an efficient tool for soil sanitation. α-HCH could be degraded to very low quantities within a period of 15 - 25 weeks under temperate field conditions, with a degradation-efficiency of >95% and 80% in slurry and moist soil respectively. However, assessment-levels for α-HCH in The Netherlands are that low (0.01 mg.kg^{-1}), that this degradation is not sufficient yet.

Moreover, toxicity potencies can increase during the total degradation process (= mineralization to CO_2 and minerals) as was shown with Endosulfan. Especially in the last part of a 12 weeks degradation period toxicity for aquatic animals increased suggesting a release (by formation of metabolites?) of toxic compounds.

7. REFERENCES

1. Anonymus: Discussienota bodemkwaliteit, 1987, Leidschendam, Ministry of Housing, Physical Planning and Environment.
2. G. Minderman, Addition, decomposition and accumulation of organic matter in forests, J. Ecol. 56, 1968, 355-362.
3. M. Alexander, Introduction to soil microbiology, Academic Press, London, 1975.
4. C.A.M. van Gestel and Wei-Chun Ma, Toxiciteit en bioaccumulatie van chloorfenolen in regenwormen in relatie tot de beschikbaarheid in de bodem, RIVM-report 718479001, RIN-report 87/12, 1987, National Institute for Human Health and Environmental Hygiene, Bilthoven and Research Institute for Nature Management, Arnhem, 65 pp.
5. E. Prins and H. Eijsackers, Bioconcentrering van organochloorverbindingen in een voedselketen op de vuilstortplaats in de Volgermeerpolder in Amsterdam, RIN-report 82/8, 1983, Research Institute for Nature Management, Arnhem, 48 pp.
6. P. Doelman, L. Haanstra and A. Vos, 1988, Microbial sanitation of soil with alpha and beta HCH under aerobic glasshouse conditions, Chemosphere (accepted).
7. P. Doelman, L. Haanstra and A. Vos, 1988 (in press), Microbial degradation by the autochthoneous soil population of alpha and beta HCH under anaerobic field conditions in temporate regions.
8. P. Doelman, M. Fredrix and H. Schmiermann, Microbiologische afbraakprocessen als saneringsmethode van met bestrijdingsmiddelen verontreinigde gronden, RIN-rapport 87/10, 1987, Research Institute for Nature Management, Arnhem, 100 pp.
9. P. Doelman, A. Vos and H. Loonen, Ecotoxicologisch onderzoek aan met Endosulfan verontreinigde grond: sanering en toxiciteit, RIN-Projectreport, 1987, Research Institute for Nature Management, 3 pp.

Further extensive information can be found in:
K. Barth and P. l'Hermite (Eds.), Scientific Basis for Soil protection in the European community, Elsevier Applied Science, London, 1987, 630 pp.

AIR POLLUTION EFFECTS IN TERRESTRIAL ECOSYSTEMS AND THEIR RESTORATION

B. ULRICH

Institute of Soil Science and Forest Nutrition
University of Göttingen, Federal Republic of Germany

1 INTRODUCTION

There are reasons to be concerned about the stability of the ecosphere. The extinction of species, the accumulation of toxic elements, the acidification of forest soils and of waters, and the forest decline indicate changes in parts of the ecosphere which are far from the centers of industrial activities. The changes mentioned have one common denominator: they can all be expressed as storage changes (e.g., of species, of nutrients, of chemical compounds, of acids and bases, of organic matter) and thus reflect changes in the balance of material of ecosystems.

2 STABILITY CONDITIONS OF TERRESTRIAL ECOSYSTEMS FROM THE POINT OF VIEW OF THE MATERIAL BALANCE

A system is defined as an integrated entirety whose properties are different from the properties of its elements. Elements of ecosystems are species, which from the point of view of the matter balance are grouped into primary and secondary producers. Primary producers are all organisms capable of photosynthesis and chemosynthesis (all autotrophic organisms), secondary producers are all heterotrophic organisms. From the point of view of the material balance, the ecosystem may be expressed in terms of storage and flow rates of chemical compounds which are aggregated in structural units like organisms or the soil. The organisms are represented in storages, their activities show up in flow rates.

The flow of matter in the ecosystem can be expressed in the following equation (ref.1):

$$\text{Photosynthesis and formation of organic substances } (\rightarrow)$$
$$\text{Respiration and mineralization of organic substances } (\leftarrow)$$
$$aCO_2 + xM^+ + yA^- + (y-x)H^+ + zH_2O + \text{Energy} \rightleftarrows$$
$$\rightleftarrows (C_aH_{2z}O_zM_xA_y)\text{org. matter} + (a+\ldots)O_2, \tag{1}$$

where a, x, y, z represent stoechiometric coefficients, M^+ cations, A^- anions of unit charge. It is accepted by many scientists that terrestrial ecosystems can approach a steady state in which the species composition and the organic matter storage in biomass and in soil show no trend. In terms of Eq.(1), steady state means that the forward reaction, caused by the activity of primary producers, is balanced by the back reaction, caused by the activity of secondary producers. Terrestrial ecosystems have thus, in the ideal state, the property of keeping the internal ion cycle closed. This property is important in maintaining the chemical state within the ecosystem and in its environment. Organisms do not have this property. Primary producers act as sinks for ions, secondary producers as sources, both produce organic waste, with the soil as the corresponding source or sink, respectively. As shown in Eq.(1), a net flow of ions may be connected with the production or consumption of protons, resulting in the acidification or alkalinization of the soil, respectively. Organisms have, therefore, the property to change the chemical state of their environment. In a metaphorical sense, man-made pollution is a consequence of this property. If terrestrial ecosystems approach the ideal (steady) state, however, the sink and source effects in respect of ions of primary and secondary producers tend to balance each other, resulting in a net effect close to zero.

3 AIR POLLUTANTS

Air pollutants may be of natural or anthropogenic origin. They can be classified as follows according to their ecochemical function:

- neutral substances: Na^+, Cl^- (sea spray);
- nutrients and essential elements: SO_4^{2-}, I^-, Mg^{2+}, etc. (sea spray); NH_4^+, NO_3^-, Ca^{2+}, trace elements;
- acidifiers: SO_2, NO_x;
- potential toxins: SO_2, HF, heavy metals, metal oxides, hydrocarbons;
- oxidants: NO_x, ozone, hydrocarbons;
- radioactive substances.

Many anthropogenic air pollutants also occur naturally and are in some cases vital to the development of ecosystems. In Europe, and particularly in Central Europe, anthropogenic emissions do, however, exceed natural emissions by a factor of around 10 in the case of most air pollutants.

Different air pollutants have differing effects on the ecosystem depending on their ecochemical function: acids deplete soils of nutrients whereas nutrients encourage plant growth. Nitrogen encourages the action of decomposer organisms in the soil in forest ecosystems with limited nitrogen supply whereas acids and heavy metals have an inhibiting effect. The effects of anthropogenic inputs in the ecosystem which have risen tenfold are hence complex and confusing, as witnessed by the current debate on acid rain.

4 DEPOSITION PROCESSES

The following deposition processes can be distinguished:

- wet deposition by rain or snow (A);
- sedimentation (gravimetric fall) of particles (B);
- impaction of aerosols (C);
- impaction of mist, fog, cloud droplets (C);
- absorption of gases (example: SO_2) (D)
  . on wet surfaces like foliage, bark, wet snow,
  . inside stomata
    .. on cell walls,
    .. on to mesophyll and palisade cell surfaces;
- re-emission.

Wet deposition of rain or snow is not influenced quantitatively by the receiving surface. Deposition of particles (sedimentation, impaction) is strongly dependent on particle size. Fog, mist, cloud droplets could be regarded as large particles and are very efficiently captured by a canopy making the rate of deposition dependent mostly on the wind speed and thus on the exposure of the receiving surface to wind action. Deposition of gases varies with kind and state of the receiving surface.

The deposition can be looked at from two different points of view: either from the deposited compound or from the receiving surface. From the point of deposition compounds, wet and dry deposition are distinguished. From the point of view of the receiving surface, it makes sense to distinguish between precipitation deposition and interception deposition. These terms are shortly defined in the comparison (see next page).

It is evident that the processes summarized under the term interception deposition depend upon size, kind and (chemical) state of the receiving surface. This has important consequences regarding the measuring techniques.

After deposition, chemical reactions can occur. The following reaction types can be distinguished:

- acid production (e.g. SO_2 reacts to H_2SO_4);
- proton buffering (e.g. cation exchange on cell walls);
- precipitation or dissolution, depending on pH (e.g. heavy metals);
- reactions with living plant cells (reactions with cell membranes, uptake into cells, assimilation in the cell metabolism);
- formation of gaseous compounds (e.g. sulphides).

After deposition, a translocation could also occur within the canopy or by leaching by rain from the outside as well as from the inside of leaves and needles. Much of the deposited material will by this process be transferred from the canopy to the ground.

Comparison

	Process	Viewpoint of depositing compound	Viewpoint of receiving surface
A	Precipitation of rain or snow "particles" with dissolved (soluble) or undissolved (unsoluble) content	wet deposition	precipitation
B	Precipitation of particles other than rain or snow according to gravity	dry deposition	deposition
C	Impaction of aerosols including fog and cloud droplets according to air or Brownian movement*	dry deposition	interception
D	Dissolution of gases on wet surfaces (with subsequent chemical reactions)	dry deposition	deposition (ref.2)

*The impaction of fog and cloud droplets is sometimes grouped under wet deposition.

Tree leaves can buffer acidity deposited on the leaf by exchange with Ca; it is assumed that Ca is exchangeable bound on acidic groups in the cell wall and that the buffering occurs by H/Ca exchange. During stomata closure this exchange may be reversed forced by an input of $Ca(HCO_3)_2$; the carbonic acid formed may be assimilated or transferred to H_2O and CO_2. If the Ca is supplied from the exchangeable pool in the soil, the amount of basicity fed in the transpiration stream is balanced by an equivalent amount of acidity formed in the soil. Proton buffering occurring in leaves is thus immediately transferred to the soil.

5 DEPOSITION OF NUTRIENTS

The nutrients being deposited accumulate in the biomass and, in connection with a decreased decomposer activity, in the forest floor. The rate of nutrient deposition may exceed considerably the amount of nutrients needed for the forest increment (Table 1).

TABLE 1

Rate of nutrient deposition from air pollutants, and of nutrient accumulation in forest increment, Solling mountain, Norway spruce

	Mg	Ca	N	S
	\multicolumn{4}{c}{$kg.ha^{-1}.yr^{-1}$}			
Deposition	4	20	40	80
Accumulation in forest increment	1.3	6.5	13	1.4

In the long-term forest ecosystem study in the Solling mountains, the amount of organic matter and of nitrogen in the humus layer (O horizons) of the forest floor have considerably increased (Table 2). This is due to a retardation in the decomposition of litter.

As a comparison with the age of the stands shows, this retardation in litter decomposition has been not a steady process during the whole stand development, but increased considerably during the last decades. It indicates a strong reduction in the activity of the decomposer organisms in the ecosystem which is due to acid deposition.

TABLE 2

	Year	Stand age (years)	Organic dry matter (kg/ha)	N (kg/ha)	C/N-ratio
Beech	1966	120	26 600	809	18.3
	1983	137	48 100	1 270	18.9
Spruce	1968	85	49 000	960	25.5
	1983	100	96 000	2 030	23.6

6 DEPOSITION OF ACIDITY

The wet deposition is independent of the receiving surface. The dry deposition of gaseous SO_2 depends upon tree species (coniferous trees have a greater absorbing surface in winter), the roughness of the canopy, and the vitality of the tree (the extent of which the acidity formed after absorption of SO_2 can be buffered and the absorbing water film can act as an infinite sink for SO_2). The deposition of particles, where fog and cloud droplets are of greatest importance, depends upon the exposition and inclination of the site, the tree species, the tree height, and the compactness of the canopy (stand closure, leaf area index).

In Table 3, data from 24 sites in Northwest Germany concerning the rate of deposition of acidity are compiled (ref.3). The rates show a considerable variation from year to year. The wet deposition can be quite low in dry/warm years, especially in rural areas and in the vicinity of cities supplying neutralizing dust.

In large closed areas, the wet deposition amounts to around 0.8 kmol $H^+ \cdot ha^{-1} \cdot yr^{-1}$, a rate which already exceeds the rate of neutralization in most soils.

The data of Table 3 are calculated from the flux balance of the canopy. The acidity due to dry deposition of SO_2 can reach high values (up to 3 kmol $H^+ \cdot ha^{-1} \cdot yr^{-1}$).

The long-term data available from the forest ecosystems investigated in the Solling mountain show that the variation from year to year can be very great, approaching zero values in some years. This variation is probably due to changes in the vitality of the stand, limiting the buffer ability in the canopy so strongly that SO_2 absorption in the water films ceases due to low pH.

TABLE 3

Deposition of acidity in Northwest Germany (values in kmol $H^+ \cdot ha^{-1} \cdot yr^{-1}$)

Wet deposition	Rural areas vicinity of cities 0.2-0.8		Large closed forest areas 0.6-1.1	
Dry deposition of SO_2	Deciduous forests		Coniferous forests	
	neutral soil 0.6-1.3	acid soil 0.3-1.0	acid soil 0.1-3.0	
	Deciduous forests		Coniferous forests	
	rural areas vicinity of cities	exposed mountains	rural areas vicinity of cities	exposed mountains
	sheltered areas		sheltered areas	
Particle deposition	0-0.3	0.5-1.8	0-0.5	0.5-3
Total H^+ deposition	0.8-1.0	1.1-2.4	1.2-2.0	3.9-5.5
H^+ buffer quota	10-87%	0-70%	30-60%	0-42%
NH_4^+ deposition	0.4-1.0	0.9-1.3	0.8-1.0	1.0-1.6
Total deposition of acidity	1.2-2.6	2.0-3.5	2.0-2.7	2.9-6.4

Damaged stands of low vitality can, therefore, show lower rates of acid deposition than prior to their damaging. On neutral soils, where the uptake of basicity may not be limiting acid buffering in the canopy, dry deposition of SO_2 is always of significance.

Particle deposition reaches high values in exposed mountain areas, higher in coniferous than in deciduous forests. It reaches also quite low values in some years, probably mainly depending upon weather conditions. In sheltered positions, e.g. also in young stands of low height sheltered by higher stands, particle deposition approaches zero. In exposed mountains, however, particle deposition may be of greatest significance. This explains the missing conformity in remote mountain areas between air quality (low concentrations of SO_2 and NO_x in air) and acid deposition (high rates of deposition).

The rates of total H^+ deposition clearly indicate the difference between sheltered and exposed sites on the one hand, and deciduous and coniferous forests on the other.

Part of the protons are buffered in the canopy. As stated already, the buffer quota may vary with the vitality of the tree and show considerable year-to-year variance. This shows up in zero values in exposed mountain areas. On the other hand, even in these regions and in other years, the buffer quota can reach 40% (coniferous stands) to 70% (deciduous stands). Since the basicity for buffering the protons stems from ion uptake from soil, the protons buffered in the canopy are balanced by an equivalent decrease of basicity in the soil close to the roots which is most strongly influenced by ion uptake. This means that in the mean, around half of the protons deposited reach the soil close to the roots. Not only the soil surface to the atmosphere, also the soil surface to roots is exposed to acid deposition. This applies especially to well buffered soils. Such soils may, therefore, show lower pH values in the subsoil than in the topsoil. This explains the observations that the die-back of old trees can be accompanied by a vital regeneration.

Also NH_4^+ must be considered as an acid, since it is completely retained in the forest ecosystem, most probably being converted to organic nitrogen. This process is connected with an equivalent release of protons. The values of total deposition of acidity include, therefore, the deposition of NH_4^+. They vary between 1.2 and 6.4, indicating that almost all forest soils in the region investigated are being acidified due to acid deposition.

7 DEPOSITION OF HEAVY METALS AND HYDROCARBONS

Table 4 gives a range of rates of deposition, accumulation on the forest floor and levels in humus for a number of forest ecosystems in Northwest Germany (ref.4).

TABLE 4

Range of rates of deposition, accumulation and levels in humus on forest floor of heavy metals

	Pb	Cd	Cu	Zn	Cr
Deposition on open land (g/ha/yr)	70-140	2-4	10-50	70-300	2-4
Deposition in forests (g/ha/yr)	130-330	3.5-5.5	50-100	300-900	7-22
Accumulation on forest floor (g/ha)	700-60000 (20 000)	0-200 (50)	200-80000 (300)	500-40000 (15 000)	1000-30000 (5 000)
Levels in forest floor organic matter (mg/kg)	100-450	0.2-1.4	10-70	50-250	10-300

The immobile (Pb) or less mobile (Cu,Cr) heavy metals occur mainly in organic compounds. This means that in forests they are occurring in increasing concentrations in the soil organic matter. In this form they could impede the breakdown of organic matter. Although there are indications that this is the case, considerable research is required in this area. Since heavy metals become less mobile as the pH rises, areas of higher pH act as sinks for them. In acid soils such areas are roots or even organisms. In the fine root system of spruce, concentrations of up to 12 mg Cd, up to 350 mg Pb, 40 mg Cr, 40 mg Cu and

0.55 mg Tl per kg dry matter have been found (ref.5). There are also indications that these heavy metals are partly responsible for root damage which is characteristic for the novel tree die-back (refs.6,7).

Owing to their low level of mobility, heavy metals will accumulate in neutral soils in the top millimetre to centimetre layer if they are not worked by tilling into a thicker layer of soil.

The fact that, with the exception of grazed grassland and ploughed arable land, the top, even only millimetre to centimetre-thick soil layers act as the sink in which heavy metals usually accumulate, will probably have growing adverse ecological effects on soil organisms and seed germination if heavy metals continue to be deposited.

There is a high level of deposition of polycyclic aromatic hydrocarbons as well as heavy metals in forests. The soil, and in particular the humus layer, act as a sink so none of these substances are washed out. The accumulated material found in Solling amounts to between 15 and 50 times the deposition rate recorded (refs.8,9).

8 THE SOIL'S FUNCTION AS A SINK/SOURCE

The soil can provide the following functions for substances deposited:

1) It can act as an inert matrix: the deposited substances leach into the groundwater through the seepage water.
   Examples:
   - Na^+Cl^- and other neutral salts ($CaSO_4$);
   - acid-soluble heavy metals (Cd, Zn and Mn) in acid soils and seepage layers;
   - nitrates in acid "nitrogen-saturated" forest ecosystems.

2) It can act as a sink: the deposited material builds up in the soil.
   Examples:
   - lead (in top organic soil horizon);
   - heavy metals in neutral soils;
   - nitrogen in "non-nitrogen-saturated" forest ecosystems;
   - hydrocarbons in acid forest soils where litter break-down is inhibited.

3) It can act as a source: the deposited material accumulates in the soil but also causes other substances to be leached with the seepage water or discharged into the air.
   Examples:
   - acids: bases equivalent to the acid input are removed in the seepage water or cation acids (e.g. Al ions) are released from the soil matrix;
   - cationic acids may accumulate to some degree but will migrate

. in an acidifying soil through the seepage layer to the aquifer,
. if acid continues to enter the soil;
- stable gaseous N and S compounds (e.g. N_2O) are released; these may contribute to the breaking down of the ozone layer in the stratosphere.

Substances for which the soil acts as an inert matrix and which are leached out in the seepage water will be concentrated in the seepage water as water is lost through the atmosphere by evaporation and transpiration in their journey through the ecosystem.

The rate of accumulation for substances for which the soil acts as a final sink can be estimated from the expected deposition rates. The accumulated acid deposition in a spruce stand in Solling since beginning of industrialization has been calculated (ref.10). Between 1969 and 1985 an average of 3.8 kmol H^+/ha was deposited each year equivalent to a total of 65 kmol H^+/ha. If it is assumed that there was a parallel trend between emissions and deposition, the accumulated acid deposition since the beginning of the last century is 200 kmol/ha. The acid deposited would require a base equivalent to 10 000 kg $CaCO_3$/ha as a buffer in the soil and seepage layer. According to available measurements, acid deposition in Central Europe is between 0.8 and over 6 kmol H^+ per hectare per year (cf. Table 3). The acid accumulated in woodland soils since the end of the last century as a result of anthropogenic air pollution is equivalent to a base of at least 2500 kg $CaCO_3$/ha.

The acidity deposited has to be related to rates of natural soil acidification in order to assess its importance for ecosystem changes.

In Table 5 the acid load (and acid buffering) for a rotation period of 100 years is compiled. Acid deposition amounts to between 50 and 75% of total acid load.

The only process which consumes protons and releases Ca, Mg and K ions is the weathering of silicates. The rate of this process depends upon the kinds and amounts of silicates present and only to a minor extent of the pH of the soil. The maximum rate of proton consumption reported in soils rich in weatherable silicates is around 2 kmol $H^+ \cdot ha^{-1} \cdot yr^{-1}$ in a soil depth of around 1 m. For most forest soils developed on magmatic and sedimentary rocks poor in silicates, the rate is around 0.5 kmol $H^+ \cdot ha^{-1} \cdot yr^{-1}$. This is in the range of soil acidification due to forest increment (cf. Fig. 1). If the harvesting is limited to the stemwood without bark, the acid production remaining in soil is around 0.3 kmol $H^+ \cdot ha^{-1} \cdot yr^{-1}$. In general one can say that for many forests the soil acidification caused by timber harvesting is balanced by silicate weathering.

This means, however, that any additional acid load leads to a decrease in exchangeable bases (Ca,Mg,K). The above calculation leads, therefore, to the

TABLE 5

Acid load and acid buffering for a rotation period of 100 years

	kmol H^+/ha
A. Acid load	
1. Accumulation of cation excess in standing biomass	20 to 40 (beech) (spruce)
2. Exported timber	around 30
3. Decrease in nitrogen storage of soil by nitrate leaching	between 0 and > 100
4. Acid deposition	between 50 and 300
Sum:	between 100 and > 400
B. Acid buffering in soil (1 m depth)	
1. Silicate weathering (release of Ca,Mg,K,Na)	between 20 and 100 (200)
2. Leaching of exchangeable Ca and Mg from cation exchange	7 to 700 (sandy soils)(clay soils)

estimate that an amount of exchangeable bases equivalent to the accumulated acid deposition (50-300 kmol H^+/ha) has been leached from forest soils or subsoils during the last century, half since 1950, one third since 1970. Assuming 100% base saturation (which is not true especially for the A horizon of many forest soils), the amount of exchangeable bases in the uppermost metre of forest soils varies between 7 (in sandy soils) and 700 (in clay soils) kmol ion equivalents (IE) per ha (see Table 5). If the initial state was 50% base saturation, the values would be 3.5 to 350 kmol/ha. The comparison in Table 5 between acid load and acid buffering during a rotation period of 100 years shows clearly that acid deposition must have led to extremely low base saturation in great soil depth in many forest soils.

From these data it can be concluded that a base amount equivalent to between 1000 and above 4000 kg exchangeable Ca has been leached from forest soils or subsoils as a consequence of acid deposition. Regional surveys have indeed shown that in the majority of forest soils in Northwest Germany, the base saturation in the rooted mineral soil down to 60-80 cm is extremely low (< 10%). The same is true for the storages of exchangeable Ca, Mg and K in mineral soil (ref.10). In the majority of these soils, the mobile nutrient stocks in mineral soil are too small to cover the demand of aggrading forests. From this it follows that aggrading stands will pass into strong nutrient deficiency and acid stress before reaching economical value.

In all these soils the acidification front, i.e. the transitional zone from base depletion to the release of toxic cation acids, now lies below the rooted soil in the percolating layer. According to the very little data available the acidification front is now between 1 and 8 m deep (in sandy soils).

9 THE EFFECT OF SOIL ACIDIFICATION ON FORESTS

Soil acidification can be the cause of serious damage to soil organisms and plant organs (roots). The concentrations and interrelationships of individual ions involved in the soil solution are excellent ecophysiological parameters which can be used to assess and analyse such damage. Examples include: H/Ca and Ca/Al ratio as a measure of proton stress and aluminium toxicity (refs.11, 12); Mg/Al ratio as a measure of aluminium-induced magnesium deficiency (ref.13). The results of laboratory or green house experiments on the meaning of ion ratios in the solution phase as ecophysiological parameters can be extrapolated to field conditions. Under field conditions the same parameters can be measured as a function of time and space. Threshold values for acid stress are compiled in Table 6.

The long-term trend of the acid stress parameters in the soil solution shows up in the seepage water collected in the forest in a depth of around 1 m (Figs. 1 and 2 for Solling and Harz). The horizontal lines represent the threshold values given in Table 6. The figures show a distinct trend of increasing acid stress with time to levels of high risk. In the Harz, Lange Bramke, needle yellowing started in spring 1982. For the spruce stand in the Solling, fine root biomass data exist since 1982. They show a significant reduction of fine roots from the subsoil, which has now reached the 10-20 cm depth layer (ref.14). Slight needle yellowing is apparent since many years. The reduction of the root system from the subsoil and the damaging of fine roots during acidification pushes increases water stress in dry periods and cold winters. The needle loss from spruce trees, which is a main feature of present forest damage, is probably the result of a systemic water stress in the tree. The water stress is the consequence of root damage, which itself is the consequence of soil acidification. The soil acidification is driven by acid deposition.

Ecosystem elasticity is reduced by soil acidification. Nutrient depletion will show up only in an advanced stage as nutrient deficiency in trees because the nutrient input from anthropogenic air pollutants covers the nutrient needs of existing stands of trees. Damage occurs in particular when growing stands take the last base and nutrient supply from the soil and/or when the external acid load is accompanied by a high acid load generated by the ecosystem itself. This is particularly the case in mull soils with a high humus content where the organic nitrogen supply of the mineral soil is broken down and the nitrogen is leached out as nitrate.

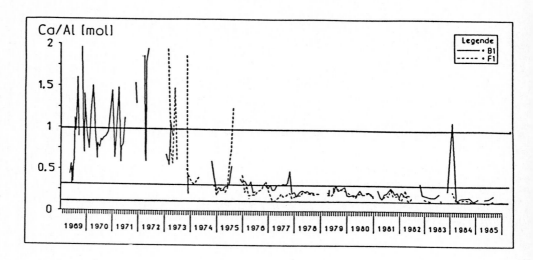

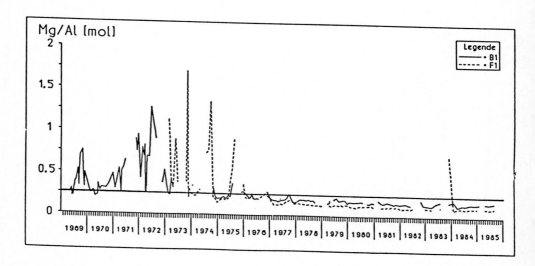

Fig. 1. Time course of Ca/Al and Mg/Al-ratios in the soil solution in 1 m depth in the Solling, European beech (B1) and Norway spruce (F1). The horizontal lines represent the threshold values given in Table 6.

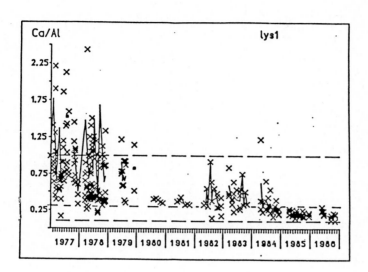

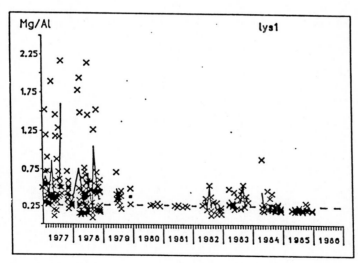

Fig. 2. Time course of Ca/Al and Mg/Al molar ratios in the soil solution in 1 m depth in the Lange Bramke watershed, Harz. The horizontal lines represent the threshold values given in Table 6.

TABLE 6

Ca/Al and Ca/H ratios in soil solution as criteria of elasticity expressed in terms of toxic effects on beech and spruce roots

Species	$\dfrac{\text{mol Ca}^{2+}}{\text{mol Al}^{3+}}$	Risk of Al-toxicity and symptoms observed in roots
Spruce	> 1	No effects, healthy roots
	0.3-1	Low risk; increased turnover of roots expected; elongation of main roots is markedly decreased
	0.1-0.3	High risk; very little root elongation observed; growth of small roots strongly restricted
	< 0.1	Very high risk; extensive injury to fine roots; new growth and elongation strongly restricted
Beech	0.1-1	Low risk
	< 0.1	High risk of Al-toxicity
Species	$\dfrac{\text{mol Ca}^{2+}}{\text{mol H}^{+}}$	
Spruce + beech	> 1	No danger of H-toxicity if pH > 3.0
Spruce	> 0.1	Less risk to H-toxicity if pH < 3.0
Beech	> 1.0	
Spruce	< 0.1	High to serious danger of H-toxicity to roots
Beech	< 1.0	
Species	$\dfrac{\text{mol Mg}^{2+}}{\text{mol Al}^{3+}}$	
Spruce	<0.2-0.3	Blocking of Mg uptake including Mg deficiency with needle yellowing

10 SOIL IMPROVEMENT

Productivity is endangered by soil depletion and acidification (ref.15). Stable ecosystems of high elasticity are required. Such ecosystems require soils with at least an average base supply and a balanced nutrient supply. In theory it is possible to restore soils to this state by liming and fertilising but the soil's original ability to store nutrients can never be restored. Enormous problems are, however, posed in practice. On the one hand, forest soils are acid to very great depths, which means that lime (and if necessary fertiliser) has to be dug in to a great depth or - if measures are limited to the surface - it can take decades to reduce the acidity of the mineral soil. On the other hand, liming, fertilising and tilling may take substances out of the soil which, in turn, pollute water (nitrates and heavy metals) and the air (nitrogen oxide). Although much is already known about liming and fertilising forests, that is not enough to be able to minimise the negative environmental effects in individual cases.

Heavy metals can in the long term be converted into insoluble compounds by increasing base saturation. This will reduce their ecophysiological effectiveness.

11 COMPARISON OF THE LOAD ON DIFFERENT TYPES OF ECOSYSTEM

Much fewer air pollutants are deposited on soils on which annual crops are grown or grassland than on forest soils. Where land is worked the accumulating potentially toxic substances are mixed into the soil and the rate of increase in concentration slowed down. The acid input into soils (or produced by the ecosystem itself) can continually be offset by liming. In these soils the load resulting from fertilisation and tilling is much more serious than that arising from deposition.

In woods of all types, even in orchards, a greater percentage of the acid input may in some circumstances reach the soil via the root system; the tree retains the acid in its leaves and draws up the base it requires from the soil. The effect of this is for acid to enter the soil throughout the area around the root system. The resultant acidification is a specific problem of forest ecosystems. In principle this acidification is reversible. However, it is impossible to offset the harmful effects of acid deposition on tree root systems by liming or fertilisation. The only alternative is to drastically reduce emissions of air pollutants, in particular SO_2, NO_x, heavy metals and hydrocarbons.

Air pollution means that the environment of the ecosphere is changed. A change in its environment leads compulsorily to a change in the ecosystem. The question is how rapid the change becomes visible (or measurable), and wether the change is of advantage or disadvantage for man. Man is not able to destroy nature, but he is able to destroy the ecosystems which give him support (e.g. as renewable resources of high productivity and product quality), and he may be able to destroy himself. The existing knowledge about air pollution quantifies the adverse effects which are exerted especially on terrestrial ecosystems containing long-lived species and on waters. The risks encountered are so high that all measures should be taken to diminish air pollution as rapid as possible.

REFERENCES

1   B. Ulrich, Stability, elasticity and resilience of terrestrial ecosystems under the aspect of the matter balance, Ecol. Studies (Springer), 61 (1987) 11-49.
2   B. Ulrich, R. Mayer and P.K. Khanna, Deposition von Luftverunreinigungen und ihre Auswirkungen in Waldökosystemen im Solling, Schriften forst. Fak. Univ. Göttingen, 58, 1979, 291.
3   B. Ulrich, in C. Troyanowski (Editor), Pollution and Plants, VCH Verlagsges., Weinheim, 1985, pp.149-181.
4   R. Schultz, Vergleichende Betrachtung des Schwermetallhaushalts verschiedener Walsökosysteme, Diss. Univ. Kassel, Ber. Forschungszentrum Waldökosysteme, Univ., Göttingen, 32, 1987.

5   G. Büttner, N. Lamersdorf, R. Schultz, B. Ulrich, Deposition und Verteilung chemischer Elemente in küstennahen Waldstandorten, Ber. Forschungszentrum Waldökosysteme, Univ. Göttingen, B1, 1987, pp.172.
6   D.L. Godbold and A. Hüttermann, Environmental Pollution (Series A), 38 (1985) 375-381.
7   D.L. Godbold and A. Hüttermann, Water, Air, Soil Pollution, 31 (1986) 509-515.
8   E. Matzner, D. Hübner and W. Thomas, Z. Pflanzenernähr. Bodenk., 144 (1981) 283-288.
9   E. Matzner, Water Air Soil Pollution, 21 (1984) 425-434.
10   B. Ulrich, in G. Glatzel (Editor), Möglichkeiten und Grenzen der Sanierung immissionsgeschädigter Waldökosysteme, FIW, Univ. für Bodenkultur, Wien, 1987, pp.1-33.
11   K. Rost-Siebert, Untersuchungen zur H- und Al-Ionen-Toxizität an Keimpflanzen von Fichte (Picea abies Karst.) und Buche (Fagus sylvatica L.) in Lösungskultur, Diss. Univ. Göttingen, Ber. Forschungszentrum Waldökosysteme, Univ. Göttingen, 12, 1985, pp.219.
12   U. Junga, Sterilkultur als Modellsystem zur Untersuchung des Mechanismus der Aluminium-Toxizität bei Fichtenkeimlingen (Picea abies Karst.), Diss. Univ. Göttingen, Ber. Forschungszentrum Waldökosystem, Univ. Göttingen, 5, 1984, pp.1-173.
13   A. Jorns, Ch. Hecht-Buchholz, Allgem. Forstzeitschr., 41 (1985) 1248-1252.
14   D. Murach, E. Matzner, The influence of soil acidification on root growth of Norway spruce and European beech, in Woody Plant Growth in a Changing Chemical and Physical Environment, IUFRO Workshop, Vancouver, in press (1987).
15   B. Ulrich, Forstwiss. Centralbl., 105 (1986) 421-435.

AIR POLLUTION EFFECTS ON AQUATIC ECOSYSTEMS AND THEIR RESTORATION

A. HENRIKSEN, Norwegian Institute for Water Research, Box 33 Blindern, 0313 OSLO 3 (Norway).

1. INTRODUCTION

A growing awareness of acid rain and its environmental effects has led to reports of acidified fresh waters, and damage to fish and other aquatic organisms, from an increasing number of areas in Europe. What was long viewed as a problem limited to Norway and Sweden has now emerged as a phenomenon affecting nearly every country in Europe.

The very first reports of acidification in Europe came as long ago as the early 1900s from Norwegian fisheries inspectors. Fish kills of Atlantic salmon (<u>Salmon salar</u>) were reported from southwestern Norway as early as 1911 (ref. 1). Several fish hatcheries had already installed limestone filters by the 1920s to treat acid water. Brown trout (<u>Salmo trutta</u>) began disappearing from mountain lakes in the 1920s and 1930s and by the 1950s barren lakes were reported from many regions in southernmost Norway. By this time, also the salmon had essentially disappeared from 7 major rivers in southernmost Norway (Fig. 1), while, for 68 rivers from other parts of the country (Fig. 1), the catches even increased from the 1940s.

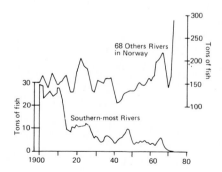

Fig. 1. Average salmon catch for 7 rivers in southern Norway that receive acid precipitation are compared with average for 68 other rivers that do not receive acid precipitation (ref. 2).

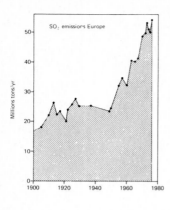

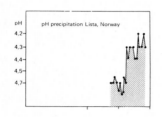

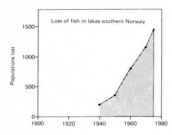

Some of these trends were due to changes in salmon management and exploitation practices. However, the decline in inland fisheries in southernmost Norway was apparently limited to acid waters. The cause of the acidity remained unexplained until the late 1950s, when the results from the European precipitation chemistry network were published (ref. 3). Dannevig (ref. 4) suggested that there was a link between acid precipitation, acid fresh waters, and the loss of fish populations. The increase in precipitation acidity and increase in sulfur emissions in Europe parallels the increased use of fossil fuels (Fig. 2a). The doubling in sulfur emissions over the 1950-1970 period resulted in an approximate doubling of the acidity in precipitation measured at one of the European precipitation chemistry network stations on the south coast of Norway over the same period (Fig. 2b). During the same period the number of fish populations that had disappeared increased tremendously (Fig. 2c). These lakes having lost fish populations were found mostly in granitic areas.

Fig. 2. Increase in consumption of fossil fuels (Fig. 2a) brings about increased acidity in precipitation (Fig. 2b) and increased fish disappearance in southern Norway (Fig. 2c) (ref. 5).

2. FOSSIL FUELS PRODUCE ACID RAIN

Acidic precipitation is coupled to the emissions of the pollutants sulphur dioxide and nitrogen oxides mainly formed during the burning of fossil fuels such as coal and oil or from the smelting of ores containing reduced sulphur. These compounds are oxidized in the atmosphere and dissolve in rain or snow forming acids which fall as acidic precipitation. Some of the sulphate and nitrate is also received from the atmosphere as dry deposition. Oxides of sulfur and nitrogen occur naturally in small quantities in the atmosphere. Rain water unaffected by man normally has a pH between 5 and 6. In areas where acidification problems occur, the yearly weighted average pH of precipitation is 4.0-4.5, while episodes with pH as low as 2.4 have been observed. (The hydrogen ion concentration of any solution is reported on a pH-scale. This scale is numbered from 0 to 14. A pH value of 7 is neutral. Values lower than 7 are acid, values above 7 are alkaline. The pH scale is a logarithmic measure; that is, each change of one pH unit - say, from 5 to 4 - represents a tenfold increase in acidity.)

Oxides of sulphur and nitrogen can stay in the air for many days. In the end, however, they return to the ground (Fig. 3), either by being washed out by rain (wet deposition) or by depositing on vegetation and moist surfaces (dry deposition). The oxides may remain long enough in the atmosphere to be transported hundreds to thousands of kilometers from the point of emission. During such transport, the gaseous oxides may be oxidized to acids (sulphuric and nitric), which ultimately are washed out to give acid rain (Fig. 3).

3. CHEMICAL EFFECTS OF ACID PRECIPITATION

The extent to which acid precipitation can change an aquatic system is determined by bedrock geology, the nature of the overburden, and the hydrodynamics of the system. These factors determine the capacities of the soil and water to neutralize acids and resist pH changes in surface waters.

The composition of precipitation is modified by a number of processes when passing through a watershed. The total sum of all the various processes occurring in watersheds determines the chemistry of the runoff. We can use the resultant water chemistry as an indication of the ability of a watershed to neutralize acidity.

Areas having appreciable amounts of calcareous rocks are not sensitive to acid precipitation, because the rocks generate large amounts of alkalinity during weathering. Small amounts of limestone in a drainage basin exert overwhelming influences in terrains that would otherwise be highly vulnerable to acid precipitation. The most sensitive watersheds are commonly those with granitic or gneissic bedrock and thin soils.

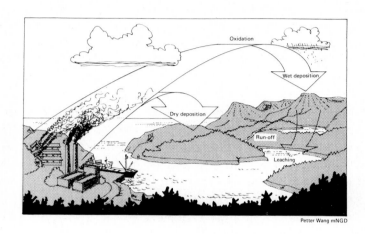

Figure 3. Sulphur dioxide and nitrogen oxides are emitted from industrial complexes and towns. A part of the oxides are dry-deposited near or at a distance from the emission source. The longer the oxides remain in the air, the more likely they will be oxidized to sulphuric and nitric acids which are carried down by precipitation (wet deposition).

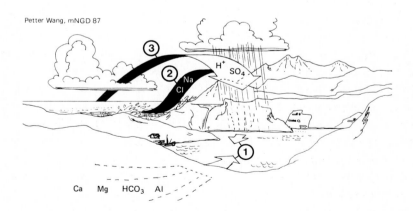

Fig. 4. The chemical composition of poorly buffered lakes lacking local pollution is largely determined by contribution from three sources.

The chemical composition of dilute lakes without direct influence by human activities is mainly derived from three sources: 1) ions that are derived from weathering-, biological- and ion exchange processes in watersheds (calcium, magnesium, bicarbonate and aluminium); 2) sea salts from sea spray (sodium and chloride principally); and 3) ions such as hydrogen, sulphate, nitrate and ammonium associated with contaminated precipitation (Fig. 4). Nitrate and ammonium are normally consumed in the watershed by biological processes. Under certain conditions nitrate can exceed the requirements of soils and vegetation and move from the watershed to the lake in the same manner as sulphate. Dissolved and particulate organic carbon (TOC) is also derived from watersheds, some of which may be comprised of organic acids.

It is generally considered that sea salts are transported with precipitation in proportions similar to those in sea water and that chloride in lake water is only derived from sea salts in precipitation. Then, sea salt contributions to lakewater can be subtracted for all of the ions based on their ratios in sea water. In this way we can estimate the ions supplied from sea-salts, air pollutants and watershed processes.

Waters in sensitive areas not receiving acid precipitation contain approximately equal amounts of calcium + magnesium and bicarbonate. Calcium and magnesium ions account normally for about 90% of the total alkalinity. The hydrogen-ion concentration is low in such waters, so that pH depends largely on the bicarbonate concentration. Bicarbonate alkalinity acts as a buffer against incoming hydrogen ions. In areas having calcareous rocks or soil, the bicarbonate concentration will be high and yield a pH of 7 to 8. In areas with quartz-bearing rocks, waters will have low concentrations of bicarbonate, and natural pH-values for such waters will be around 6.

The influence of acid precipitation on runoff water is mainly a function of the acidity of the precipitation and the geologic and hydrologic characteristics of the watershed. A watershed's ability to neutralize acidity is related to its ability to "produce" alkalinity (bicarbonate). Watersheds with thin soils and siliceous bedrock such as granites or granitic gneisses normally produce little alkalinity. Simply stated, when the acidity of precipitation exceeds the ability of the watershed to produce alkalinity, the runoff water will be acid.

In the areas of Scandinavia containing granites or granitic gneisses, alkalinity production ranges from about 15-60 keq/km^2/yr. When the deposition of acid components exceeds about 15 keq/km^2/yr, runoff water will be acid in the most sensitive areas; those with the lowest concentrations of base cations in the runoff water. This deposition of acidic components is equivalent to an annual average pH in precipitation of about 4.8 in the areas of Norway receiving 800-1000 mm of precipitation annually depending on the amount of dry deposition.

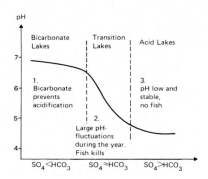

Fig. 5. As the deposition of sulphate increases, the chemical balance in the affected lakes follows the full line from left to right until the loss of alkalinity is complete and the lakes have a low but stable pH. In areas where lakes have high buffering the chemistry changes little and the lakes never reach an acidic state. During acid episodes the pH will decline along the curve and return in much the same way after the episode is over.

When a watershed receives acid precipitation, sulphuric acid (and nitric acid) can influence water quality. At constant concentrations of base cations and organic anions, there will be a decreasing concentration of bicarbonate with an increasing concentration of sulphate. When bicarbonate decreases, pH also decreases. As runoff water is acidified, pH will decline as shown in Fig. 5. The excess sulphate in acidic waters is associated with hydrogen ions and aluminium, such as is found in acidic lakes. Waters with low concentrations of ions will be near or in the zone of transition lakes shown on Fig. 5. These lakes will be sensitive to episodic acidification associated with snowmelt or storm runoff, while the chronically acidic lakes will have low pH throughout the year.

Poorly-buffered, low ionic-strength soft waters are generally located in environments dominated by carbonate-free, highly-siliceous overburden and soil. Such areas can, in turn, be found under 3 sets of circumstances, namely, (1) glaciated areas on granitic or other highly-siliceous bedrock, with overburden and soil derived from these materials; (2) areas with thick overburden of siliceous sands; and (3) areas with relatively old, well-weathered and well-leached soils. All three types of areas are found in Europe (Fig. 6). Most of the lakes scattered over the granitic areas in Scandinavia and Scotland are of the first category. Lakes on sand plains in Denmark and Holland are of the second category. Streams in several areas of Germany belong to the third category. In areas having thin soils, the portions expected to contain sensitive fresh waters can usually be located with the aid of bedrock geology maps. This method applies to most of Scandinavia and northern Great Britain, as well as to mountainous areas elsewhere in Europe.

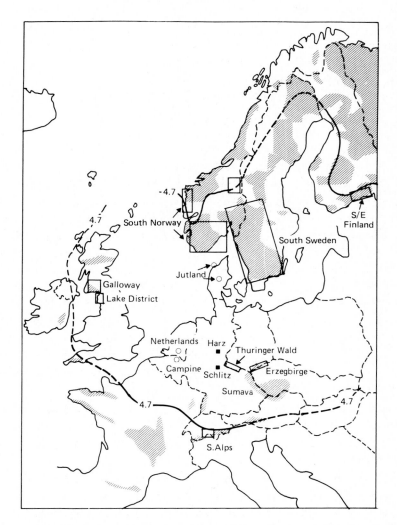

Fig. 6. Map of Europe areas sensitive to fresh-water acidification (from bedrock geology). Areas with the pH 4.7 isoline receive precipitation of acidity exceeding the "threshold" for ecological effects in the most sensitive waters. Areas from which acid fresh waters (pH less than 5) have been reported lie on sensitive bedrock (squares) or siliceous sands (circles) (after ref. 5).

3.1 Dose/response

The concept of dose-response is widely used in connection with water pollution (ref. 6). Any lake system subjected to an external dose (load) of pollutants will have an internal resistanse (or buffer capacity) to the change. The response of the lake system will depend on the relative magnitudes of the dose and resistance parameters.

The dose-response concept is used in many mathematical and empirical water quality models. Empirical models of nutrient pollution of lakes (ref. 6) and the empirical ion balance model of lake acidification (ref. 7) (Fig. 7) are both based on this principle. There are, however, major differences in the models. The Vollenweider model is developed from log-log plots of lake characteristics, phosphorus loading and mean depth. By contrast, the empirical acidification model is built from simple linear relationships with the important constraint of electroneutrality of the solution. Thus, this model is based on theoretical considerations. It does not suffer from the inherent variability in the Vollenweider model, and it incorporates basic elements of water chemistry. The slopes of the lines in the eutrophication model are given as phosphorus concentrations (areal loading divided by mean depth), while the slopes in the acidification model are in hydrogen ion concentration (pH), both of which generally describe the conditions of lakes in meaningful ways. The polluted condition (acid or eutrophic) is placed above the unpolluted condition (bicarbonate or oligotrophic) in each case. The acidification model estimates the pH of lakes as a function of sulphate (or H^+) in precipitation, as reflected in the concentrations of base cations and sulphate in lakes, based on ion balance considerations. The model is simple, but it is based on empirical water chemistry and its predictions must obey ion balance.

4. BIOLOGICAL IMPACTS OF FRESHWATER ACIDIFICATION

The acidification of freshwaters had profound impacts on aquatic life, and all trophic levels are affected.

4.1 Effects on fish

The best documented biological aspects of fresh-water acidification are the decline and loss in fish populations. Acidification has by now eliminated fish populations in lakes in Norway extending over an area of 13,000 km^2 and is affecting an additional area of 20,000 km^2 (ref. 9). Findings from the Norwegian Monitoring Program for Long Range Transported Air Pollutants indicate that the area of affected fish populations continues to increase

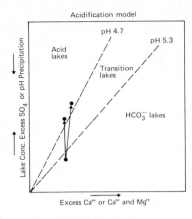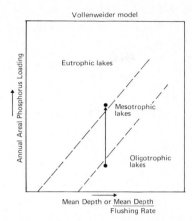

Fig. 7. Comparison of two empirical dose-response models, the empirical model of acidification (ref. 7) and the Vollenweider (ref. 6) model of eutrophication. The different paths of response of the lake to increased acid (sulphate) loading is determined by the amount of increase in the weathering of calcium and magnesium (F-factor, ref. 8).

during the 80s, and that in addition to the new areas, the losses continue in the "old" affected areas (ref. 10). The situation in southern Norway provides an excellent illustration of the basic concepts of fresh-water acidification (fig. 8).

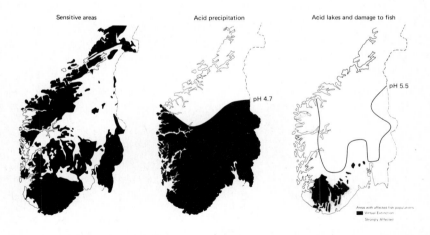

Fig. 8. Acid lakes (pH < 5.5) having affected fish populations are found in areas with sensitive bedrock (mainly granitic) receiving acid precipitation

lakes (pH < 5.5) are found only in areas having sensitive bedrock (mostly granitic) receiving acid rain (pH < 4.7). When a lake starts to show a sparse fish population, acidification has usually been exerting its effect on the fish population for a number of years.

The responses of fish to changes in water chemistry depend on the species, age, size and maturity (ref. 11). In most lakes the cause of fish loss appears to be the loss of reproduction (ref. 12), either by excessive roe mortality between fertilization and hatching, or mortality to fry or among spawners. The Atlantic salmon is the most sensitive species among salmonids in Norway, and the smolt stage (the period of migration to sea) is the most sensitive life stage (refs. 13 and 14). Observations from field sites and from well-designed laboratory experiments indicate strongly that concentrations of calcium and monomeric inorganic aluminium together with pH are the most important chemical factors in determining the fishery status of lakes and streams. Acid streams can usually support a trout population if the calcium concentrations are sufficiently high. In waters containing organic acids some of the aluminium is bound in organic complexes which are much less toxic to fish.

The direct causes of fish kills depend on the developmental stage of the fish. Eggs appear to be most sensitive to pH (H^+) alone (ref. 15), while labile aluminium combined with pH is toxic mainly at stages after hatching (ref. 16).

The toxic effect of the combination of pH (H^+) and labile aluminium seems to be a combined, i.e., a synergistic effect leading to an impaired ion balance over the gills (refs. 13,14,17), but H^+ and aluminium seem to act differently. H^+-ions appear to act directly on the ability of the cell membrane to retain the ions (i.e., they increase permeability) (ref. 18). Labile aluminium acts more directly by reducing the enzyme system that control the active ion uptake over the gills (ref. 19). The ameliorating effect of calcium on critical H^+/aluminium concentrations has been demonstrated (refs. 20,21). It appears that calcium reduces the permeability of ions over the cell membrane (ref. 18), thus reducing the loss of plasma ions and preventing fish kills.

As already pointed out, the toxicity of aluminium depends on the ionic species present. It is not possible to specify absolute levels of toxicity, but salmonid populations fail to survive if the pH stays below 5.2 for long periods, the concentration of labile aluminium exceeds 50 µeq/l and the calcium concentration falls below 1 mg/l.

Recent investigations show that Alantic salmon are killed during short acidic episodes, when pH decreases and the concentrations of ionic aluminium increase rapidly (ref. 22). Fish kills are documented in controlled experiments combined with continous monitoring of pH and daily water sampling (Fig. 9). During a two-week period, four episodes with pH drops from 5.9 to

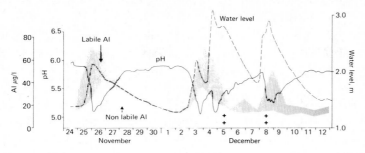

Fig. 9. Variations in pH. water level, and labile aluminium (shaded area) in River Vikedalselva (southwestern Norway) Nov. 24 to Dec. 12, 1983. pH-curve is plotted on readings every second hour from a continuous pH monitor. Water levels are from 4 h readings of limnigraph records. Aluminium data are from daily samples. Times of death of salmon presmolts (1+) are indicated (+ = 1 fish) (ref. 22).

5.1 coincided with increased water flow due to rainfall and snowmelt, accompanied by dilution of calcium and substantial changes in aluminium-speciation (Fig. 9). The concentration of ionic Al increased from 0 to 50 µeq/l during the pH-drops. These rapid changes in Al-speciation could be due to dissolution of previously precipitated Al on the river bed by episodic flushing of the river from acid snowmelt and storm events. Of the three year-classes of salmon (eggs to presmolts), only the presmolts died, illustrating the higher sensitivity of salmon during smoltification (refs. 13,14). Continuous monitoring of water chemistry and fish behaviour in waters subjected to acid episodes are considered a key to the understanding of biological responses to the acidification process.

4.2 Effects on other aquatic organisms

Acidification of water affects all categories of organisms in the aquatic ecosystem (ref. 23). Effects on fish are the most visible as fish will disappear completely from highly acidified lakes and streams.

Acidification also has comprehensive direct and indirect effects on primary producers, decomposers and consumers, resulting from the disappearance of fish (refs. 24,25,2).

4.2.1 Primary producers

Acidification of rivers and lakes affects their planktonic and benthic algae communities. There appear to be clear tendencies toward a reduction in number of species, and green algae (Chlorophyceae) seem to be seriously affected (refs. 24,26,27). Although the number of phytoplankton species is reduced, no reduction in biomass and production in acid water relative to a more neutral environment has been observed, provided that the concentration of phosphorus remains the same (refs. 24,28).

Diatom populations shift toward more acid-tolerant forms, and diatom analyses are presently used in the study of sediments as a method of demonstrating the historical development of the acidification process. The use of diatoms in reconstructing pH of lakes has evolved rapidly during the the last 10 years and has reached a point when the technique is well tested (ref. 29). It offers today the only means whereby the pH-history of a lake can be estimated, both in terms of trends and terms of past pH-values. The sediments of lake Gårdsjöen near the west coast of Sweden (ref. 30) (Fig. 10) and the Round Loch of Glenhead in the Galloway area in south west Scotland (ref. 31) (Fig. 11) provide excellent examples of the use of the diatom technique.

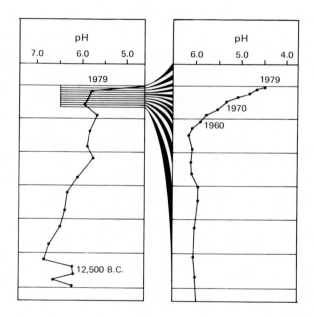

Fig. 10. The sediment in Lake Gårdsjön on the west coast of Sweden has been studied with respect to the composition of diatoms back to the last Ice Age. The figure covers along time perspective (rather more than 10,000 years B.C. and down to our own time), and also a segment of the time from the beginning of the 1900s to end of the 1970s. The changes in the siliceous-algae composition become discernable as early as the 1950s, but the tendency is most dramatic during the 1960s and 1970s. (Adapted from ref. 30.)

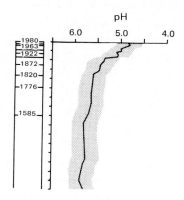

Fig. 11. Diatom diagram for Round Loch of Glenhead (Galloway, Scotland) shows that the pH of the lake has declined by about 1 pH unit (from ~ pH 5.5 to 4.5) during the last 130 years. (Ref. 31.).

Expected patterns are now emerging from diatom studies carried out in many countries: lakes in sensitive bedrock areas within the regions of high acid deposition have have been acidified, whereas similar sites outside these regions have not. All the data so far available show a time and space pattern that is consistent with the hypothesis of acid deposition producing acidic lakes in sensitive areas.

Rapid growth (increased biomass) of filament algae and mosses is a typical acid water phenomenon (refs. 23,32). The number of species comprising the periphytic algal population is reduced. Increased biomass is most easily explained by a decrease in feeding by invertebrates, parallel with a reduction of microbiological decomposition. Investigations demonstrate that the photosynthetic activity of algae is profoundly reduced by reduced pH (ref. 23), indicating that increased biomass does not imply increased primary production.

Among macrophytes, a pronounced increase in peat moss (Sphagnum) has taken place at the expense of grasses (Lobelia and Isoetes) in some places. The most significant increases appear to have taken place in Swedish lakes (ref. 33), but the phenomenon has also been observed in the U.S. (ref. 32) and in isolated cases in Norway (ref. 34).

4.2.2 Decomposers

Many acid lakes fall under the definition "clearwater lakes", a term suggesting great clarity. Data from certain areas suggest that depth visibility will increase during the acidification process (ref. 26). One reason may be reduced primary production, because the phosphorus content in the water tends to decrease. Another cause may be a change in the optical characteristics of the humus, as well as the fact that less humus will remain in suspension when the pH level falls (ref. 35). At the same time there seems to be an increase in non-decomposed organic material. This phenomenon has lead to a hypothesis, suggesting that acidification contributes to an oligotrophic situation (ref. 33).

An increase in non-decomposed organic material (leaves, twigs and other allochtonous material) indicates a reduction in microbiological decomposition, i.a. of leaves, with falling pH levels (ref. 36). Microbiological decomposition also appears to be reduced in sediment cores with a trend away from from bacteria in favour of fungi as the metabolizer (ref. 32). These circumstances may lead to reduced recirculation of limiting materials for primary producers (e.g. phosphorus), a condition which can have a direct limiting effect on energy turnover.

4.2.3 Consumers

There is a clear correlation between increasing acidification and a reduction in the number of zooplankton species in water (refs. 23,24,27,36,37) Species reduction is seen in all major categories of zooplankton; however, it is more evident in certain groups than others. The genus Daphnia, a group belonging to the Cladocera (water fleas), is particularly affected (Fig. 12).

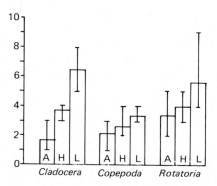

Figure 12. Average number of zooplankton species of the groups Cladocera (water fleas), Copepoda and Rotatoria in 3 types of water from western Norway. A: Acid water; H: Humus water; L: Less acid water (adapted from ref 36).

Marked changes in species dominance, in parallel with a reduction in the number of species, are often observed at certain times of the year. A combination of biotic and abiotic factors, such as altered predation patterns for insects and fish and different physiological tolerance to variable water qualities, is probably responsible for these changes.

4.2.4 Insects

Hatching is one of the most critical periods in the life of an insect (refs. 38,39) Insect survival in acid water depends upon water quality at the time of hatching. Because many species hatch simultaneously with spring thawing, serious consequences may arise when runoff water is acid. A reduced number of insect species has been found in water affected by increasing acidification. Certain groups, such as midges (Chironimidae) and Mayflies (Ephemeroptera) appear to be more sensitive than others (refs. 40,41,42). The Mayfly species Baetis rhodani is especially vulnerable, and is presently used as an indicator when monitoring acidification and tracing early acidity before effects on fish are apparent.

While some species vanish, others appear able to take advantage of changes in the natural competitive situation. A typical example is found among the waterboatmen (Hemiptera), where the Corixidae increase significantly in number and importance as fish food with increasing acidification (refs. 12,43). This is partly caused by reduced predatory pressure from diminishing fish populations during increasing acidification, and partly because of Corixids high physiological tolerance to acid water (ref. 44).

4.2.5 Crustaceans, snails and bivalves

Benthic invertebrates are an important source of food for fish. As with other groups described, they show a decreasing number of species present as acidification progresses (Fig. 13) (refs 40,45).

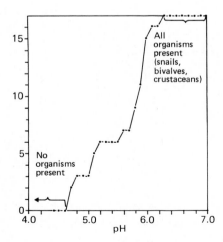

Fig. 13. Distribution of 17 common species within the categories snails, mussles and crustaceans as a function of water pH. (Adapted from ref. 45).

Among the most sensitive crustaceans are (<u>Lepidurus arcticus</u>) and amphipods (<u>Gammarus lacustris</u>) (refs. 46,47), whilst <u>Asellus aquaticus</u> is more tolerant. While the first two seem to dissappear at pH < 6 and pH < 5.5 respectively, <u>Asellus aquaticus</u> has been found as low as at pH 4.8. Adult freshwater crayfish (<u>Astacus astacus</u>) appear to tolerate fairly acid water, but it has been shown that osmoregulatory disturbances appear when crayfish are exposed to acid water with a high aluminium content (ref. 48). Disturbance of the reproductive process resulting from increased acidity, is however most likely responsible for the disappearance of crayfish (ref. 49).

At pH below 6.0, the presence of freshwater snails with calcareous shells and small bivalves (Sphaeridae) can be drastically reduced. This is particularly observed among those species which do not live in benthic sediments (ref. 43,40,50).

5. THE FUTURE, WHAT CAN BE DONE?

Most scientists now accept that the change in chemical composition of precipitation over the last decades is the major cause of the acidification of fresh waters in Scandinavia, North America, and other exposed regions in the world having similar geology. Natural acidification processes may lead to runoff waters characterized by lower alkalinity, thus rendering them more sensitive to acidification by external acid inputs, rather than to large numbers of acidic lakes.

Forest damage in Europe is extensive, accelerating, and of immense economic significance. Heavy damage to historical monuments and artwork has been recorded in many European countries. Continued acid precipitation may lead to further damage to material, forests, soil, and water. More research is always needed; however, the scientific work carried out so far offers a sufficient basis for action today.

As with most environmental problems mitigation of acidification can take two forms (1) - treatment of the causes, such as by reducing emissions and (2) -treatment of the symptoms, for example by liming.

5.1 Reducing emissions

As a first step for the mitigation of acidification the Scandinavian countries have initiated measures for implementing a 30 percent reduction in national sulfur emissions (or their transboundary fluxes) by 1993. The levels of emissions in 1980 were used as a basis for the calculation of reductions. This important step in reducing emissions was taken by 34 countries (almost all of the countries in Europe together with the USA and Canada) signing a convention on long-range transport of air pollutants across international boundaries. The convention was held in March 1983 and a joint resolution was drafted, which was thereafter ratified by most of the countries. In July 1983, Canada and 20 countries in Europe committed to a 30% reduction in emissions by 1993.

It is frequently discussed whether the acidification process is reversible, that is: will there be complete recovery with a 100% reduction in levels of sulphate. Such a reduction will not necessarily result in lakes returning to the same conditions of water chemistry and biology as they had prior to receiving acid rain if changes in soil chemistry have occurred. It is most important, however, that lakes should recover to levels that are acceptable to fish, and that fish should survive in areas where they previously occurred.

Observations from Sudbury, Ontario in Canada, indicate that water quality has improved rapidly with decreased deposition of sulphur. Here, emissions of SO_2 declined from an average of $1.45 \cdot 10^6$ tonne yr^{-1} in 1973-78 to $0.68 \cdot 10^6$ tonne yr^{-1} in 1979-85. As a result, sulphate concentrations in lakes in the area have decreased, and the pH of each of the acidic lakes that was studied has increased (fig. 14). Aluminium concentrations have also decreased (ref. 51).

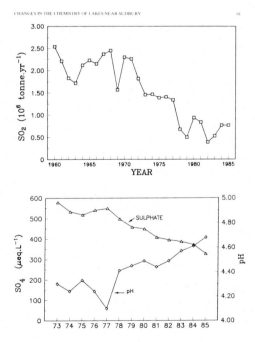

Fig. 14. a) SO_2-emissions (10^6 tonne yr^{-1}) in the Sudbury basin, 1960-1985. b) SO_4 concentration (µeq/l) and pH in clearwater lake 1973-1985. Results are whole-lake annual averages (ref. 51).

The RAIN-project (Reversing Acidification in Norway) (ref. 52) has shown that when acid rain is excluded from a small catchment, and unpolluted rain is distributed instead, a rapid response is observed (fig. 15). After three years of treatment the sulphate in runoff was reduced by about 50% relative to the sulphate runoff occurring in the reference catchment.

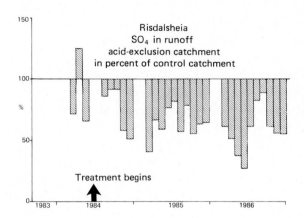

Fig. 15. Monthly average concentrations of sulphate in runoff in acid-excluded catchment relative to control catchment in percent. Concentrations in experimental catchment have decreased by about 50% during the first 2 1/2 years of treatment (ref. 54).

Data from monitored rivers and catchments in southern Norway likewise indicate declining sulphate concentrations since 1980 (ref. 53). This probably reflects the reduced discharges of sulphur to the atmosphere that have taken place in western Europe during the most recent years.

A recent analysis of data from 786 lakes in southern Norway indicates that a 30% reduction in sulphur deposition would reduce the number of acid lakes by about 20%, while a 50% reduction would reduce the number of acid lakes by about 35%. Substantial reductions in sulphur deposition will be necessary to achieve conditions acceptable to fish in most Norwegian lakes, as well as in other areas in the world strongly affected by acid deposition.

The data available so far strongly indicate that measurable reductions in acidic deposition should give a quick response in water chemistry. This should result in improved conditions for fish and other aquatic organisms.

5.2 Liming

The two forms of mitigation of acidification, reduced emissions and liming, are not mutually exclusive, and it has been argued that large-scale liming can in fact actually inhibit vigorous international efforts required to obtain reductions in emissions of the acid precursors SO_2 and NO_x. Nevertheless, extensive liming of acid waters is now being carried out in Scandinavia, particularly in Sweden.

Lime treatment has been used for several decades in hatcheries in southernmost Norway (ref. 55). As early as in 1926 intake water to hatcheries previously plagued by fish mortality was successfully limed. About 4000 lakes have by now been limed in Sweden since their liming program started in 1976. The yearly spending is about 17 mill. US dollars. The activities are more modest in Norway: the yearly spending is about 1.5 mill. US. dollars.

The lakes limed in Sweden normally have retention times of several years so that the effects of one lime application lasts for many years. In Norway most of the acidified lakes have very short retention times, mostly less than half a year. This difference in retention times is largely due to smaller lake size, higher annual runoff and thinner soil cover in Norway than in Sweden.

The use of calcium carbonate containing substances, such as powdered limestone is a safe method for liming lakes and rivers. Overdosage does not result in pH-levels which are potentially dangerous for fish and other aquatic organisms, and no serious negative effects of liming on biota have been observed in Scandinavia (ref. 56).

Liming is, however, not a cure, it is like taking aspirin for a headache. Acidification is a disease that must be cured at the source. Liming is only a temporary remedy, and must not be conducted at the expense of efforts necessary to eliminate the major environmental problem of acidification.

6. SUMMARY

One of the major environmental effects of acid rain is the acidification of lakes and rivers and the resulting decline in and finally disappearance of fish populations. Fish kills were reported in Norway as early as the 1910s. The acidification problem did not, however, receive serious attention until the end of the 1960s, when Scandinavian scientists were able to link the increasing acidity of precipitation with the increasing number of fishless lakes and rivers.

Poorly-buffered, low-ionic strength soft waters are most vulnerable to acid inputs. Such waters are in general located in environments dominated by carbonate-free, highly-siliceous overburden and soil. Such areas are found in many countries in Europe, particularly in southern Scandinavia. The chemical composition of precipitation will be modified by a number of processes when passing through a watershed. The ionic composition of the runoff water reflects, however, the net integrated effects of these watershed processes. Runoff water chemistry can thus be used as an indicator of the buffering capacity existing in the watershed and its susceptibility to acidification. Waters in areas not receiving acid precipitation (precipitation pH > 4.7) contain approximately equivalent amounts of calcium + magnesium and bicarbonate. These ions account normally for about 90 percent of the total ion content.

In areas receiving acid precipitation, the strong acids will replace an approximately equivalent amount of bicarbonate. If the amount of added acid is larger than the amount of bicarbonate originally present, the runoff water will turn acid. The acidification of fresh water may thus be considered as analogous to a large-scale titration of a bicarbonate solution with a strong acid.

Acid clear-water lakes have elevated concentrations of aluminium. Depending on the pH, inorganic ionic aluminium may be toxic to fish. The direct cause of fish kills depends on the developmental stage of the fish. Eggs appear to be most sensitive to pH alone, while labile aluminium combined with pH appears to be toxic during post-hatching stages.

Aquatic ecosystems under acidification show both reduced production and reduced decomposition. The invertebrate fauna shows reduced diversity during acidification. This applies both to zooplankton, larger crustaceans, insects, snails and bivalves. However, fish in acid waters do not disappear from the lack of food. Fewer food species survive, but those with a particular advantage in acid water will dominate. This makes the ecosystem more vulnerable to changing conditions and may mean an additional stress on the top predators, as food will not always be easy available.

It is now scientific consensus that the change in chemical composition of precipitation over recent decades is the major cause of acidification of fresh waters in exposed regions. Although liming of acidic waters have been successful, liming of acidified soils, lakes, and rivers offer no regional solution to the acidification problem. It is not the symptoms, it is the causes of acidification that must be attacked. Thus, the Scandinavian countries have initiated measures for implementing a 30 percent reduction of national sulphur emissions by 1993. By now, 20 European countries and Canada have joined the so-called "30 percent club". International cooperation across the boundaries is the only way to solve one of the major ecological disasters of the twentieth century.

may still be used today in very small mines and was used everywhere until about 1920. This was the process of jigging. The mixture of rock and ore was placed in a compartmented wooden box (jig) through which water was sluiced causing the heavier ore to sink downwards leaving the lighter rock at the top. The efficiency of jigging is poor and the effluent water from the jig contains both metal solutes and suspended ore particles. Until the late nineteenth century in Britain the effluent was passed direct into water courses where the pollutants were transported as solutes or suspended sediment. Wherever the current velocity decreased the suspended sediment was deposited and this process was particularly important after overbank flooding since a toxic silt was left behind which coated herbage and contaminated the surface of alluvial soils. In the middle decades of the nineteenth century farmers in several river valleys in Wales, notably the Ystwyth and Rheidol rivers, were vociferous in their complaints of lead pollution of crops and stock. Subsequently, the (British) Rivers Pollution Act of 1876 forced mine operators to pass their jig effluent into settling lagoons before discharging the water to rivers.

Modern mines use a different, very efficient technology called froth flotation (ref. 3). Various organic chemicals are added to a slurry of ore and rock and air is forced upwards through the mix. The many small bubbles carry the ore particles to the surface where they are skimmed off. By varying the mix of organics one ore type at a time can be separated.

The final stage is ore smelting. Here sulphide ores are heated in the presence of carbon, originally charcoal, and the metal sulphide reduced to molten metal with the evolution of SO_2. Old smelters caused much environmental damage: forests were felled to provide charcoal and the SO_2 and metal oxide fumes devastated the remaining vegetation. Accelerated soil erosion was often a consequence. In modern smelters the sulphides are first calcined to produce oxides and the sulphur oxides are trapped to generate sulphuric acid. Air quality in and around the works is monitored continuously and an unacceptable rise leads to a temporary shut down of the sinter plant. The oxide sinter is then smelted in a blast furnace. Finally, the impure metal must be treated to remove its impurities some of which may be valuable, e.g., Ag or Cd.

3.0 THE PROCESS OF POLLUTION

From this brief description of the mining and smelting process it can be seen that there are numerous opportunities for heavy metals to be released to the aqueous environment as solutions, colloids and suspended sediment or, to the atmosphere as gases and aerosols from smelter stacks (fallout) or as fugitive dusts from ore stock piles and tailings (i.e., fine rock waste) lagoons. After release they follow normal geochemical pathways until they reach sinks such as sediments, soils or biota.

Flowing water is the dominant agent in landscape evolution and weathered and eroded rocks provide dissolved and suspended materials which rivers transport. But rivers are not simple chutes and sediment is transported from its source to the sea by a complex process. The suspended sediment load is deposited as floodplain alluvium which weathers into soil. As the meander configuration of a river alters this soil is eroded by bank collapse, resuspended and later redeposited. The composition of the alluvial matrix is meanwhile enriched by deposition, adsorption and precipitation processes. River basins are therefore zones of geochemical enrichment which enhances the fertility of their soils (the classical example is the Nile valley before the construction of the Aswan Dam) but it also makes them unusually susceptible to contamination.

In mining areas, soils on slopes become contaminated through fallout and by loss of materials from waste heaps uphill. Along roads and railways there may be significant losses of ore concentrates during transportation. Areas downwind of fine grain waste can be affected by blowing dust.

The result of these various processes is a heterogeneous spread of contaminants in the landscape leading locally to very high concentrations in sediments or soils and hence in the biota.

4.0 THE PATTERN OF SOIL CONTAMINATION

Figure 1 shows the distribution of lead in surface layers (0-15 cm) of soils in Wales. It was plotted using analytical data derived from soils sampled on a regular grid at 5000 m intervals (ref. 5). The plotting intervals are derived from a cumulative frequency distribution curve of data from 845 samples after a $\log_{10}$ transformation and are the 80, 90, 95, 99 and 99.9 percentiles. When comparison is made with the distribution of metal mines (Figure 2) a close correspondence is observed. In addition, the influence of industrial south Wales is marked. Thus, a general contamination of the terrestrial ecosystem is evident at a regional level despite the coarse spacing of the sampling grid.

When considering contaminated ecosystems it is essential to know what is 'normal', that is, what ranges of metal concentrations could be expected arising from natural processes. For Welsh soils the 10, 50 and 90 percentiles were chosen as the lower, median and upper limits and are presented in Table 1 for selected metals.

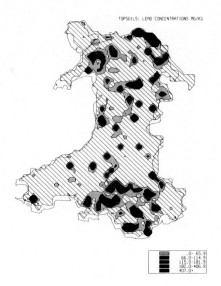

Figure 1

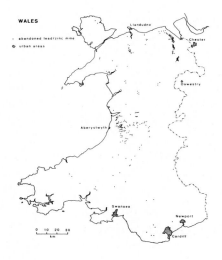

Figure 2

Table 1.   Calculated normal concentrations of metals in Welsh surface soils (ref. 5).

	mg/kg dry soil		
Metal	lower value	median value	upper value
Pb	12	29	115
Zn	19	53	115
Cu	3.6	9.3	20
Cd	0.05	0.3	1.0

Within the regional distribution shown in Figure 1 it is possible to resolve the areas of metal accumulation into different landscape segments. As would be expected from the earlier discussion, alluvial soils downstream of metal mines are frequently contaminated. The data in Table 2 demonstrate the difference between slope and alluvial soils and mining and non-mining areas.

Table 2.   Metal contents of different soils in west Wales (ref. 6).

	Range and mean (mg/kg dry soil)		
	Pb	Zn	Cu
Slope soils			
Mine area	33-1680 (270)	20-1900 (193)	5-25 (16)
No mines	17-57 (30)	35-157 (82)	3-15 (10)
Alluvial soils			
Mine area	90-2900 (1420)	95-800 (455)	17-42 (30)
No mines	24-56 (42)	80-171 (129)	14-31 (19)

The alluvial soils can be resolved further by subdivision according to sediment type and age of deposition. The former is illustrated (Table 3) by representative concentrations for Pb and Hg in soils from the Ystwyth floodplain in west Wales.

Table 3. Representative Pb and Hg concentrations in soils/sediments differentiated according to type (ref. 7).

Sediment type	mg/kg dry soil	
	Pb	Hg
Active channel sediment	1593	1.43
Old channel sediment (active during mining)	3423	1.78
Alluvial flood loam	1024	0.46
Valley side (slope)	104	0.08

Work within a developing meander loop in the neighbouring Rheidol valley (ref. 8) illustrated an inference to be drawn from Table 3, that sediments deposited during the mining period contain higher metal concentrations than modern sediments. The modern sediments can still be contaminated since they have incorporated material derived from bank collapse of contaminated riverine soils but the introduction of fresh sediment implies generally lower concentrations. The areas within the meander loop were dated from air photographs and maps of different ages. The data in Table 4 for Pb reveal a clear age dependency and the organic contents (% ignition loss at 430°C) illustrate the maturation of the soils with time. But there is little differentiation of zinc contents since the solubility of Zn is higher than for Pb and it will be noted later that Zn pollution of Welsh streams is still active.

Table 4. Soil data for dated sections of a developing meander area in the Rheidol valley of west Wales (ref. 8).

Age zone	% ignition loss	mg/kg dry soil	
		Pb	Zn
1845-1886	7.2	1500	485
1886-1904	4.6	1011	364
1904-1951	6.5	785	368
1951-1971	2.2	503	324
1972	1.4	368	243

The above data suggest that over a long period of time (many centuries) soils in valley bottoms will recover from contamination through the process of floodplain renewal. But the fate of metals deposited on stable slopes or level plateaux is unknown. Studies of the distribution with depth of metals in soil profiles lead to the conclusion that the downwards movement is generally slow over a period of about a century. The data in Table 5 are for two, pedologically similar soils (ref. 9). Both are dystrochrepts formed on Silurian shale: profile I was sampled in a non-contaminated area and profile II was situated about 100 m away from an abandoned nineteenth century mine. There is a marked accumulation of silver in the surface layers of profile II but below 40 cm the two profiles are indistinguishable. Lead concentrations are also high in the surface of profile II but there has been some movement downwards. There is therefore little evidence to suggest that leaching losses are likely to have any significance in reducing surface soil contamination in the short time in contrast with metals such as calcium or potassium where losses are appreciable and farmers have continuously to make good these losses through fertilizer applications. Nor do we know whether the contaminating metals are likely to become fixed, that is unavailable for plant root absorption, over time. One recent example from archaeological survey work in Greece (B.E. Davies and J. Bintliff, unpublished) suggests that where soils are stable, lead contamination persists for millennia and that sites of abandoned, long-vanished farmsteads are identifiable by a halo of residual lead accumulation in the soils around the site.

Table 5. Distribution of lead and silver in two profiles; I: non-contaminated and II: contaminated (ref. 9).

cm depth	I ng/g Ag	ug/g Pb	cm depth	II ng/g Ag	ug/g Pb
0-5	38	39	0-10	5300	4300
5-10	33	43	10-20	250	2500
10-15	24	43	20-30	90	495
15-20	18	43	30-40	61	270
20-25	16	49	40-50	32	180
			50-60	33	190
			60-70	32	140

5.0 CONTINUING CONTAMINATION PROBLEMS

Although metal mining has ceased and rivers are no longer burdened by fine ground waste active contamination occurs sporadically in the old mining fields. Two sources are involved. One is drainage from abandoned mines and the other is movement of mine spoil.

Before the invention, and widespread adoption in the nineteenth century, of the steam engine the only effective way of draining a mine was to extend working levels until they emerged at a hillside, thus forming a drainage adit. Some adits have been plugged but most continue to drain old workings and thereby introduce pollutants to river systems. An example is the Cwm

Rheidol mine near Aberystwyth. This mine produced sphalerite with galena in a matrix containing much marcasite (FeS_2). These minerals continue to dissolve. The adit drainage becomes aerated after it leaves the mine and oxidation causes a ferric hydroxide (ochre) to be precipitated as a coating on the channel bed. This is a widespread process in metal and coal mining areas and is called acid mine drainage:

$$4FeS_2 + 14H_2O + 15O_2 = 8H_2SO_4 + 4Fe(OH)_3$$

Metals, acidity and the suffocating coating of ochre make acid drainage a serious hazard for aquatic ecosystems. An attempt has been made to remedy the problem by passing the drainage water through limestone filter beds but the following data indicate that this is not particularly successful. The filter beds are interposed between sampling stations C and D. Water pH was raised slightly but metal contents were not significantly reduced.

Table 6. Mean metal contents and pH of filtered water samples taken on 5 occasions between November, 1981 and April, 1982 from acid mine drainage in Wales (ref. 10).

SITE	pH	Pb (ug/l)	Zn (ug/l)
A	3.2	540	6500
B	2.7	75	65000
C	3.8	450	17000
D	4.1	330	14000

It is notable in Table 6 that zinc concentrations are much higher than lead concentrations reflecting the higher solubility of zinc compounds. Zinc contents of rivers draining mineralized areas have been reported (ref. 11) as an order of magnitude greater than elsewhere and lead concentrations were raised by factors of 2-12.

When fine textured mine spoil dries it becomes liable to deflation. In Wales, although the prevailing winds are from the west and south these are generally accompanied by rain whereas easterlies are drying winds. Davies and White (ref. 12) have provided a case study of the hazards of blowing lead-rich waste. This particular mine site has since been levelled and revegetated in order to reduce the dangers to local children.

6.0 CONSEQUENCES FOR RIVER ECOLOGY

In the eighteenth century the rivers of west Wales were prized for their salmon catches but mining rendered the Ystwyth almost sterile and seriously damaged other rivers. Scientific investigations did not begin until after the effective cessation of mining in 1920. A survey of the lower reaches of the Ystwyth in 1919 (ref. 13) revealed that fauna was restricted to 9 species, mostly insects and water lead concentration was around 0.4 mg/l. Dissolved lead caused mucus to form on fish gills which then suffocated them (ref. 14). By 1922, the number of species had increased to 26, again mostly insects and the lead concentration had fallen below 0.1 mg/l. By 1939 the number of species had increased to 63 (but no fish) and lead concentrations were below 0.05 mg/l (refs. 15, 16). By 1975 fish had returned to most of the river but were still absent from a 3 km stretch below the Cwm Ystwyth mine, the highest in the river. The river Rheidol recovered much faster (ref. 17) and trout were first caught again in 1952, but salmon were not seen until 1970. In north Wales, tributaries of the river Conwy drain a group of old mines and Elderfield et al. (ref. 18) proposed that dissolved metals caused the intermittent failure of oyster larvae at a hatchery in the estuary some 18 km below the mines.

There is evidence that metal tolerance has emerged in the aquatic flora of these rivers. McLean and Jones (ref. 19) found that certain organisms appeared to be resistant to dissolved heavy metals. These included filamentous algae, especially Hormidium rivulare and the bryophyte Scapania undulata. Burton and Peterson (ref. 20) also found evidence of tolerance and metal accumulation in several aquatic bryophytes in the same area, namely Philonitis fontana, Scapania undulata and Solenostoma crenulata. Later, Jones et al. (ref. 21) concluded that Scapania undulata could be used as a biological monitor of silver pollution.

In addition to the effects on river ecosystems coastal communities may also be affected when polluted river water enters the sea and is diverted along the coast by long shore drift. Littoral and sublittoral animals are influenced by dissolved zinc coming from polluted rivers and high levels of zinc were found in barnacles (Balanus balanoides) in Cardigan Bay (ref. 22).

On land, earthworms have been shown to accumulate metals from soils contaminated by metal mining, e.g., a good correlation between soil lead and tissue lead was demonstrated for Dendrobaena rubida (ref. 23) and for the slug Arion ater (ref. 24). Small mammals and invertebrates living in the vicinity of abandoned mines also accumulate lead in their tissues (ref. 25).

7.0 CONCLUSIONS

Metal mining using simple technologies and in the absence of environmental protection laws led to a widespread contamination of ecosystems in Wales and elsewhere in Great Britain. The major period of mining activity was 1830-1870 and most mines ceased to operate by 1920. Soil contamination is still widespread. Alluvial soils, especially those formed on river channel

TABLE 1.1 - (cont.)

Element	Mean crust mg X/ kg (ref. 2)	Mean soil range mg X/ kg (ref. 2)	Freshwater range µg X/l (ref. 2)	Sea water range µg X/l (ref. 2)	Geochemical mobility (ref. 3)
Hf	5.3	0.5 - 34	0.005- 0.13	0.007	
Hg	0.05	0.01 - 0.5	0.0001- 2.8	0.01 - 0.22	*
In	0.049	0.7 - 3		0.00011	
Ir	0.000003?				
Mn	950	20 -10000	0.02 -130	0.03 -21	+
Mo	1.5	0.1 - 40	0.03 - 10	4 -10	+++
Nb	20	6 - 300		0.01 - 0.015	
Ni	80?	2 - 750	0.02 - 27	0.13 -43	+++
Os	0.0001?				
Pb	14	2 - 300	0.06 -120	0.03 -13	+
Pd	0.0006?				
Pt	0.001?				
Re	0.0004			0.004 - 0.0084	
Rh	0.0002?				
Ru	0.001?				
Sb	0.2	0.2 - 10	0.01 - 5	0.18 - 5.6	+++
Se	0.05	0.01 - 2	0.02 - 1	0.052 - 0.2	++
Sn	2.2	1 - 200	0.004- 0.09	0.002 - 0.81	
Ta	2	0.4 - 6	<0.002	0.002	
Te	0.005?				
Ti	5600	150 -25000	3 - 18	1	
Tl	0.6	0.1 - 0.8		0.019	
V	160	3 - 500	0.01 - 20	0.9 - 2.5	
W	1	0.5 - 83	<0.02 - 0.1	0.001 - 0.7	
Zn	75	1 - 900	0.2 -100	0.2 -48	+++
Zr	190	60 -2000	0.05 - 20	0.03	

+++ high, ++ less high, + low, * very low

It is evident that only for very mobile elements, like As, the concentration in ocean waters is higher than that in continental waters, whereas for all other elements freshwater usually has a higher content than sea water.

2.1 ECOSYSTEM POLLUTION

Local conditions, however, can differ significantly from those shown in this picture; in fact, every year many millions tons of metals like Cr, Mn, Cu, Zn, Pb are extracted; but even the rarest elements (Hf and Ru excluded), are mined in a considerable quantity despite their relative scarcity because of their use in several human activities. As a consequence, the geologic weathering now accounts only for a part of the metal supply reaching the hydrographic network; additional sources are represented by mining effluents, effluent resulting from the industrial processing of ores and metals, domestic effluents and urban stormwater runoff (ref. 4), leaching from solid waste dumps, ... (Table 2.1).

As an example, the estimated balance of heavy metal loads in the Ruhr catchment (ref. 13) shows that for lead, copper, zinc, nickel, chromium and cadmium, only 45 % of the load has a geochemical origin and comes from the 4500 km^2 of the drainage basin, while the remaining part is discharged from municipal (54 %) and industrial (1 %) waste treatment plants that treat wastewater from 2.2 million people and 2 million industrial population equivalents.

The metal inputs to lakes and rivers are both in particulate and dissolved forms (complexed and free metal species), yet a substantial part of the load soon becomes associated with the sediments (ref. 14) and suspended particulates: Martin and Meybeck (ref. 15) estimated on a world-wide basis the "Dissolved Transport Index (DTI)", that is the ratio between dissolved to total transport in rivers, finding that most metallic elements are carried preferentially by the suspended particles (Table 2.2). Three main mechanisms are involved in the enrichment of the solid phase (refs. 16-18): physico-chemical adsorption from the water, biological uptake by bacteria and algae, and the sedimentation and physical entrapment of the particulate matter.

These processes thus play a major role in the biogeochemical cycling of the elements, as the solid phase represent the internal reservoir of metals: but they are not completely lost to the ecosystem, for a variation in environmental conditions can made them available again (ref. 5).

TABLE 2.2 - Dissolved Transport Index in rivers: percentage of elements transported in dissolved form over total transport (ref. 15).

%	Elements
90-50	Br, I, S, Cl, Ca, Na, Sr
50-10	Li, N, Sb, As, Mg, B, Mo, F, Cu, Zn, Ba, K
10- 1	P, Ni, Si, Rb, U, Co, Mn, Cr, Th, Pb, V, Cs
1- 0.1	Ga, Tm, Lu, Gd, Ti, Er, Nd, Ho, La, Sm, Tb, Yb, Fe, Eu, Ce, Pr, Al

For instance, the influence of changing redox conditions on the solubility of iron and manganese is well known: both elements usually occur as insoluble compounds in the higher oxidation state, whereas in the lower one they are more soluble and relatively free from complexation. The sediments then can release metals when the water bodies receive acidic precipitations, acid mine drainage, or even organic complexing agents such as EDTA or NTA. Even microbial activities can mobilize metals from sediments by destroying organic matters and forming lower molecular weight compounds with increased complexing capacity and solubility, by changing physical properties of the medium (pH and Eh conditions), or by converting inorganic compounds into organometallic forms. Finally, major changes are usually observed in the estuarine mixing zone, where the salinity gradient affects the sorption/complexation equilibria (ref. 19).

The man-induced alterations in the normal biogeochemical cycles are of environmental concern, for it is well known that the metals can interact with the biota in many ways; for instance, some of these elements are "essential" to life, that is animals and/or plants cannot live without a minimum supply of them (ref. 20); on the other hand, practically all metals produce undesirable, noxious or even lethal effects when supplemented in too high an amount to a living being (ref. 21).

Wuhrmann and Eichenberger (ref. 22) pointed out that sometimes even a minimum increase in the metal supply can induce an unwanted ecosystem response: this is the so-called metal-induced eutrophication observed when river pollution enhances the availability of essential metals to the benthic phytocoenoses. But usually the concept of metal pollution is bound to the disappearance of animal or vegetal species that are unable to survive in the environmental conditions created by the excessive input of those elements.

Different organisms, however, have different sensitivities to the same metal, and the same organisms may be more or less damaged by different metals. In aquatic environments, the concentrations of metallic elements in the organisms are usually far higher than in the water; the ability to accumulate these elements, known as bioaccumulation, does not imply that active uptake mechanisms are necessarily involved. Two different routes of element absorption contribute to the bioaccumulation processes: the bioconcentration, or accumulation from water, and the biomagnification, or the building up of residues via food chains (ref. 23).

Obviously, primary producers cannot profit from the second mechanism, which on the contrary assumes for consumers a different relevance according to element type and organism species, or even for the same organism in different metabolic conditions and for the same element in different environments. Finally, the uptake of an element from the surrounding medium is seldom exactly proportional to the amount present in the water, for often only a part of this quantity is bioavailable, that is available for incorporation into living matter (Fig. 2.1). In fact, bioaccumulation occurs by a modification of inorganic biochemistry and cell physiology that results in a regulation of intracellular metal concentrations (ref. 24).

As regards the effect on the biota then, in general terms it can be said that, for a given metal and organism, both the health status of the organism itself and the environmental conditions may significantly change the extent of interference with the normal metabolism; the factors that influence the toxicity of metals in aquatic environments, in fact, have been summarized by Bryan (ref. 25) as shown in Table 2.3.

Despite the difficulty of clearly attribute a change in the aquatic population to a specific toxicant immission, often the "biological monitoring" has been suggested as a diagnostic tool for assessing water quality. For instance, during the International Seminar on "The use of biological tests for water pollution assessment and control", organized by OCDE and held at the

Joint Research Centre of Ispra in May 1986, the field-related methods used to assess water quality degradation have been subdivided between two categories, structural biomonitoring and functional biomonitoring. The first refers to the evaluation of the biota stress by means of: species diversity indexes of various type; studies of population structure above, at and downstream a toxic effluent discharge; biotic indexes (that is, a numerical scale that correlates organism presence and water quality); "field monitors", such as artificial substrate mimicking the real environment and providing a convenient surface for colonizing species, and caged animals, fish or molluscs, kept in the water bodies to be monitored.

Nevertheless, it is not easy to assess the extent of a contamination from an examination of species check lists, either for species abundance and species presence/absence data, for the animal and vegetal communities respond to the whole set of physical, chemical and biological conditions existing in the surronding environments, and this produces countless possible natural situations against which the observed one must be compared; however, it seems that macroinvertebrate community structure shows a predictable, graded response to heavy metal pollution (ref. 26), with control (usually upstream, or in very similar water bodies), impact and recovery (= downstream) zones differing for species richness and relative invertebrate abundance (ref. 27).

Following this line, Sheehan and Knight (ref. 28) proposed a multilevel approach for the assessment of the ecotoxicological effects in a heavy metal polluted stream, specifically intended to evaluate if the absence of species is a reliable indication of toxicity and if the tolerant species, too, can be used for detecting signs of stress.

Waterhouse and Farrell (ref. 29) suggest that the identification of pollution related changes in biotic community composition on the basis of the diversity indexes at the generic or family level are often adequate, but emphasise that these indexes are generally sensitive to a few dominant species; on the other hand, species identification is necessary for identifying indicator organisms according to Resh and Unzicker (ref. 30).

For their ability to survive metal pollution, the indicator species are often selected among insects (refs. 26, 31): chironomids (refs. 29, 32), Ephemeroptera (ref.33), especially when studying river ecology.

However, sometimes bacteria are preferred (ref. 34), or protozoan (ref. 35), plankton (refs. 36-37), algae (refs. 38-39), and aquatic bryophytes (refs. 40-42), but even fish (refs. 43-44) or molluscs (refs. 45-47) and aquatic plants (refs. 48-49) can be used to relate presence/absence or body content to the existence of polluting metal sources. Obviously, the requisite for any kind of meaningful assessment is the knowledge of the background level for the type of sample selected (ref. 50).

The alternative way to study the impact of pollutants on aquatic environments, that is functional biomonitoring, is based on the monitoring of toxicant induced physiological, biochemical, cytological, histological and behavioral changes within test organisms, as compared with control groups. The methods consists in exposing a selected species, or several species, to different concentrations of a toxicant or to different dilutions of a water sample, and observing if this produces sublethal or lethal effects.

Bioassays have the great advantage of submitting the conditions of exposure to the will of the observer, thus facilitating an understanding of the mechanisms involved; however, it is usually difficult to make generalizations on the basis of the results obtained, and their environmental significance is often questionable.

Parameters that can be monitored, sometimes _in situ_, for instance by using caged animals, but, more often, in aquaria (refs. 51-52), vary according to the organism species: just to mention few examples, using fish it can be studied swimming behaviour, ventilation rates or other physiological changes, even at sub-cellular level; with plants, gross modification or photosynthetic rate; with bacteria, bioluminescence, nitrification of ammonia or genotoxicity (Ames test).

La Point _et al_. (ref. 27) studied the benthic community structural response to pollution in 15 streams, and concluded that the measure of the effect can be estimated only by combining bioassays and biomonitoring: onsite bioassays, in fact, reflect the toxicity under ambient field conditions and are influenced by the physical and chemical characteristics of the water, while a seasonal monitoring regime should make it possible to distinguish the normal variations in species abundance from a pollution effect (ref. 53).

TABLE 2.4 Classification of elements according to toxicity and availability (ref. 26).

Non critical	Very toxic and relatively accessible	Toxic but very insoluble or very rare
Na K Mg Ca H O N	Be Co, Ni Cu, Zn Sn	Ti Hf Zr W Nb Ta Re
C P Fe S Cl Br	As Se Te Pd Ag Cd Pt	Ga La Os Rh Ir Ru Ba
F Li Rb Sr Al Si	Au Hg Tl Pb Sb Bi	

With so many different factors which can modify the effective extent of toxic action, it is evident that a generalized classification of the potential hazard of elements, from the environmental viewpoint, will depend on the criteria used: Wood (ref. 54), for instance, on the basis of the toxicity and availability of elements produced the results shown in table 2.4.

Bowen (ref. 2), on the other hand, compared the rates of mining with the natural rates of cycling and considered as potential pollutants those elements for which the first exceed the second by a factor of ten (element in bold letters):

Ag, As, **Au**, B, Bi, **C**, **Cd**, Co, **Cr**, **Cu**, Fe,

Hg, **Mn**, Mo, **N**, Ni, P, **Pb**, Sb, Sn, U, **W**, Zn

Finally, Li (ref. 55) examined the chemical composition of marine organisms and plants, and comparing them with that of clay, sea water and soil found that living organisms apparently separate elements into biophile and biophobe group.

3.1 ECOSYSTEM PROTECTION AND RECOVERY

According to the literature (refs. 2, 5-6, 56), among the heavy metals the most frequently reported pollutants are cadmium, copper, mercury, lead, and zinc; however, point sources of other metals are often detected in relation with specific industrial activities.

After some catastrophic episodes of metal poisonings (for a review, see ref. 5), the ecotoxicological properties of the most common heavy metals have been carefully investigated, and the maximum allowable concentrations that should be reached in aquatic environments have been fixed (refs. 57-63) according to different objectives:

- for the protection of human health, limits have been set for bathing water, drinking water, water to be used in agriculture and farming or in food industries, and considering the potential hazard due to metal transfer along a trophic chain (ref. 23), also the concentrations in aquatic animals or plants that are used as food have been fixed by laws and regulations, at least for elements like Hg or As;

- for the protection of the human economy, water to be used for specific industrial purposes must be of a minimum quality;

- for the protection of freshwater and marine biota, domestic and industrial effluents must not exceed given concentration limits, and even for the receiving water bodies maximum allowable levels have been fixed.

These protective measures are intended to preserve the water quality, and surely in the future their application and the effort in improving waste water treatments will diminish the hazard to water bodies; at present, the technique most used for controlling the waste water concentration of metals is the chemical precipitation of these pollutants as oxides, hydroxides or sulfides (ref. 6); as an alternative, the metals can be adsorbed on alumino silicates, activated carbon, fly ash (refs. 64-65), or removed from the solution by ion-exchange, reverse osmosis, electrodialysis and so on (ref. 66). Unfortunately, these treatments are sometimes not sufficient to reach safety levels, and as a side effect they produce huge amounts of heavily polluted sludges.

In addition, metal pollution may also derive from diffuse sources, to which the conventional treatments cannot be applied. This is the case, for instance, of metal increase in water bodies as a consequence of the acidification due to acid precipitations: according to the review of Campbell and Stokes (ref. 67), between pH 6 and 5 a number of metals are mobilized, and this could affect the aquatic biota that can tolerate this acidity level. In a survey of 34 shield lakes, Stokes et al. (ref. 68) pointed out that in acid-stressed lakes the bioavailability of zinc, lead, aluminum and mercury increases, controversial results are shown by manganese, and nickel and copper were of concern only where point-sources were present.

Another metal source is represented by metal mines and tailings (refs. 69-70). The best strategy would be the stabilization of tailings in situ, for instance by planting vegetation (refs. 71-72), and in this way at least the erosion due to the wind is controlled; but in addition the drainage water should be collected and treated in order to lower the metal content (refs. 73-74).

The control of inputs is not sufficient for restoring already polluted places, for the heavy metals are know to be "persistent" pollutants, that is once they have entered the water body they are not destroyable, but remain inside the system and can become available again upon a change in environmental conditions (refs. 75-76) and sediment bioturbation due to macroinvertebrates (refs. 77-79).

As an example, ten years after a sharp reduction in mercury loading from a chlor-alkaly plant Parks et al. (ref. 80) found that the water and sediments of Wabigoon River were still heavily polluted, and the elevated mercury levels were maintained mainly because of a chemical and biological remobilization from sediments. Koppe (ref. 81) investigated the fate and effects of heavy metals in the impounded Ruhr River, and concluded that more research on self-purification from hazardous substances in rivers is necessary to permit improvements in the treatment techniques of industrial wastes.

As already seen, particulate matter plays a critical role in determining the amount and the effect of heavy metal in aquatic environments; in fact, the fate of these elements depends mostly on the heterogeneous equilibrium established between dissolved and suspended compounds (refs. 82-83), and the

sediments act as a sink for most pollutants.

For this reason, the restoration of the quality of polluted environments must involve the inactivation of this internal source. Many experiences in this sense have been gathered in trying to control the eutrophication process, as even for phosphorus the sediments may represent for a lake a notable internal source. In fact, along with cosmetic treatment, such as the use of herbicides and algicides to lower excessive primary production, and the aeration of the hypolimnion, the two most popular solutions (refs. 84-85) have been dredging and capping.

The same technical solutions (mechanical, chemical and biological) have also been suggested for dealing with bottom sediments containing toxic substances (ref. 86). The selection of appropriate management technologies will depend on the chemical (inorganic or organic) and physical state of the contaminants (liquid, solid or gaseous), as well as on the horizontal and vertical extension of the contaminated zone (ref. 87).

4.1 MECHANICAL METHODS

4.1.1 DREDGING

Dredging is a process which has been long used for mining sand and gravel or for removing sediments from navigable streams, rivers, lakes, coastal waters, harbors. To mention a few data, Brannon et al. (ref. 88) estimate that every year the U.S.Army Corps of engineers alone dredge 290 million cubic metres of sediments from 30,580 km of waterways and 500 harbors throughout the United States; Foerstner et al. (ref. 89) indicate that about 20 million m^3 have to be dredged from Rotterdam harbor and an other 10 million m^3 from the rivers Scheldt, Weser and Elbe. The sediment resulting from the dredging activities of The Netherlands, Belgium and West Germany amounts to a dozen times the total suspended matter carried by the river Rhine.

According to Adams and Darby (ref. 90), 80 % of the dredged materials are finer than 2 mm, and one-third is composed of silts and clay-sized particles (<0.062 mm); during the dredging operations, both hydraulic and mechanical, it is possible that the sediment resuspension can physically affect the biota, but the extent of the perturbation is small if compared with the normal resuspension due to wind-wave action (ref. 91). Peddicord (ref. 92), reviewing the available information on the direct effects of suspended sediments on aquatic organisms, concluded that it is unlikely that most dredging or disposal operations result in an ecological degradation, even if fluid muds have a definite effect on relatively immobile benthic infauna and epifauna.

However, especially if they come from harbors, the sediments may contain organic materials, oil and grease, metals and other contaminants in high amounts. In this case, the exposure of the dredged material to oxygenated water may produce the oxidation of reduced substances and possibly an oxygen depletion (ref. 93). In addition, the variation of redox and pH conditions may change the chemical species compositions of metals, possibly increasing their bioavailability and damaging the benthic biota (ref. 94). Peddicord (ref. 92) in fact noticed that the dispersion of contaminated sediments usually results in a tissue accumulation of pollutants, but even so lethal conditions are seldom reached during most typical dredging operations.

Moreover, as the resulting material must be discharged on land or in open water, pollution from the sediment could affect the environmental quality of disposal sites. Engler (ref. 95) reviewed the guidelines and criteria in use in the U.S. since the late 1960s (Table 4.1): at that time it seemed enough to evaluate the bulk characteristics of the dredged material, but this measure proved to be ineffective in protecting environmental characteristics and biota, so that by now the official protocols for the discharge of dredged material into inland waters or dumping into ocean waters prescribe leaching tests, toxicity and bioaccumulation tests with various aquatic organisms as well as a general ecological evaluation of the proposed disposal sites (refs. 97-98).

In a detailed study on the release of heavy metals from dredged sediments, Lee et al. (ref. 99) noticed that only manganese was released in an amount potentially significant for the water quality, whereas Zn, Cu, Ni, Pb, Cd, Cr, Hg were sorbed or released in such small quantities that the concentrations in water were not affected; however, they recommanded improving the benthic bioassay tests to study possible chronic effects and/or food web accumulation.

For Lake Trummen (ref. 100), the mercury concentration did not increase in lake water during dredging, and only slightly and temporarily in sedimentation ponds; however, when drying dredged sediments on land a 30 % decrease of mercury was observed for the rapid formation of volatile dimethylmercury. According to the results of some simulations, a problem may arise for the lake biota from the increase of methylmercury concentration in water due to the stirring and turbulence of the sediment; in fact, shortly after dredging the methylmercury should increase in pike, but after one year the concentrations should begin to decrease and in eight years a new, lower steady state should be reached.

Sometimes it is possible to change the characteristics of the dredged material: for instance, Gambrell et al. (ref. 96) studied the sediment composition of City Park Lake, a highly eutrophic water body, and found that they were lower than acceptable limits (Table 4.1) for Mn, Cu, Cr, Ni, close to the limit for Hg, and over the limits for Fe, Pb, Zn and Cd. For this reason, to minimize adverse effects on crops they suggest increasing the pH of the dredged material by liming it before disposing of the slurry on agricultural soils.

TABLE 4.1 EPA Sediment Criteria[a] and screening level[b] for open-water disposal of dredged material (concentrations in mg/kg dry weight).

		Sediment Criteria					Screening level
		Non Polluted		Moderately Polluted		Heavily Polluted	
Pb	<	40		40 - 60	>	60	50.0
Zn	<	90		90 - 200	>	200	75.0
Hg	<	1.0			>	1.0	1.0
Fe	<	17000		17000 - 25000	>	25000	20000
Ni	<	20		20 - 50	>	50	50.0
Mn	<	300		300 - 500	>	500	512
As	<	3		3 - 8	>	8	
Cd					>	6	2.0
Cr	<	25		25 - 75	>	75	100
Cu	<	25		25 - 50	>	50	50.0

(a) EPA Region V Criteria (ref. 95)
(b) EPA Region VI (ref. 96)

Also Berrow and Cheshire (ref. 101) showed that incorporation of peat in contaminated soil was effective in reducing bioavailability of copper for lucerne (Medicago sativa) and ryegrass (Lolium perenne), when the acidification effect of the peat is neutralized by liming.

Ziegler and Krenkel (ref. 102) suggested stabilizing mercury in dredged material by adding sulfides and clay. Alternatively, it might be possible to "mine" the contaminated spoils for metals, for instance by sieving and separating the most polluted size fraction; other suggested methods are flotation, pyrometallurgy, leaching with acids or by hypochlorination.

4.1.2 CAPPING

To accelerate the rate of burial and prevent the release of toxicants from polluted sediments, they could be isolated from the water column by covering them with a layer of unpolluted sediment-like material. This can be done with a minimum of disturbance by merely dredging bottom sediments from a cleaner area of the same waterbody and redispersing them over the contaminated zone. During settling, the particles could even adsorb any contaminants remaining in water column: according to the experiments of Rudd et al. (103), the resuspension of sediment in enclosures in fact reduces the bioaccumulation of mercury and selenium by the biota and results in the formation of a sediment layer with a lower metal concentration.

The segregation of the contaminated sediments can also be obtained by depositing layers of uncontaminated sand or gravel (ref. 104), or with a mixture of iron and sand (ref. 105). The possibility of using clay-rich material also, instead of real sediments has been considered. In this case, the fact that some fish, fouling organisms and sandy bottom epifauna can be relatively sensitive to suspended clay mineral (ref. 106) must be taken into account.

The effectiveness of capping in isolating contaminated dredged material from biota and the overlying water depends either on the nature and thickness of the capping material: in fact, with time the diagenetic and diffusive processes may end up by releasing the contaminants if the coverage is not deep enough, and obviously the size and chemical composition of the particles have an influence on the rates of chemical, physical and biological processes. On the other hand, for the economy of the operation it is necessary to keep the depth of capping to a reasonable minimum: for this reason Branson et al. (ref. 107) suggest the use of large- and small- scale laboratory reactor units for monitoring the biological uptake of contaminants with clams and polychaetes.

Jernelov and Asell (ref. 100), by means of aquaria experiments, found that a substantial reduction (70-95 %) of methylmercury in fish was obtained when the contaminated bottom sediments were covered with a layer of fine sand or ground silica; in Lake Garlangen and Lake Mellanfryken the contaminated area was then covered with mine tailings and sand, with a covering efficiency of around 95 %.

Other proposed methods include the use of cotton net covered with sulfide (ref. 108), of wool (ref. 109), or of films of nylon polymer or polyethylene (ref. 110), but these pose technical problems for keeping the film in position against water movements and gas formation in the anaerobic layer beneath the artificial boundary; the film can be formed in situ by distributing alcohol-mixed organic film, but in this case a severe drop in oxygen concentration may be expected.

4.2 CHEMICAL METHODS

The restoration of a polluted water body may in theory also be achieved by modifying its chemistry: again on analogy with eutrophication restoration projects, for instance, the oxidation of the hypolimnion might be considered, either by aeration of the water (ref. 111), or by treatment of sediments with nitrate (ref. 112).

Unfortunately, these techniques may be effective only in the unlikely case of pollution from iron or manganese, for these two elements become less soluble in oxygenated waters (ref. 113); other elements, on the contrary, can actually be mobilized, for the treatment transforms less soluble sulfides into more soluble compounds.

This suggests that, in dealing with heavy metal pollution, the opposite measures should be taken. Jernelov and Asell (ref. 100), knowing that mercuric sulfide has a much lower solubility (about 10^{-53}) and is much more difficult to methylate than divalent inorganic mercury (about 1000 times more sulfide than divalent inorganic mercury is needed to obtain the same yield of methylmercury), performed a number of aquarium experiments to test different in situ method for converting divalent inorganic mercury in the sediment to mercuric sulfide: addition of glucose to create anaerobic conditions, direct addition of sulfide ions, covering of sediments with FeS or FeS_2. All the treatments significantly decreased methylmercury formation and its uptake by fish. Of course these techniques cannot be proposed as a restoring option for a whole polluted lake, but it is interesting to note that the natural generation of sulfide ions during the winter period of anoxic conditions is more effective in reducing mercury bioavailability than capping the sediments with a clean fill.

Feick et al. (ref. 114) investigated the possibility of using chemical binding agents, or scavengers, to clean up water bodies and hold mercury in the bottom sediments: laboratory experiments with long-chain alkyl thiols, chemicals not toxic to fish, proved that these scavengers are very effective in binding mercury, and they can also be expected to bind copper and lead, and to a lesser extent even cadmium and zinc. According to these Authors, a possible method of treating a contaminated lake in this way would be that of spreading sediments or sand impregnated with thiols, that are oily liquid with a density > 1.

Another reclaiming techniques that now is being widely used is the liming of acidified environments, both lakes (refs. 112, 115) and rivers (refs. 116-117). As we have seen, both acidified ecosystems and acid mine water may contain a high level of metals, thus the restoration of the buffer capacity of the acidified environment also affects the availability of metals. According to Lindstroem et al. (ref. 118), a successful recover of acid high mountain lakes must also take into account of the level of elements like aluminum, for the survival of fish is greatly enhanced when the lake water contains less then 50 ug Al/l. It is well known that the solubility of metals is pH sensitive; in fact, Dillon et al. (ref. 119) found that the addition of base ($Ca(OH)_2$, $CaCO_3$) to acidic lakes in Ontario, Canada, resulted in a substantial reduction of Cu, Ni, Zn, Mn, Fe and Al, probably by a precipitation mechanism. Also Bengtsson et al. (ref. 120) observed that liming acid lakes in Sweden reduced the toxic effects of metal, in addition to providing circumneutral pH.

4.3 BIOLOGICAL METHODS

For the adsorptive characteristics of biogenic particles, Stumm and Baccini (ref. 121) consider the internal production of biomass to be an important feed-back mechanism controlling the heavy metal balance in lakes: in fact, the increase of phytoplankton due to the eutrophication process has the beneficial effect of increasing the sedimentation of the metals, but when the pollutant load reaches the toxic threshold the biomass begins to reduce and the metal residence time increases. In addition, when high trophic conditions are established, the hypolimnetic waters usually become anoxic and the reduced state leads to the formation of metal sulfides, thus accelerating the sedimentation rate of elements like Cu, Zn, Cd, Pb, Hg. But, again, in these conditions there could be a risk of biomagnification along the trophic chain due to the formation of alkylated species, which lead to a new mobilization of the toxicants.

Wangersky (ref. 122) too recognised that the dominant mechanism controlling the removal of metal from water is the adsorption on biologically produced particulate matter, and experimental evidence is provided, among other methods, by microautoradiography (ref. 123) and X-ray microanalysis (refs. 124-125) of algae. According to Skowronski and Przytocka-Jusiak (ref. 126), the green alga Stichococcus bacillaris at constant biomass density of 1g dry weight/l can remove 60-80 % of the cadmium from a solution containing as much as 4 mg Cd/l, while Simon et al. (ref. 127) thought of using strains of Alcaligenes eutrophus, a facultative chemolitotroph bacteria resistant to Cd, Co, Hg, Ni and Zn, as a microbial device for metal recovery.

Indeed, Allan (ref. 83) speculated on the possibility of recovering metal polluted environment by the addition of nutrients, for the increased productivity would reduce soluble concentrations of toxicants; there is however the risk of increasing the flux of the same toxicants along the trophic chain. On the basis of enclosure experiments, Rudd and Turner (ref. 128) concluded that stimulation of primary production cannot be devised as an ameliorating measure for mercury contaminated environments: in fact, they observed that the addition of nutrients resulted in increased primary production, as expected, but also in a high concentration of mercury in the whole body and muscle of fish, for the uptake exceeded the dilution expected, as indicated by the enhanced body growth rates.

As an alternative, Turner and Rudd (ref. 129) tried to reduce the bioavailability of mercury by introducing, as a competing ion, sodium selenite at 1, 10, and 100 ug Se/l: using radioisotope techniques, they observed a marked reduction of the mercury uptake at the various levels of the food chain, and concluded that the spiking with selenium might be of benefit in the treatment of aquatic environment facing a moderate pollution produced by atmospheric deposition or point sources.

For eutrophication control, the possibility of harvesting or culturing aquatic macrophytes to remove nutrient from water and/or sediments has also been considered (refs. 131-132).

In the same way, several authors investigated the use of plants as biological filters for removing heavy metals: in fact, some aquatic plants can live in highly stressed ecosystems, and are still able to remove heavy metal from the surrounding medium (refs. 133-134). As in growing they concentrate the metals as much as thousands of times, it may be possible, by carefully planning the harvesting, to use these biological absorbers for cleaning moderately polluted environments. Ozimek (ref. 135) for instance compared the role of helophytes (Schoenopletus lacustris, Typha latifolia and Glyceria maxima), pleustonic plants (Lemna minor and L.gibba), and elodeids (Potamogeton lucens and Ceratophyllum demersum) in the cycling of heavy metals in ponds supplied with effluents from sewage treatment plants; Haritonidis (ref. 136) studied the uptake of Cd, Zn, Cu and Pb by marine macrophyceae Ulva lactuca, Enteromorpha linza, and Gracilaria verrucosa under culture conditions; Staves and Knaus (ref. 137) proposed using duckweeds for the removal of chromium from water, and in their experiments Spirodela polyrhiza (L.) grown in water containing 10 mg Cr/l reached a concentration as high as 8.6 g Cr/kg dry weight of plant!

These alternative biological treatment techniques, however, have the major disadvantage of producing a high quantity of heavily polluted material, which cannot be used to feed animals and must be disposed of in some way.

5.1 THE BIOTA RESPONSE

When a polluted environment no longer receives an excessive load of metals, the biotic communities start to change in order to adapt to the new conditions; however, the recovery processes can require long periods of time. According to the experiments of Eichenberger et al. (ref. 138), which used eight artificial river channels to study the recover of the algal community subjected to "metal stress" with copper and a mixture of cobalt, copper and zinc, the first response consisted in a vigorous expansion of the species already present, but the species composition did not change for months, remaining fundamentally different from the controls.

Chadwick et al. (ref. 139) monitored for ten years the recovery of benthic invertebrate communities in a creek, after an improved metal mine wastewater treatment had dramatically reduced the metal concentrations; they detected the first signs of recovery only three years after the improvements in the water quality, and they were restricted to the furthest downstream station. Subsequently, colonization was established even at upstream stations after a period of eight years; every time, the earliest colonizers were chironomids, empidids and oligochaetes. Even after ten years, the results indicate stressed invertebrate communities; in the specific case, the Authors concluded that the delayed recovery may have initially been due to a pH stress of high pH, rather than metals, and it is very likely that the slow rate of colonization depends on the lack of an upstream source of invertebrates, which appear to come from tributary streams and aerial colonization.

Norris (ref. 140) studied the effectiveness of remedial works at Captain Flat Mining area: the Molonglo River had received trace metals from mine working since 1939, the main sources of pollutants being erosion and leaching of mine waste dumps, water flowing from the mine itself, and river bed sediments; as a consequence of the lethal and sublethal effects of zinc, Crustacea, Mollusca and Oligochaeta were absent. In 1974 the water courses were rerouted to prevent them entering the mine, and tailings were capped with impervious material to avoid surface erosion and runoff, then stabilized with shallow vegetation, but a study conducted between 1978 and 1982 revealed little or no discernible change in the benthic fauna, as the mine seeps and the river bed sediments act as a major source of toxic metals.

Benthic colonization after sediment removal by dredging seems to be more rapid, but may take as long as 5 years (refs. 141-142). The progressive structuration of the community depends on the immigration distance and on the stock of colonizing species, but through different mechanisms, it is usually the number of species which recovers faster than abundance and biomass.

At the already mentioned International Seminar on "The use of biological tests for water pollution assessment and control", organized by OCDE and held at the Joint Research Centre of Ispra in May 1986, it was estimated that the recovery times of damaged ecosystems were in the order of years for both fresh and marine water (Table 5.1). Providing that no new pollution episodes do happen during the generation time of the affected organisms, obviously the recovery times depend on the size of the impacted environment; but the most

TABLE 5.1 Estimated ecosystem recovery times.

FRESH WATER ENVIRONMENTS		
Lotic systems:	Fast flowing streams	3- 5 years
	Slow flowing rivers	5-10 "
Lentic systems:	Small ponds	10 "
	Large lakes never recover their original state without human intervention	
MARINE ENVIRONMENTS		
Intertidal shore:	Sand beach	1- 2 years
	Rocky shore	5-10 "
	Tidal flats	5-10 "
Intertidal wetlands:	Marshes	10-20 "
	Mangrove swamps	20-80 "
Subtidal systems:	Seagrass systems	50 "
	Coral reefs	10-20 "
	Soft bottom benthos	10 years minimum

(presented at the joint OECD/EEC International Seminar on "The use of biological tests for water pollution assessment and control", Ispra, 24-26 June 1986)

important condition is the opportunity for recruiting replacement species, that is the nature and proximity of the area from which the species must emigrate, and the absence of immigration impediments of whatsoever physical, chemical or biological kind.

6.1 CONCLUSIONS

Up to now, most of the experiences in aquatic environment restoration have referred to eutrophication or acidification problems: for these subjects, in fact, a recent updated bibliography reported 215 citations of lake restoration projects (ref. 143). The theory and practice of the different technological interventions are quite well understood, so that limnologists can predict the outcome of the various management options and are thus able to convince the public of the need for restoration plans (ref. 144).

On the other hand, there are few references in literature to the rehabilitation of metal polluted environments, so that a general statement regarding the feasibility of such a recovery is impossible; however, as is already required in projects of lake restoration from eutrophication (refs. 145-146), after every type of treatment (even the reduction of inputs alone), it should be mandatory to monitor the limnological characteristics, for only in this way can the appropriate corrective measures be applied before irreversible changes appear in the ecosystem.

REFERENCES

1. C.C. Burrell, Atomic spectrometric analysis of heavy metal pollutants in water, Ann Arbor Sci., Ann Arbor, 1974.
2. H.J.M. Bowen, Environmental Chemistry of the Elements, Academic Press, London, 1979.
3. M. Dall'Aglio, R. Marchetti and E. Sabbioni, in M. Beccari, A.C. Di Pinto, D. Marani, M. Santori and G. Tiravanti (Editors), I metalli nelle acque: origine, distribuzione, metodi di rimozione, IRSA, Roma, Quaderni 71, 1986, pp. 3-78.
4. P. Liston and W. Maher, Environ.Contam.Toxicol., 36 (1986) 900-905.
5. U. Foerstner and G.T.W. Wittmann, Metal Pollution in the Aquatic Environment, Springer-Verlag, Berlin Heidelberg, 1979.

6   M. Beccari, A.C. Di Pinto, D. Marani, M. Santori and G. Tiravanti (Editors), I metalli nelle acque: origine, distribuzione, metodi di rimozione, IRSA Quaderni 71, Roma, 1986.
7   J. Marsalek, Sci.Tech.Eau, 17 (1984) 163-167.
8   E.E. Angino, L.M. Magnuson and G.F. Stewart, Water Resour.Res., 8 (1972) 135-140.
9   H. Muntau, Mem.Ist.Ital.Idrobiol., 38 (1981) 505-529.
10 D.P.H. Laxen and R.M. Harrison, Water Res., 11 (1977) 1-11.
11 A.J. Rubin, Aqueous-environmental chemistry of metals, Ann Arbor Sci.Publ., Ann Arbor, 1974.
12 J.A. Davis, III and J. Jacknow, Jour.WPCF, 47 (1975) 2292-2297.
13 K.R. Imhoff, P. Koppe and F. Dietz, Prog.Wat.Tech., 12 (1980) 735-749.
14 L.J. Warren, Environ.Pollut.Ser.B, 2 (1981) 401-436.
15 J.M. Martin and M. Meybeck, Mar.Chem., 7 (1979) 173-206.
16 B.T. Hart, Hydrobiologia, 91 (1982) 299-313.
17 G. Gunkel, Arch.Hydrobiol., 105 (1986) 489-515.
18 G. Gunkel and A. Sztraka, Arch.Hydrobiol., 106 (1986) 91-117.
19 U. Foerstner, Proc. 4th Int.Wadden Sea Symposium "The role of organic matter in the Wadden Sea", 1-3 November 1983, Texel, The Netherlands. Netherlands Inst.Sea Res. Publ.Series No 10, 1984, pp. 195-209.
20 J.E. Zajic, Microbial biogeochemistry, Academic Press, NY & London, 1969.
21 D.H.K. Lee, Metallic Contaminants and Human Health, Academic Press, NY & London, 1972.
22 K. Wuhrmann and E. Eichenberger, Verh.Internat.Verein.Limnol., 19 (1975) 2028-2034.
23 R. Baudo, Mem.Ist.Ital.Idrobiol., 43 (1985) 281-309.
24 K. Simkiss and G.H. Schmidt, Mar.Environ.Res., 17 (1985) 188-191.
25 G.W. Bryan, in R. Johnston (Editor), Marine Pollution, Academic Press, NY, 1976, pp. 185-302.
26 R.W. Winner, M.W. Boesel and M.P. Farrell, Can.J.Fish.Aquat.Sci., 37 (1980) 647-655.
27 T.W. LaPoint, S.M. Melancon and M.K. Morris, Jour.WPCF, 56 (1984) 1030-1038.
28 P.J. Sheehan and A.W. Knight, VerhInternat.Verein.Limnol., 22 (1985) 2364-2370.
29 J.C. Waterhouse and M.P. Farrell, Can.J.Fish.Aquat.Sci., 42 (1985) 406-413.
30 V.H. Resh and J.D. Unzicker, Water Qual.Monit., 49 (1975) 9-19.
31 H.V. Leland, Verh,Internat.Verein.Limnol., 22 (1985) 2413-2419.
32 M. Yasuno, S. Hatakeyama and Y. Sugaya, Verh.Internat.Verein.Limnol., 22 (1985) 2371-2377.
33 M. Cotta Ramusino and G. Pacchetti, Verh.Internat.Verein. Limnol., 22 (1985) 2383-2384.
34 M.J. Gauthier, J.P. Breittmayer, R.L. Clement and G.N. Flatau, Environ. Technol. Lett., 7 (1986) 335-340.
35 Y.-F. Shen, A.L. Buikema, Jr., W.H. Yongue, J.R. Pratt and J. Cairns, Jr., J.Protozool., 33 (1986) 746-751.
36 R. Oliveira, Hydrobiologia, 128 (1985) 51-59.
37 R. Oliveira, T. Monteiro, C. Cabecadas, C. Vale and M.J. Brogueira, Verh. Internat.Verein.Limnol., 22 (1985) 2395-2404.
38 A. Austin and N.Munteanu, Environ.Pollut.Ser.A, 33 (1984) 39-62.

39 G. Gorbi and G. Campanini, Verh.Internat. Verein.Limnol., 22 (1985) 2385-2389.
40 C. Mouvet, Verh.Internat.Verein.Limnol., 22 (1985) 2420-2425.
41 K.C. Jones, J.Geochem.Explor., 24 (1985) 237-246.
42 C. Mouvet, E. Pattee and P. Cordebar, Acta Oecol., 7 (1986) 77-91.
43 K. Essink, Neth.J.Sea Res., 19 (1985) 177-182.
44 J. Skurdal, T. Qvenild and O.K. Skogheim, Environ.Biol.Fish., 14 (1985) 233-237.
45 K. Julshamn, Fiskeridir.Skr.(Ernaering), 1 (1981) 161-182.
46 N. Crescenti, S. Martella and G. Martino, Nova Thalassia, 6, Suppl. (1983-1984) 113-121.
47 Ya. Shalanky, Zh.Obshch.Biol., 46 (1985) 743-752.
48 O. Sortkjaer, Ecol.Bull., 36 (1984) 75-80.
49 K.H. Bowmer, Aust.J.Mar.Freshwat.Res., 37 (1986) 297-308.
50 B. Wachs, in J. Salanki (Editor), Heavy metals in water organisms, Symp.Biol.Hung., 29 (1985) 179-190.
51 E.R. Christensen, Schweiz.Z.Hydrol., 46 (1984) 100-108.
52 P. Couture, S.A. Visser, R. van Coille and C. Blaise, Schweiz.Z. Hydrol., 47 (1985) 127-158.
53 C.E. Hornig, Macroinvertebrate inventories of the White River, Colorado and Utah: Significance of annual, seasonal, and spatial variation in the design of biomonitoring networks for pollution detection, EPA-600/7-84-063, 1984.
54 J.M. Wood, Science, 183 (1974) 1049-1052.
55 Y.-H. Li, Schweiz.Z.Hydrol., 46 (1984) 177-184.
56 W. Salomons and U. Foerstner, Metals in the Hydrocycle. Springer- Verlag, Berlin Heidelberg, 1984.
57 NAS, Water Quality Criteria 1972, National Academy of Sciences, EPA, Ecol.Res.Ser. EPA R373033, 1973.
58 EPA, Quality Criteria for Water, Office of Water and Hazardous Materials, U.S.Environmental Protection Agency, Washington D.C., 1977.
59 R. Baudo, R. de Bernardi, G. Giussani, E. Grimaldi and L. Tonolli, Quality objectives for aquatic life in European fresh waters. Mem.Ist.Ital.Idrobiol., 36 Suppl., Pallanza, 1978.
60 R.E. Train, Quality criteria for water, Castle House Publ.Ltd., 1979.
61 AFS, A review of the EPA Red Book: Quality Criteria for Water, American Fisheries Society. Water Quality Section. Bethesda, Maryland, 1979.
62 EIFAC, Water Quality Criteria for Freshwater Fish, Butterworths, London, 1980.
63 IRPTC, IRPTC Legal File 1983. 2 Volls, International Register of Potentially Toxic Chemicals, United Nations Environment Programme, Geneva, 1983.
64 H.A. Elliott and C.P. Huang, Water Res., 15 (1981) 849-853.
65 K.K. Panday, G. Prasad and V.N. Singh, Water Res., 19 (1985) 869-873.
66 J.W. Patterson and R.A. Minear, in P.A. Krenkel (Editor), Heavy metals in the aquatic environment, Pergamon Press, NY & London, 1975, pp. 261-272.
67 P.G.C. Campbell and P.M. Stokes, Can.J.Fish.Aquat.Sci., 42 (1985) 2034-2049.
68 P.M. Stokes, R.C. Bailey and G.R. Groulx, Environ.Health Perspect., 63 (1985) 79-87.

69 R.D. Johnson, R.J. Miller, R.E. Williams, C.M. Wai, A.C. Wiese and J.E. Mitchell, Proc.Int.Conf.Heavy Metals in the Environment, Toronto, Ontario, Canada, Oct.27-31, 1975, Vol. II, Part 2, pp. 465-485.
70 J. Ford, Water quality problems caused by abandoned metal mines and tailings, EPA 440-5-85-001, 1985, pp. 344-345.
71 R.P. Gemmell, Proc.Int.Conf.Heavy Metals in the Environment, Toronto, Ontario, Canada, Oct.27-31, 1975, Vol. II, Part 2, pp. 579-598.
72 A.D. Bradshaw, in Proc.Int.Conf.Heavy Metals in the Environment, Toronto, Ontario, Canada, Oct.27-31, 1975, Vol. II, Part 2, pp. 599-622.
73 R.D. Andrews, in Proc.Int.Conf.Heavy Metals in the Environment, Toronto, Ontario, Canada, Oct.27-31, 1975, Vol. II, Part 2, pp. 645-676.
74 M.H. Bates, J.N. Veenstra, J. Barber, R. Bernard, J. Karleskint, P. Kahn, R. Pakanti and M. Tate, in J.M. Bell (Editor), Proceedings of the 39th Industrial Waste Conference, Purdue Univ., West Lafayette, Indiana, May 8-10, 1984, Butterworth Publ., Stoneham, Ma (USA), 1985.
75 J.M. Brannon, R.H. Plumb, Jr. and I. Smith, Jr., in R.A. Baker (Editor), Contaminants and Sediments, Ann Arbor Sci.Publ., Ann Arbor, 1980, Vol. 2, pp. 221-266.
76 G. Matisoff, J.B. Fisher and P.L. McCall, Geochim.Cosmochim.Acta, 45 (1981) 2333-2347.
77 G. Krantzberg and M. Stokes, Can.J.Fish.Aquat.Sci., 42 (1985) 1465-1473.
78 G. Matisoff, J.B. Fisher and S. Matis, Hydrobiologia, 122 (1985) 19-133.
79 W.M. Starkel, Can.J.Fish.Aquat.Sci., 42 (1985) 95-100.
80 J.M. Parks, J.A. Sutton and A. Lutz, Can.J.Fish.Aquat.Sci., 43 (1986) 1426-1444.
81 P. Koppe, Z. Wasser-Abwasser-Forsch., 19 (1986) 14-19.
82 C.R. Olsen, N.H. Cutshall and I.L. Larsen, Mar.Chem., 11 (1982) 501-533.
83 R.J. Allan, The role of particulate matter in the fate of contaminants in aquatic ecosystems: Part I. Transport and burial. Part II. Bioavailability, recycling and bioaccumulation, CCIW, Burlington, Ontario, Canada, National Water Research Institute Contribution No 84-18, 1984.
84 G. Barroin, Rev. Fr.Sci.Eau, 3 (1984) 295-308.
85 A.J. White, Diss.Abst.Int.Pt.B - Sci.& Eng., 46 (1985) no. 2.
86 Army Corps of Engineers, Commencement Bay nearshore/tideflats superfund site, Tacoma, Washington, remedial investigations. Evaluation of alternative dredging methods and equipment, disposal methods and sites, and site control and treatment practices for contaminated sediments, EPA/910/9-85/134f, 1985.
87 T.R., Patin, Management of bottom sediments containing toxic substances, Proceedings of the U.S./Japan Experts Meeting, Oct.30-31, Kyoto (Japan), 1985.
88 J.M. Brannon, R.M. Engler, J.R. Rose, P.G. Hunt and I. Smith, Selective analytical partitioning od sediments to evaluate potential mobility of chemical constituents during dredging and disposal operations, Tech. Report D-76-7, U.S.Army Engineer Waterways Experiment Station, Vicksburg, MS, 1976.
89 U. Foerstner, W. Ahlf, W. Calmano and M. Kersten, in G. Kullenberg (Editor), The Role of the Oceans as a Waste Disposal Option, D.Reidel Publ.Co., 1986, pp. 597-615.
90 D.D. Adams and D.A. Darby, in R.A. Baker (Editor), Contaminants and Sediments, Vol.1, Ann Arbor Sci.Publ., Ann Arbor, 1980, pp. 373-392.

91 P.G. Sly, in H.L. Golterman (Editor), Interactions between Sediments and Fresh Waters, Junk Publ., The Hague, 1977.
92 R.K. Peddicord, in R.A. Baker (Editor), Contaminants and Sediments, Ann Arbor Sci.Publ., Ann Arbor, 1980, Vol. 1, pp. 501- 536.
93 N.M. Burns and C. Ross, in N.M. Burns and C. Ross (Editors), Project Hypo, CCIW Paper 6, Burlington, 1972, pp. 120-126.
94 J.G. Seeyle, R.J. Hesselberg and M.J. Mac, Environ.Sci.Technol., 16 (1982) 459-464.
95 R.M. Engler, in R.A. Baker (Editor), Contaminants and Sediments, Ann Arbor Sci.Publ., Ann Arbor, 1980, Vol. 1, pp. 143-169.
96 R.P. Gambrell, C.N. Reddy and R.A. Khalid, Jour.WPCF, 55 (1983) 1201-1210.
97 W.S. Davis, in R.A. Baker (Editor), Contaminants and Sediments, Ann Arbor Sci.Publ., Ann Arbor, 1980, Vol.1, pp. 289-296.
98 R.W. Alden, III, A.J. Butt, S.S. Jackman, G.J. Hall and R. Young, Jr, Comparison of microcosm and bioassay techniques for estimating ecological effects from open ocean disposal of contaminated dredged sediments, NTIS AD-A165 057/1/gar, 1985.
99 G.F. Lee, J.M. Lopez and G.M. Mariani, Proc.Int.Conf.Heavy Metals in the Environment, Toronto, Ontario, Canada, Oct.27-31, 1975, Vol. II, Part 2, pp. 731-764.
100 A. Jernelov and B. Asell, in P.A. Krenkel (Editor), Heavy metals in the aquatic environment, Proc.Int.Conf.Nashville, Tennessee, Dec.1973, Pergamon Press, 1975, pp. 299-309.
101 M.L. Berrow and M.V. Cheshire, in T.D. Kekkas (Editor), International Conference Heavy Metals in the Environment, Athens - Sept.1985, Vol. 2, pp. 397-399.
102 F.G. Ziegler and P.A. Krenkel, in P.A. Krenkel (Editor), Heavy metals in the aquatic environment, Proc.Int.Conf.Nashville, Tennessee, Dec. 1973, Pergamon Press, 1975, pp. 311-314.
103 J.H.W. Rudd, M.A. Turner, A. Furutani, A.L. Swick and B.E. Townsend, Can. J. Fish.Aquat.Sci., 40 (1983) 2206-2217.
104 L.H. Bongers and N.M. Khattak, Sand and gravel overlay for control of mercury in sediment, Water Pollution Control Series 16080 HTD, 1972.
105 G. Feick, E.E. Johanson and D.S. Yeaple, Control of mercury contamination in fresh water sediments, EPA-R2-72-077, 1972.
106 V.A. McFarland and R.K. Peddicord, Arch.Environ.Contam.Toxicol., 9 (1980) 733-741.
107 J.M. Branson, R.E. Hoeppel, T.C. Sturgis, I. Smith, Jr., and D. Gunnision,Effectiveness of capping in isolating contaminated dredged material from biota and the overlying water, AEWES, Vicksburg, MS (USA), WES/TR/D-85-10, 1985.
108 J.D. Suggs, D.H. Petersen and J.D. Middlebrook, Jr., Mercury pollution control in stream and lake sediments, EPA 16080 HTD, 1972.
109 M. Friedman, C.S. Harrison, W.H. Ward and H.P. Lundgren, Sorption behavior of mercuric and methylmercuric salts on wool, Proc.ACS Division of Water, Air and Waste Chemistry Meeting, Los Angeles, CA, March 29-April 2, 1972.
110 M.A. Widman and M.M. Epstein, Polymer film overlay system for mercury contaminated sludge-phase I, Water Pollution Control Series 16080 HTZ, 1972.

111 D.J. McQueen and D.R.S. Lean, Water Pollut.Res.J.Can., 21 (1986) 205-217.
112 W. Ripl, Oxidation of lake sediments with nitrate - A restoration method for former recipients, Inst.Limnol., Univ.Lund, LUNBDS/(NBLI-1001)/1-151/(1978), 1978.
113 W. Davison, Hydrobiologia, 92 (1982) 463-471.
114 G. Feick, E.E. Johanson and D.S. Yeaple, in P.A. Krenkel (Editor), Heavy metals in the aquatic environment, Proc.Int.Conf.Nashville, Tennessee, Dec. 1973, Pergamon Press, 1975, pp. 329-332.
115 G.K. Lindmark, Hydrobiologia, 92 (1982) 537-547.
116 B.I. Andersson, I. Alenaes and H. Hultberg, Rep.Inst. Freshwat.Res., Drottningholm, 61 (1984) 16-27.
117 H.U. Sverdrup, P.G. Warfvinge and J. Frase, Vatten, 41 (1985) 155-163.
118 T. Lindstroem, W. Dickson and G. Anderson, Rep.Inst.Freshwat.Res., Drottningholm, 61 (1984) 128-137.
119 P.J. Dillon, N.D. Yan, W.A. Scheider and N. Conroy, Arch.Hydrobiol., 13 (1979) 317-336.
120 B. Bengtsson, W. Dickson and P. Nyberg, Ambio, 9 (1980) 34-36.
121 W. Stumm and P. Baccini, in A. Lerman (Editor), Lakes Chemistry Geology Physics, Springer - Verlag, New York, 1978.
122 P.J. Wangersky, Mar.Chem., 18 (1986) 269-297.
123 C. Glueck and K.H. Lieser, Naturwissenschaften, 69 (1982) 391.
124 M. Pedersen, G.M. Roomans, M. Andren, A. Lignell, G. Lindahl, K. Wallstrom and A. Forsberg, Scanning Electron Microscopy, 1981/II, (1981) 499-509.
125 G. Lindahl, K. Wallstroem, G.M. Roomans and M. Pedersen, Botanica Marina, 26 (1983) 367-373.
126 T. Skowronski and M. Przytocka-Jusiak, Chemosphere, 15 (1986) 77-79.
127 C. Simon, J. Remacle and M. Mergeay, in T.D. Lekkas (Editor), International Conference on Heavy Metals in the Environment, Athens, Sept. 1985, Vol. 1, pp. 57-59.
128 J.H.W. Rudd and M.A. Turner, Can.J.Fish.Aquat.Sci., 40 (1983) 2251-2259.
129 M.A. Turner and J.W.M. Rudd, Can,J,Fish.Aquat.Sci., 40 (1983) 2228-2240.
130 G.D. Cooke, in J. Taggart and L. Moore (Editors), Lake restoration, protection and management, EPA 440-5-83-001, 1983, pp. 257-266.
131 J.D. Jackson, Diss.Abst.Int.Pt.B-Sci.& Eng., 44 (1984) no. 8.
132 Anonymous, Die Bedeutung von Makrophyten fuer die Gewaesseroekologie. Vegetationsentwicklung. Einfluesse auf den Stoffhaushalt in Gewaessern, Tech.Univ.Muenchen (FRG),1985.
133 J.H. Rodgers, Jr., D.S. Cherry and R.K. Guthrie, Water Res., 12 (1978) 765-770.
134 J.R. Clark, J.H. Van Hassel, R.B. Nicholson, D.S. Cherry and J. Cairns, Jr., Ecotoxicol.Environ.Saf., 5 (1981) 87-96.
135 T. Ozimek, in J. Salanki (Editor), Heavy metals in water organisms, Symp.Biol.Hung., 29 (1985) 41-50.
136 S. Haritonidis, Mar.Environ.Res., 17 (1985) 198-199.
137 R.P. Staves and R.M. Knaus, Aquatic Botany, 23 (1985) 261-273.
138 E. Eichenberger, F. Schlatter, H. Weilenmann and K. Wuhrmann, Verh. Internat.Verein.Limnol., 21 (1985) 1131-1134.
139 J.W. Chadwick, S.P. Canton and R.L. Dent, Water, Air, and Soil Pollution, 28 (1986) 427-438.
140 R.H. Norris, Aust.J.Mar.Freshwat.Res., 37 (1986) 147-157.

141 E. Bonsdorff, Oceanol.Acta, (1983) 27-32.
142 C. Hily, Oceanol.Acta, (1983) 113-120.
143 NTIS, Lake restoration, 1977-May 1986 (citations from the Selected Water Resources Abstracts Database), NTIS, Springfield, VA (USA), PB86-864832/GAR, 1986.
144 L.L. Klessig and N.W. Bouwes, Sr., in J. Taggart and L. Moore (Editors), Lake restoration, protection and management, EPA 440-5-83-001, 1983, pp. 267-270.
145 R.E. Pine, R.A. Soltero and R.D. Riley, in J. Taggart and L. Moore (Editors), Lake restoration, protection and management, EPA 440-5-83-001, 1983, pp. 135-140.
146 L.E. Keup, Environ.Monit.Assess., 5 (1985) 339-360.

SUBJECT INDEX

Abiotic, 96, 97, 115, 195, 245, 248, 252
-, components 1, 8
-, control 115
-, factors 304
-, interaction 6, 245
Abundance, 161, 188
Accumulation
-, rates 281
Acclimation, 167, 171
Acid, 226, 276, 277, 298
-, buffering 283, 284
-, cations 282
-, lakes 303, 308, 335, 341
-, leaching cations 190
-, load 283, 284, 285, 306
-, organic 295, 300
-, precursor 308
-, stress 284, 285
-, sulphuric 235
-, wastes 90
-, waters 321
Acid deposition, 55, 66, 76, 90, 91, 95, 115, 131, 170, 190, 261, 276, 279, 284, 285, 289, 291, 293, 294, 296, 303, 307, 308, 329, 335
Acid mine, 55, 321
-, drainage 321, 329
-, water 341
Acidification, 73, 81, 95, 96, 102, 164, 170, 173, 185, 186, 235, 236, 238, 276, 277, 281, 282, 283, 285, 288, 289, 291, 292, 293, 295, 296, 298, 299, 300, 301, 303, 304, 305, 306, 307, 308, 309, 339, 341, 346
-, indicators 305
Acidity, 154, 156, 164, 279, 281, 288, 292, 295, 306, 321
Activity
-, heterotrophic 239, 240
-, metabolic 127, 188
-, photosynthetic 169
Adaptation, 191, 245, 246, 247, 253, 270
Adenosine-triphosphates, 63
-, diphosphates 63
-, monophosphates 63
Adsorption, 171, 231, 264, 267, 271, 316, 326, 342
-, coefficient 272
Aerobic conditions, 267, 269, 271
Aerosols, 315
Agriculture, 113, 142, 164, 338
Agroforestry, 23
Air pollution, 131, 148, 275, 280, 283, 285, 289, 291, 295, 307, 315
Algae, 44, 46, 59, 60, 64, 68, 72, 73, 76, 87, 92, 94, 96, 99, 102, 109, 110, 111, 176, 177, 179, 196, 197, 198, 200, 203, 204, 207, 208, 209, 220, 223, 233, 236, 240, 241, 301, 303, 322
-, biomass 198, 220

-, community 200, 203, 208, 209, 301, 344
-, density 197, 204
-, growth 196
-, production 197
Algicides, 219, 336
Aliphates, 261, 267
Alkalinity, 84, 89, 95, 104, 106, 154, 156, 163, 219, 220, 236, 293, 295, 306
Alkanes, 267
Alkenes, 267
Alkil-phiols, 341
Aluminium, 86, 89, 91, 92, 96, 110, 115, 164, 169, 170, 174, 175, 185, 187, 190, 232, 233, 234, 282, 285, 296, 300, 306, 307, 334, 335, 341,
-, distribution 174
-, effects 232
-, hydroxide 89, 90
-, speciation 300, 301
-, sulphate 89, 90, 109
Aminoacid, 151
Ammonium, 109, 167, 172, 175, 181, 182, 183, 188, 295, 332
Amphipods, 176, 306
Anaerobic conditions, 93, 100, 110, 267, 269, 271, 340
Animals, 12, 123, 126, 131, 173, 217, 220, 245, 257, 329
-, aquatic 272, 322, 334
-, caged 331, 332
-, development 125
-, population 220
-, production 173
-, terrestrial 272
-, tissues 323
Anion
-, organic 256
Anoxic, 198, 342
Antagonism, 73, 165, 261
Anthropogenic, 276
Aromates, 261, 267
Arsenic, 314, 334
Arthropods, 249
Atmosphere, 131, 137, 138, 139, 140, 144, 148, 153, 164, 281, 283, 293, 308, 315, 343
Autotrophs, 67
Aquaculture, 64
Availability, 267
Avoidance, 245, 253, 254, 257

Bacillariophyta (diatoms), 96, 101, 104, 106, 115, 176, 179, 198, 237, 301, 303
Bacteria, 64, 68, 110, 126, 179, 198, 223, 235, 239, 240, 252, 253, 304, 328, 332
-, chemolitotroph 342
-, cyanobacterial 101, 109, 114
-, growth 100
-, heterotrophic 100
-, marine 240

-, mercury resistant 240
-, mercury tolerant 240
-, planktonic 224
Bacteriostatic effect, 240
Barite, 314
Barnacle, 322
Base, 275, 284, 288
Basicity, 277, 280, 281
Bedrock
-, gneissic 294
-, granitic 294
-, siliceous 295, 296
Behaviour, 168
Benthos, 34, 37, 51, 84, 87, 101, 105, 161, 162, 163, 164, 165, 167, 169, 170, 171, 174, 176, 177, 180, 181, 182, 183, 185, 188, 189, 190, 224, 242, 305, 324
-, biossay 337
-, colonization 345
-, community 332
-, density 161
-, migration 190
-, phytocoenoses 330
-, population 169, 242
Bentonite, 110
Benzene, 266, 277
Bicarbonate, 295
Binding mechanisms, 253
Bioaccumulation factor (BCFs), 270, 272
Bioassays, 167, 168, 189, 332
-, field 167, 168, 189
-, laboratory 189
Bioavailability, 264, 330, 335, 337, 339, 343
Biocides, 60, 220, 239, 241
Biocoenosis, 126, 203, 209
Bioconcentration, 264, 265, 266, 272, 330
BOD 182, 183
Biodegradation, 264, 266, 267, 268, 269, 271
Biogeography, 19, 22
Biogeochemical cycle, 328, 329
Biological
-, absorbers 343
-, conditions 331
-, control 195, 201, 211, 219
-, degradation 261
-, evolution 5
-, filters 343
-, impact 298
-, injury 314
-, interactions 190, 195
-, model 19
-, monitor 322, 330
-, organization 8

-, responses 301
-, surveillance 188
-, treatment 343
Bioluminescence, 332
Biomagnification, 330
Biomanipulation, 91, 128, 167, 168, 185, 261, 264, 266, 272, 330
Biomass, 33, 60, 61, 62, 64, 65, 69, 72, 74, 75, 84, 94, 107, 114, 115, 132, 138, 139, 142, 144, 148, 203, 220, 238, 240, 242, 276, 277, 301, 303, 342
-, production 113
-, recovery 240, 242
-, terrestrial 137
Biomembranes, 151
Biome, 16
Biomethylation, 253, 257
Biomonitoring, 331, 332
Biosanitation, 269
Biosphere, 137, 138, 139, 140, 148, 262
Biota, 56, 90, 96, 138, 141, 142, 148, 195, 217, 220, 239, 242, 309, 316, 329, 330, 336, 339, 344
-, aquatic 91, 335
-, marine 334
-, stress 331
Biotic score, 170
Biotope, 126
Bioturbation, 108
Bird, 19, 88, 265, 266,
Bivalve, 168, 176, 305
Bovid evolution, 12
Breakdown
-, natural processes 245
-, products 262
Bream, 107
Buffer, 137, 149, 151, 154, 277, 279, 280, 283, 295
-, capacity 90, 91, 133, 151, 298, 341
-, concentration 151
-, effect 164
Bryophyte, 322, 332

Cadmium, 73, 167, 172, 175, 191, 230, 232, 239, 253, 281, 282, 314, 328, 334, 337, 338, 341, 342, 343
Calanoida, 44, 45, 46, 198, 203
Calcite, 314
Calcium, 73, 86, 104, 106, 190, 277, 283, 284, 300, 301, 320
Carbon, 73, 84, 99, 100, 110, 137, 138, 139, 140, 142, 144, 207, 236, 238, 315, 334
-, cycle 137, 140, 142
-, dissolved organic (DOC) 91, 112
-, organic 100, 207
-, particulate 207
Carbonate, 73, 236, 238, 309
Carnivor, 138, 219

Catchment area, 82, 84, 85, 97, 110, 293, 294, 295, 296, 328
Cathabolism, 176
Cation, 296, 298
Cell, 125, 127, 149, 151, 154, 156, 176, 250, 253, 277, 300, 330
Chalk, 267, 269
Chalcopyrite, 314
Charcoal, 315
Chelation, 238, 240
Chemosynthesis, 275
Chironomid, 101, 106, 176, 177, 180, 242, 331, 344
Chloride, 163
Chlorobenzene, 265
Chlorococcales, 105, 108
Chlorophenol, 239, 264, 272
Chlorophyll, 46, 84, 85, 94, 96, 98, 99, 102, 103, 104, 106, 107, 108, 113, 114, 197, 206, 219, 229, 231, 232, 233
Chlorophyceae, 101, 102, 104, 105, 106, 109, 179, 198, 233, 342
Chloroplast, 149, 151, 156
Chromosome, 5
Chromium, 73, 168, 172, 281, 328, 337, 338, 343
Chrysophyceae, 96, 102, 115
Cianophyta, 72, 73, 84, 93, 94, 101, 102, 103, 104, 106, 108, 112, 114, 115, 198, 203, 207, 233, 236
Cirriped, 51
Cladocera, 41, 42, 44, 45, 46, 48, 68, 70, 73, 83, 105, 111, 198, 203, 271,
Clam, 340
Clay, 333, 336, 339
Climate, 6, 7, 9, 12, 13, 14, 15, 17, 18, 19, 20, 72, 84, 90, 123
Climax, 11, 61, 131, 139
Cluster, 7
Coal, 172, 181, 293, 321
Cobalt, 344
Coevolution, 3, 4, 16
Coliform, 91
Colloid, 315
Community, 4, 5, 7, 17, 19, 22, 23, 61, 64, 67, 73, 75, 91, 103, 109, 126, 127, 128, 160, 162, 163, 164, 166, 168, 169, 171, 187, 195, 196, 197, 198, 199, 200, 201, 211, 217, 218, 219, 220, 222, 226, 230, 238, 240, 241, 242, 246, 250, 252, 331, 344
Compartment, 134, 135, 137, 217
Competition, 4, 5, 60, 65, 72, 196, 197, 240
Complexation, 329
Conductivity, 232, 234, 236
Conifer, 164, 279, 280
Consumer, 62, 132, 219, 301, 304, 330
CO_2, 131, 138, 142, 144, 148, 149, 198, 262, 277
Copepoda, 41, 42, 48, 52, 105, 198, 203, 240
Copper, 73, 168, 172, 178, 227, 229, 230, 239, 240, 247, 249, 254, 281, 313, 314, 328, 334, 335, 337, 338, 339, 341, 342, 343, 344
Coprecipitation, 326
Coregonid, 198
Crayfish, 306

Creek, 344
Crops, 219, 289
Crustacea, 72, 73, 93, 101, 105, 107, 111, 174, 180, 266, 305, 344
Current waters, 219, 344
Cuticle, 149
Cyanide, 175, 261
Cyclopoida, 45, 47
Cyprinid, 198
Cysteine, 314
Cystine, 314
Cytosol, 149, 151, 154, 156

Damage, 148, 174
DDT, 175
Decomposer, 132, 277, 301, 303
Decomposition, 4, 90, 171, 261, 304
Deforestation, 131, 142, 144, 148
Degradation, 159, 169, 187, 189, 190, 219, 239, 266, 268, 269, 270, 271, 272
Degradation Time (DT), 268
Denitrification, 110
Deoxygenation, 170
Deposition
-, acid 279, 280, 281
-, dry 131, 277, 279, 280, 293, 296
-, wet 277, 279, 293
Desert, 218
Detritus, 198
Dichlobenil, 70
DCP (2,4 Dichlorophenol), 240
Dinophyceae, 96
Diphosphate, 127
Discharges, 82, 85, 90, 101, 171, 189
-, chronic 189
-, industrial 101
-, municipal 82, 85, 101
-, rural 85, 90
Dispersion, 170, 173, 174, 189
Distribution, 171, 318
Disturbance, 10, 12
Diversity, 34, 35, 36, 39, 46, 47, 48, 50, 51, 60, 61, 69, 71, 72, 73, 74, 75, 77, 101, 162, 163, 183, 189, 218
Dose, 128, 245, 249, 252, 297
Dredging, 336
Drift, 170, 176, 177
Drinking water, 110, 114, 219
Duckweed, 343
Dumping, 267
Dust, 279, 316

Earthworm, 247, 248, 249, 251, 254, 256, 264, 265, 266, 270, 271, 272
Ecology, 3, 4, 7, 9, 21, 23, 25, 159

Ecosystem, 5, 9, 10, 12, 16, 17, 19, 21, 22, 23, 25, 55, 56, 57, 58, 59, 60, 61, 62, 63, 64, 65, 66, 67, 68, 69, 70, 71, 72, 73, 74, 75, 76, 77, 81, 86, 95, 111, 113, 114, 123, 126, 127, 128, 131, 132, 133, 139, 142, 160, 165, 168, 169, 195, 196, 201, 217, 218, 219, 220, 221, 222, 226, 230, 234, 238, 245, 246, 250, 257, 261, 275, 276, 279, 281, 283, 288, 289, 313, 328, 344
-, aquatic 7, 34, 40, 55, 57, 62, 66, 68, 72, 73, 74, 76, 81, 91, 97, 115, 201, 217, 222, 291, 301, 313, 321, 343
-, degradation 217, 218, 222, 227
-, evolution 1, 2, 3, 5, 6, 7, 8, 10, 11, 12, 13, 15, 16, 19, 20, 21, 23, 217
-, management 201
-, recovery 13, 217, 220, 222, 227, 313, 325, 334
-, responses 171
-, stressed 83, 131, 316, 343
-, terrestrial 1, 6, 13, 14, 63, 76, 81, 121, 128, 131, 137, 217, 245, 257, 275, 276, 313, 316, 323
-, theory 5
EDTA, 73, 329
Effect, 175, 180, 189, 246, 282, 331, 332
-, chronic 173, 177, 337
-, lethal 329, 332, 344
-, sub-lethal 175, 332, 344
-, synergistic 300
Effluent, 3, 28, 102, 171, 172, 177, 178, 181, 219, 315, 331, 334, 343
Egg, 175, 176, 300
Elasticity, 123, 134, 246, 285, 288
Electrodialysis, 334
Electrofishing, 181
Element, 275
-, basic, 298
-, essential 276, 314
-, nutrients 276
-, trace 313, 314
Embryo, 179
Emigration, 169, 170
Endosulfan, 269, 271, 272
Energy, 3, 4, 132, 162, 164, 198
-, flow 22, 25, 32, 61, 73, 75, 77, 110, 122, 191, 195, 304
Entomology, 201
Environment, 125, 126, 127, 148, 149, 153, 195, 196, 197, 200, 201, 203, 204, 207, 209, 210, 211, 213, 219, 220, 224, 226, 230, 241, 247, 276, 289, 330, 331
-, aquatic 14, 20, 211, 220, 315, 326, 330, 332, 335, 346
-, polluted 211, 336, 342, 343, 344, 346
Environmental factors, 125, 126, 161, 241
Enzyme, 151, 154, 268, 300, 314
Ephemeroptera, 305, 331
Erosion, 72
Estuary, 177, 322
Eukaryote, 252
Euphotic zone, 87

Eutrophication, 34, 38, 39, 55, 60, 64, 65, 72, 75, 82, 85, 86, 94, 99, 105, 111, 112, 114, 196, 197, 198, 201, 203, 207, 213, 219, 336, 340, 342, 343, 346,
Evaporation, 262, 283
Evolution 5, 7, 8, 9, 10, 19, 22, 23, 217, 218, 221
Excretion, 253, 254, 257
Experimental channels, 48

Fall-out, 315
Fauna, 132, 163, 176, 182, 183
Faunal distribution, 164
Fertilization, 144, 148, 289, 320
Filter feeder, 198
Fish, 37, 46, 59, 60, 65, 71, 72, 75, 76, 84, 85, 87, 88, 91, 94, 96, 97, 108, 109, 110, 114, 160, 165, 166, 167, 169, 170, 172, 175, 176, 177, 178, 179, 180, 181, 183, 184, 185, 186, 187, 189, 197, 198, 200, 201, 202, 203, 204, 206, 207, 210, 223, 224, 266, 271, 291, 292, 298, 300, 301, 304, 305, 307, 308, 309, 323, 331, 332, 339, 340, 341, 343
-, community 198
-, distribution 180
Fishery, 75, 87, 94, 110, 187, 292
Flagfish, 175
Flora, 132, 322
Flotation, 339
Fluctuation, 72, 131, 133, 175, 249
Flux, 134, 138
Fly-ash, 256, 334
Food, 157, 197, 200, 266
Food-chain, 4, 22, 23, 95, 96, 110, 114, 138, 166, 195, 197, 198, 200, 201, 254, 261, 266, 323, 343
Food-web, 9, 16, 56, 82, 167, 190
Forest, 16, 35, 128, 131, 142, 148, 149, 185, 187, 218, 252, 262, 275, 276, 277, 279, 280, 281, 282, 283, 284, 285, 288, 289, 306, 315
Fossil fuel, 55, 140, 141, 142, 148, 292, 293
Freshwater, 31, 36, 38, 168, 291, 297, 327, 337
-, protection 334
Frost, 128, 131, 149
Fungus, 179, 252, 304

Galena, 314, 321
Gas, 277, 315
Genetics, 1, 6, 7, 8, 22, 125, 126, 162, 167, 171, 189, 191, 217
Genotoxicity, 332
Geochemical cycle, 62, 316, 326
Geological weathering, 328
Geomorphology, 218
Glutamic acid, 270
Gneiss, 295
Granite, 295
Grassland, 16, 251, 282, 285
Grazer, 67, 177, 198, 228, 241
Groundwater, 262

Growth rate, 63, 64, 109, 126, 141, 167

Habitat, 164, 167
Hardness, 73, 164, 165, 170, 185
Harvesting, 74, 283
Hatchability, 175
HCH, 267, 269, 270, 272
Heat, 123
Heavy metal, 55, 60, 61, 71, 73, 87, 91, 115, 167, 168, 170, 172, 174, 175, 178, 183, 184, 191, 220, 226, 227, 229, 230, 232, 233, 234, 239, 240, 242, 245, 247, 248, 249, 251, 252, 253, 254, 256, 257, 261, 276, 277, 281, 282, 288, 289, 314, 315, 322, 325, 331, 335, 337, 340, 342, 343
Helophyte, 343
Hemiptera, 305
Herbicide, 70, 131, 132, 175, 336
Herbivore, 138, 198, 219
Heron, 266
Heterotrophy, 67, 138
Hexachlorobenzene, 266
Homeostasis, 75, 123, 124, 125, 196, 238
Homogenization, 270
Homothermy, 93
Hooke's law, 121
Hormone, 123, 125, 126
H_2S, 151
Humus, 138, 139, 141, 142, 144, 148, 281, 304
Hydrocarbon, 239, 261, 267, 276, 281, 282, 289
-, chlorinated 261
-, polycyclic aromatic 282
Hydrogen, 293, 295, 296, 298
-, sulphide 104, 149
Hydrography, 109, 328
Hydrology, 218, 293, 295
Hydrosphere, 326
Hydroxides, 334
Hypochlorination, 339
Hypolimnion, 89, 93, 94, 110, 113, 115, 336, 340, 342
Hysteresis, 121, 123, 134

Index
-, biotic 160, 162, 163, 167, 180, 182, 183, 187, 188, 219, 331
-, Carlson's 85, 219
-, Chandler 160, 163
-, diversity 77, 132, 160, 161, 163, 180, 187, 190, 237, 251, 331
-, evenness 250
-, functional 164
-, leaft area 279
-, saprobic 162, 219
-, Shannon-Weaver 101, 161, 162, 180
-, Trend biotic 162, 180
-, trophic state 105, 219

Inertia, 14, 133, 172, 246
Insect, 128, 131, 149, 167, 169, 170, 200, 249, 254, 271, 305, 322, 331
Insecticide, 72, 131
Invertebrate, 111, 128, 169, 170, 176, 179, 180, 183, 184, 185, 189, 190, 303, 322
Immigration, 172, 176, 177, 189
Ionic strength, 296
Iron, 86, 90, 91, 104, 112, 172, 319, 326, 329, 338
Isopoda, 247, 254

Lagoon, 51, 315, 316
Lake, 38, 39, 40, 42, 44, 46, 55, 59, 60, 64, 65, 68, 71, 72, 75, 81, 82, 83, 84, 86, 87, 88, 89, 90, 91, 92, 93, 94, 95, 96, 97, 98, 99, 101, 102, 104, 106, 107, 108, 110, 111, 113, 114, 115, 195, 196, 197, 198, 199, 200, 201, 210, 211, 213, 217, 218, 219, 220, 221, 222, 226, 291, 292, 296, 297, 298, 300, 309, 326, 335, 336, 338, 341, 342
-, eutrophic 90, 91, 94, 104, 109, 195, 198, 200, 215, 218, 221, 226
-, hypertrophic 75, 93, 106, 109
-, mesotrophic 109, 111, 113, 221
-, oligotrophic 38, 39, 44, 46, 59, 100, 101, 109, 110, 197, 203, 218, 221, 226
Landscape, 5, 16, 25, 316
Land-use, 185, 186, 187, 190, 282, 289
Lead, 172, 178, 183, 239, 253, 281, 282, 314, 316, 318, 320, 321, 322, 323, 328, 334, 337, 338, 341, 342, 343
Leaching, 190, 277, 320, 333
Leaf, 148, 149, 156, 262, 277, 289
Leeches, 176, 180
Lethal dose, 73
-, concentration 166
-, LC 50, 175
-, time 175
Life time, 126, 166, 188
Light, 75, 84, 90, 96, 104, 164, 196
Limestone, 95, 293, 321, 329, 391
Liming, 82, 95, 96, 178, 185, 190, 288, 289, 306, 308, 309, 338, 341
Lindane, 175, 269
Litter, 137, 138, 144, 279, 282
Loading, 101, 106, 108, 113, 114, 213
Locust, 201

Macroinvertebrate, 163, 167, 180, 331, 335
Macrophyte, 52, 68, 75, 84, 85, 87, 88, 89, 90, 95, 100, 114, 303, 332, 343
-, marine 343
Magmatic, 283
Magnesium, 73, 104, 190, 284
Mammal, 14, 64, 127, 265, 266, 270, 272, 322
Man, 123, 127, 131
Manganese, 282, 328, 329, 335, 337, 338, 340, 341
Mangrove community, 132
Marcasite, 321
Maturity, 34, 35
Mayfly, 176

Membrane, 149
Mercury, 73, 239, 240, 314, 318, 334, 335, 337, 338, 339, 340, 341, 342, 343
Mesophyll, 277
Metabolism, 148, 176, 268, 313, 330
Metabolite, 149, 261, 269, 270, 272
Metal, 55, 169, 171, 172, 175, 179, 183, 184, 185, 190, 231, 232, 257, 314, 315, 316, 319, 320, 321, 322, 323, 325, 326, 328, 329, 330, 331, 332, 334, 335, 337, 341, 342, 344, 376
-, excretion 257
Metallothioneine, 253
Metazoan, 167
Methoxychlon, 175
Microclimate, 7, 8, 16
Microcosm, (Enclosure, Limnocorral, Mycroecosystem, Mesocosm), 33, 40, 56, 69, 70, 164, 165, 168, 180, 188, 197, 203, 211, 217, 222, 223, 224, 226, 227, 228, 230, 231, 232, 234, 235, 237, 238, 239, 240, 241, 242, 339,
Microfauna, 262
Microflora, 262
Microorganism, 91, 110, 126, 128, 131, 137, 162, 165, 176, 177, 189, 248, 249, 252, 253, 261, 262, 267, 268, 269, 270, 271, 272, 303, 329
Micropollutant, 220, 226, 261
Migration, 94, 168, 176, 181, 183
Mine (mining), 183, 184, 189, 190, 313, 314, 315, 316, 318, 319, 320, 333, 335, 336, 339
Mineral, 61, 262, 272, 326
Mineralization, 218, 262, 268, 269, 272
Mite, 176, 256
Mixing zone, 174
Model, 56, 62, 63, 67, 74, 75, 76, 85, 88, 98, 109, 113, 114, 142, 148, 149, 151, 154, 160, 186, 187, 188, 190, 208, 298
Moisture, 267
Mollusc, 51, 73, 180, 331, 332, 344
Monitoring, 188, 332, 340
Monophosphate, 127
Moorland, 164, 185, 186, 187
Mortality, 108, 165, 168, 169, 170, 175, 178, 180, 185
Moss, 73, 179, 303
Multivariate analysis, 190, 251
Mushroom, 271
Mussel, 176
Mutualism, 3, 4, 16

Na, 104
Naidid, 180
Nannoplankton, 101
Natural catastrophy, 131
Nickel, 73, 172, 328, 335, 337, 338, 341
Nitrate, 91, 110, 181, 282, 285, 288, 293, 295, 340
Nitric acid, 293, 296
Nitrification, 93, 251, 332
Nitrogen, 58, 60, 64, 73, 75, 84, 85, 92, 96, 99, 100, 102, 103, 106, 109, 110, 163, 181, 198, 251, 276, 281, 285

-, oxyde 288, 293
NTA, 329
Nucleotide, 5
Nutrient, 3, 4, 7, 8, 21, 68, 72, 74, 75, 81, 82, 84, 85, 86, 87, 88, 89, 92, 93, 94, 96, 97, 98, 99, 102, 104, 113, 114, 115, 123, 131, 134, 168, 169, 197, 198, 199, 200, 203, 204, 220, 221, 224, 232, 241, 242, 245, 262, 269, 275, 276, 277, 284, 285, 342, 343

Ocean, 137, 138, 144
Oil, 71, 170, 220, 241, 267, 293, 337
Oligochaeta, 344
Ore, 293, 313, 314, 315, 316, 328
Organic matter, 73, 74, 89, 91, 100, 102, 104, 114, 134, 137, 171, 218, 219, 245, 248, 261, 262, 265, 267, 275, 304, 329, 337
-, degradation, 226, 238, 261
-, production 238
-, sedimentation rate 242
Organochlorine, 269
Organomethal, 329
Osmoregulation, 306
Ostracoda, 176
Outdoor channel, 189
Over-fishing, 183
Oxydation, 91, 92, 115, 151, 156, 253, 329, 340
Oxydes, 293, 334
Oxygen, 38, 44, 48, 68, 73, 82, 84, 86, 92, 93, 94, 104, 110, 113, 164, 165, 170, 171, 177, 178, 181, 188, 198, 219, 229, 238, 241, 337, 340
Oxygenation, 337, 340
Oyster, 322
Ozone, 283

Paleoclimatic change, 12
Palinology, 6, 12
Parasite, 200
Particulate matter, 89, 100, 179, 188, 189, 220, 234, 277, 335, 339, 342
Partition coefficient, 151
PCB, 71, 87
Perch, 110, 200
Periphyton, 169, 170, 224, 239
Persistance, 131, 134, 190, 269
Perturbation, 127
Pest control, 200
Pesticide, 23, 71, 169, 174, 175, 262, 269
pH, 48, 73, 75, 82, 84, 89, 90, 91, 92, 95, 96, 99, 106, 109, 149, 154, 156, 164, 165, 169, 170, 171, 172, 175, 178, 190, 198, 203, 220, 232, 233, 234, 235, 236, 237, 241, 248, 264, 267, 268, 277, 279, 281, 283, 293, 295, 296, 298, 300, 301, 304, 306, 309, 321, 329, 335, 341, 344
Phenol, 175, 262, 272
Phosphorus, 58, 60, 64, 73, 75, 76, 81, 84, 85, 86, 87, 89, 90, 91, 92, 93, 94, 96, 99, 102, 104, 106, 109, 110, 111, 113, 114, 115, 197, 206, 219, 231, 232, 233, 234, 298, 301, 303, 304, 336
-, external loading 111

-, internal loading 92, 108
-, release 100, 109, 114
-, removal 89, 91, 99, 101, 108, 112, 113
Photosynthesis, 74, 75, 131, 165, 198, 275
-, activity 108, 109, 138
-, pigments 33
-, rate 220, 240, 332
Phyletic radiation, 15
Phylogeny, 17, 19, 20, 22
Physical barrier, 183
Physiological processes, 190
Phytohormone, 126
Phytoplankton, 37, 38, 41, 46, 64, 65, 68, 71, 73, 75, 84, 90, 92, 94, 96, 101, 102, 103, 104, 108, 111, 114, 137, 198, 200, 203, 206, 207, 220, 224, 226, 228, 229, 231, 232, 233, 234, 235, 236, 237, 240, 241, 242, 301, 342
Pike perch, 110
Plankton, 4, 34, 36, 37, 41, 42, 51, 52, 81, 89, 90, 103, 165, 196, 202, 203, 224, 332
Planktophagous, 35, 75, 76, 200
Plant, 12, 19, 21, 125, 126, 128, 142, 148, 154, 156, 179, 180, 217, 245, 262, 276, 277, 285, 320, 329, 332, 343
-, communities 131
-, population 220
Plecopteran, 182, 183
Pollution (pollutant), 23, 55, 60, 61, 62, 65, 71, 72, 81, 87, 88, 94, 97, 110, 113, 114, 115, 128, 131, 132, 148, 149, 153, 154, 156, 159, 160, 161, 162, 163, 164, 165, 166, 167, 168, 169, 170, 171, 172, 173, 174, 175, 176, 177, 181, 187, 188, 189, 190, 191, 217, 218, 219, 220, 221, 222, 224, 226, 229, 230, 232, 238, 239, 241, 242, 245, 246, 250, 256, 261, 271, 276, 313, 315, 320, 328, 330, 332, 335, 336, 337, 342
-, abatement 172, 177
-, air 76, 245, 276
-, control 159, 172, 177
-, monitoring 190
Polyacrylamide, 109
Polychaete, 340
Pond, 56, 110, 197, 201, 210, 338, 343
Pool, 218
Population, 3, 4, 5, 6, 8, 12, 16, 22, 23, 126, 128, 131, 164, 167, 176, 177, 195, 196, 236, 237, 239, 247, 250, 270, 331
-, density 177, 206, 223, 229, 234, 241
-, fluctuation 246
Potassium, 283, 284, 320
Precipitation, 277, 295, 296, 326
Predation, 6, 60, 65, 72, 74, 76, 85, 93, 96, 110, 196, 197, 200, 201, 203
Predator, 168, 196, 198, 240, 304, 305
Prey, 196, 240
Primary productivity, 19, 33, 36, 37, 38, 51, 53, 60, 64, 70, 71, 73, 75, 101,

134, 138, 139, 142, 144, 148, 198, 210, 219, 303, 336, 342, 343
Producer, 200
-, primary 85, 132, 140, 196, 275, 276, 301, 304, 330
-, secondary 275, 276
Production, 65, 72, 75, 142, 218, 240, 276, 301
-, rate 220, 238
Prokaryote, 252
Protection, 334
Protein, 314
Proton, 149, 156, 276, 277, 280, 281
Protozoan, 332

QSAR's (Quantitative Structure Activity Relation), 248, 264

Rainfall, 190, 277, 293, 301, 321
Rainforest, 16, 21
Recolonization, 176, 180, 247, 256
Recovery (restoration), 65, 68, 70, 75, 82, 83, 84, 85, 86, 87, 88, 89, 91, 92, 94, 95, 96, 99, 100, 101, 102, 104, 106, 107, 108, 109, 110, 111, 112, 113, 114, 115, 131, 134, 159, 162, 169, 171, 172, 173, 174, 175, 177, 178, 179, 180, 181, 183, 184, 187, 189, 190, 201, 209, 211, 213, 217, 221, 230, 231, 233, 238, 241, 242, 246, 247, 256, 275, 291, 307, 313, 323, 331, 341, 344, 345, 346
Regulation, 3, 22
Renewal time, 95
Reproduction, 63, 64, 126, 167, 175, 241, 251, 300, 306
Reservoir, 82, 85, 86, 91, 109, 110, 111, 112, 113, 115, 181, 185
Resilience, 4, 7, 8, 17, 56, 69, 70, 73, 77, 99, 106, 115, 160
Resistance, 20, 21
Respiration, 62, 73, 74, 138, 139, 249, 251, 252, 270
Response, 128, 148, 167, 168, 169, 170, 177, 189
-, behavioural 128, 167, 189
-, biochemical 128, 189
-, biological 169
-, physiological 189
-, sub-lethal 168
Restoration, 238, 336, 340, 346
Retention time, 82, 99, 139, 174, 309
Ricefield, 41, 42
River, 92, 159, 163, 164, 165, 166, 167, 168, 170, 171, 172, 173, 174, 175, 176, 177, 179, 181, 183, 185, 187, 188, 189, 190, 218, 219, 291, 301, 309, 315, 316, 320, 321, 322, 326, 330, 331, 336, 341, 344
-, continuum concept 190
-, monitoring 308
Root, 281, 282, 285, 289
Rotifera, 38, 41, 43, 44, 45, 46, 47, 48, 49, 50, 51, 72, 84, 101, 105, 107, 198, 203, 233
Run-off, 177, 179, 293, 295, 296, 305, 308, 309, 328, 344

Salinity, 218, 329

Salmon, 176, 181, 183, 185, 187, 198, 291, 292, 300, 301, 321, 322
Salt marsh, 64
Sand, 296, 297, 336, 339
Scavenger, 341
Sea, 36, 52, 53, 316, 322, 327, 337, 345
-, water 295, 333
Sea-trout, 181
Seepage, 282, 283
Sediment, 64, 65, 68, 73, 82, 84, 86, 87, 89, 90, 91, 92, 93, 94, 95, 96, 98, 99, 100, 101, 102, 109, 115, 169, 171, 174, 179, 198, 219, 220, 221, 223, 224, 226, 227, 229, 230, 231, 232, 236, 239, 266, 271, 301, 304, 306, 316, 318, 319, 323, 326, 328, 329, 335, 336, 337, 338, 339, 340, 341, 344
-, dredging 87, 88, 89, 96, 100, 101, 339, 345
-, removal 87, 88, 89, 108, 115, 345
-, suspended 169, 315, 316, 336, 337, 339
Sedimentation, 86, 88, 111, 234, 277, 342
Selenium, 314, 339, 343
Self-purification, 335
Seston, 67, 68, 84, 87, 103, 104, 107, 112, 233, 239, 326, 328, 336
Silicate, 283, 314
Silt, 336
Silver, 314, 320, 322
Silviculture, 246
Slurry, 269, 270
Snail, 168, 305, 306
Snow, 277, 293
Snowmelt, 301
Soil, 8, 16, 17, 21, 131, 132, 138, 148, 171, 190, 217, 245, 246, 247, 248, 249, 251, 252, 253, 254, 256, 257, 261, 262, 263, 265, 266, 267, 269, 270, 271, 272, 275, 276, 277, 279, 280, 281, 282, 283, 284, 285, 288, 289, 293, 294, 306, 313, 314, 316, 318, 319, 320, 322, 339
-, erosion 315
-, type 245, 257, 263, 264, 271, 272
SO_2, 71, 128, 148, 149, 151, 153, 279, 280, 289, 293, 315
Sphaeridae, 306
Sphalerite, 314, 321
Spawner, 179
Spawning, 175, 183
Spillage, 173, 174
Springtail, 254, 256
Spruce, 128
Stability, 4, 12, 21, 34, 35, 36, 69, 73, 132, 133, 137, 139, 162, 190, 247, 275
Stage, 131, 166, 177, 187, 300
Stand, 285
Stonefly, 182, 187
Stress, 55, 56, 57, 58, 59, 60, 61, 62, 63, 64, 65, 67, 68, 69, 70, 71, 72, 73, 75, 76, 77, 81, 95, 101, 121, 122, 123, 124, 126, 128, 131, 132, 133, 149, 159, 160, 218, 234, 238, 246, 247, 249, 261, 285
-, biochemical stress index 63
-, Selye concept 123, 124, 127, 128
Succession, 1, 3, 7, 12, 13, 16, 19, 171

Sulphate, 102, 151, 156, 293, 295, 296, 298, 307, 308
Sulphide, 189, 277, 314, 315, 339, 340
Sulphite, 149, 334
Sulphur, 149, 151, 154, 156, 283, 293, 306, 307, 308, 314, 315
Sulphuric acid, 190, 293, 296, 315
Surveillance, 188
Survival, 178, 187, 247
Synergism, 148

Taiga, 16
Temperature, 58, 60, 61, 64, 65, 68, 72, 74, 75, 84, 94, 99, 163, 164, 165, 172, 175, 196, 226, 267, 268, 269
Titanium, 326
Tin, 314
Thermal stratification, 36, 68, 93, 94
Thermal effluent, 55
Thermocline, 68, 137
Thermodynamics, 7, 8, 23
Threshold, 128, 133
Tolerance, 149, 167, 195
Toxic, 73, 81, 87, 88, 89, 115, 156, 164, 165, 178, 179, 184, 189, 221, 222, 234, 241, 254, 272, 275, 300, 333, 342
Toxicity, 96, 165, 166, 168, 172, 175, 178, 187, 189, 219, 220, 232, 238, 239, 240, 264, 269, 271, 272, 285, 300, 331, 333
-, test 165, 166, 167, 168, 169, 174, 178, 179, 188, 332, 337
Toxicology, 128, 174, 179
Toxin, 180
Transparency, 44, 46, 87, 93, 99, 103, 104, 108, 109, 110, 111, 197, 203, 204, 210, 219, 229, 232, 236, 241
Transpiration, 293
Trap, 176
Treatment plant, 86, 178, 213, 261, 335
Tree, 15, 149, 280, 281, 282, 285, 289
Tributaries, 171, 174, 176, 179, 181, 182, 183, 189
Trichoptera, 180
Triphosphate, 127
Trophy, 9, 22, 32, 50, 75, 76, 77, 81, 98, 107, 109, 113, 114, 115, 138, 162, 190, 195, 196, 197, 199, 200, 219, 220, 226, 238, 298, 342
Trophic pyramid, 197
-, chain 201, 334, 342
-, web, 4
Trophodynamic, 195
Tropical reef, 35
Trout, 178, 179, 181, 184, 185, 186, 300, 322
-, brown 166, 178, 184, 291
-, rainbow 90, 166
Tubificid, 101, 180, 242
Tungsten, 314
Tundra, 16
Turbidity, 68, 87, 91, 104, 112
Turbulence, 12, 72, 224
Turnover, 9, 62, 138

Vacuole, 149, 151, 156
Variability, 16, 159, 161
Vegetation, 12, 17, 142, 144, 148, 295, 315, 335, 344
Vertebrate, 128
Volcanic eruption, 72, 131, 144
Vole, 272
Vulnerability, 10

Waste, 55, 109, 111, 172, 173, 189, 190, 245, 265, 269, 276, 344
-, leaching 328, 344
-, treatment 113, 328, 334
Water, 131, 132, 149, 166, 167, 168, 169, 170, 171, 172, 173, 175, 177, 179, 180, 181, 182, 183, 185, 187, 190, 220, 221, 224, 226, 231, 236, 262, 266, 271, 275, 277, 295, 296, 306, 326, 328, 330, 335, 340
-, chemistry 293, 298, 300, 301, 307, 308
-, ecosystem 114, 219, 220, 222, 224, 236, 238, 241, 329, 331, 334, 335, 339
-, pollution 111, 162, 163, 298, 331, 345
-, quality 162, 163, 164, 166, 167, 168, 169, 170, 171, 172, 173, 179, 180, 181, 182, 183, 185, 187, 190, 195, 201, 213, 229, 296, 304, 305, 307, 331, 334, 337, 344
-, sediment interference 342
-, uses 334
Weathering, 190
Wind, 72, 277, 327, 335
Woodland, 17, 218, 246, 283

Zinc, 170, 172, 175, 178, 183, 184, 239, 249, 282, 314, 319, 321, 322, 328, 334, 335, 337
Zoobenthos, 52, 65
Zooplankton, 35, 36, 37, 41, 44, 46, 47, 48, 51, 52, 59, 60, 65, 68, 71, 72, 74, 75, 76, 84, 85, 94, 95, 101, 103, 107, 108, 109, 111, 114, 137, 198, 200, 201, 202, 203, 204, 207, 228, 234, 235, 236, 241, 304
-, grazing 109, 197